OSCILLATION THEORY FOR DIFFERENCE AND FUNCTIONAL DIFFERENTIAL EQUATIONS

Oscillation Theory for Difference and Functional Differential Equations

by

Ravi P. Agarwal
Department of Mathematics,
National University of Singapore,
Singapore

Said R. Grace
Department of Engineering Mathematics,
Cairo University,
Orman, Giza, Egypt

and

Donal O'Regan
Department of Mathematics,
National University of Ireland,
Galway, Ireland

SPRINGER-SCIENCE+BUSINESS MEDIA, B.V.

A C.I.P. Catalogue record for this book is available from the Library of Congress.

ISBN 978-90-481-5447-0 ISBN 978-94-015-9401-1 (eBook)
DOI 10.1007/978-94-015-9401-1

Printed on acid-free paper

Contents

Preface

This monograph is devoted to a rapidly developing area of research of the qualitative theory of difference and functional differential equations. In fact, in the last 25 years *Oscillation Theory* of difference and functional differential equations has attracted many researchers. This has resulted in hundreds of research papers in every major mathematical journal, and several books.

In the first chapter of this monograph, we address oscillation of solutions to difference equations of various types. Here we also offer several new fundamental concepts such as oscillation around a point, oscillation around a sequence, regular oscillation, periodic oscillation, point-wise oscillation of several orthogonal polynomials, global oscillation of sequences of real-valued functions, oscillation in ordered sets, $(f, \mathbb{R}, \leq)$-oscillate, oscillation in linear spaces, oscillation in Archimedean spaces, and oscillation across a family. These concepts are explained through examples and supported by interesting results.

In the second chapter we present recent results pertaining to the oscillation of n-th order functional differential equations with deviating arguments, and functional differential equations of neutral type. We mainly deal with integral criteria for oscillation. While several results of this chapter were originally formulated for more complicated and/or more general differential equations, we discuss here a simplified version to elucidate the main ideas of the oscillation theory of functional differential equations. Further, from a large number of theorems presented in this chapter we have selected the proofs of only those results which we thought would best illustrate the various strategies and ideas involved.

We hope this monograph will fill a vacuum in the oscillation theory of difference and functional differential equations and will be a stimulus to its further development.

It is impossible to acknowledge individually colleagues and friends to whom we are indebted for assistance, inspiration and criticism in writing this monograph. We must, however, express our appreciation and thanks to Sadhna for her careful typing of the entire manuscript.

Ravi P Agarwal
Said R Grace
Donal O'Regan

Chapter 1
Oscillation of Difference Equations

1.1. Introduction

The theory of difference equations (recurrence relations), the methods used in their solutions and their wide applications have advanced beyond their adolescent stage to occupy a central position in Applicable Analysis. In fact, in the last few years, the proliferation of the subject is witnessed by hundreds of research articles and several monographs [2,25,28,51,129, 136,159,167,201,202,214,215,266]. In particular, oscillation of solutions of difference equations has attracted many researchers. The purpose of this chapter is to offer several new fundamental concepts in this fast developing area of research. These concepts are explained through examples and supported by interesting results.

The plan of this chapter is as follows: In Section 1.2, we shall consider scalar difference equations and introduce concepts such as *oscillation around a*, *oscillation around the sequence*, *regular oscillation*, and *periodic oscillation*. In Section 1.3, we shall prove the *point-wise oscillation* property of several orthogonal polynomials, namely, Chebyshev polynomials of the first and second kind, Hermite polynomials, and Legendre polynomials. Here an oscillation theorem for second order difference equations is also established. In Section 1.4, we define the *global oscillation* of sequences of real-valued functions. Here, for a second order nonlinear continuous–difference equation, sufficient conditions to ensure global oscillation of solutions are provided. In Section 1.5, oscillation in ordered sets is considered. Here, the concept $(f, I\!R, \leq)$*-oscillate* is introduced, and two interesting examples are given. Section 1.6 deals with the oscillation in linear spaces, and the ideas are elaborated in two simple theorems. In Section 1.7, we consider oscillation in Archimedean spaces, and prove a result for a third order difference equation. In Section 1.8, for the partial difference equations we define *oscillation across the family* $\gamma = \{\gamma_k : k \in I\!N\}$ *of each other disjoint path arguments*, and prove two theorems. Oscillation of system of equations is the subject matter of Section 1.9. Here for the system of two linear difference equations with constant coefficients oscillation is established by using some geometric ideas. In Section 1.10, oscillation between sets is introduced. Here several examples illustrating the basic ideas are included. In Section 1.11 the oscillation of continuous–discrete difference equations is studied. Here two interesting results dwelling upon the importance of the concepts are proved. In Section 1.12 we shall obtain sufficient conditions for the oscillation of all solutions, as well as necessary conditions for the existence of nonoscillatory solutions for second order quasilinear difference equations. In Section 1.13 we shall establish sufficient conditions for the oscillation of all solutions of even order difference equations by comparing with certain difference equations of lower order whose oscillatory character is known. In Section 1.14 we shall present explicit conditions for the *almost oscillation* of certain odd order difference equations. Sections 1.15 and 1.16, respectively, furnish sufficient conditions for the oscillation of neutral and mixed type higher order difference equations. In Section 1.17 we shall offer criteria which guarantee for the nonoscillatory solutions of difference equations involving quasi–differences to be bounded and/or ultimately tend to zero. Here several interesting examples which dwell upon the importance of the results are also illustrated. In Section 1.18 we shall develop nonexistence criteria for eventually positive (negative) solutions of certain inequalities involving distributed deviating arguments. These criteria give rise to oscillation results of the related ordinary as well as very general partial difference equations subject to two different types of boundary conditions. In Section 1.19 we shall address the oscillation of all bounded solutions of higher order systems of difference equations. Finally, in Section 1.20 we shall provide sufficient conditions for all solutions of a partial difference equation with continuous variables to be oscillatory.

Throughout this monograph we shall use the following notations: $\mathbb{N} = \{1, 2, \cdots\}$ represents the set of positive integers. For integers $a, b\ (> a \geq 0)$, we shall denote the discrete intervals $\mathbb{N}(a,b) = \{a, a+1, \cdots, b\}$ and $\mathbb{N}(b) = \{b, b+1, \cdots\}$. For $j \in \mathbb{R}$ and $k \geq 0$ integer we define the usual factorial expression $j^{(k)} = j(j-1)\cdots(j-k+1)$ with $j^{(0)} = 1$. Further, $\mathbb{R} = (-\infty, \infty)$, $\mathbb{R}_0 = [0, \infty)$, $\mathbb{R}_+ = (0, \infty)$, and $\mathbb{R}_- = (-\infty, 0)$ are the usual sets of real, nonnegative, positive, and negative numbers, respectively. Finally, $\mathbb{R}^*$ denotes the extended real line, i.e. $\mathbb{R}^* = \mathbb{R} \cup \{\infty, -\infty\}$.

1.2. Oscillation of Scalar Difference Equations

Let $y = y(n) = (\{y_n\}_{n=1}^{\infty}) : \mathbb{N} \to \mathbb{R}$, i.e. y is a real–valued sequence (function). For this y the difference operators Δ, Δ_a, $a \in \mathbb{R}$, and Δ^k, $k \in \mathbb{N}$ are defined as follows:

$$\Delta y_n = y_{n+1} - y_n, \quad \Delta_a y_n = y_{n+1} - a y_n, \quad n \in \mathbb{N}$$

and

$$\Delta^k y_n = \Delta^{k-1} y_{n+1} - \Delta^{k-1} y_n, \quad k > 1, \quad \Delta^1 = \Delta.$$

Definition 1.2.1. The sequence y is said to be *oscillatory around* a $(a \in \mathbb{R})$ if there exists an increasing sequence of integers $\{n_k\}_{k=1}^{\infty}$ such that

$$(y_{n_k} - a)(y_{n_k+1} - a) \leq 0 \quad \text{for all } k \in \mathbb{N}. \tag{1.2.1}$$

If in (1.2.1) the strict inequality holds then y is called *strictly oscillatory around* a. Further, if $(y_{n+1} - a)(y_n - a) < 0$ for all $n \in \mathbb{N}$ then y is said to be *quickly oscillatory around* a.

Thus, the oscillation of y is the same as the sequence $\{y_n - a\}_{n=1}^{\infty}$ has infinite number of *generalized zeros*. (We recall that the sequence y has a generalized zero at n provided $y_n = 0$ if $n = 1$ and if $n > 1$ then either $y_n = 0$ or $y_{n-1}y_n < 0$.)

Usually in literature oscillatory behavior is studied only around 0, whereas in applications mostly we need oscillation around some other point, e.g. at a stable (stationary) point.

Definition 1.2.2. The sequence y is said to be *strictly oscillatory around* 0, or strictly oscillatory, or simply oscillatory if there is no ambiguity, if for every $n \in \mathbb{N}$ there exists a $k \in \mathbb{N}$ such that $y_n y_{n+k} < 0$. Further, y is said to be *nonoscillatory* if it is eventually of constant sign.

In Definitions 1.2.1 and 1.2.2 the oscillation of y is from the phase space point of view, whereas the definition of nonoscillation looks at the sequence as the chain of points on the enlarged space.

Definition 1.2.3. The sequence $y = \{y_n\}_{n=1}^{\infty}$ is called *oscillatory around the sequence* $x = \{x_n\}_{n=1}^{\infty}$ if there exists an increasing sequence of positive integers $\{n_k\}_{k=1}^{\infty}$ such that

$$(y_{n_k} - x_{n_k})(y_{n_k+1} - x_{n_k+1}) \leq 0 \quad \text{for all } k \in \mathbb{N}.$$

From the above definition it is clear that if the sequence $\{y_n\}_{n=1}^{\infty}$ oscillates around the sequence $\{x_n\}_{n=1}^{\infty}$ then $\{x_n\}_{n=1}^{\infty}$ oscillates around the sequence $\{y_n\}_{n=1}^{\infty}$. Moreover, every sequence $\{y_n\}_{n=1}^{\infty}$ oscillates around itself. However, the relation 'oscillation around' does not have the transitivity property. Further, the sequence $\{y_n\}_{n=1}^{\infty}$ is oscillatory around the number a (the number 0) if it is oscillatory around the constant sequence x such that $x_n = a$, $n \in \mathbb{N}$ (around the sequence θ such that $\theta_n = 0$, $n \in \mathbb{N}$). We also note that if the sequence $\{y_n\}_{n=1}^{\infty}$ oscillates around the sequence $\{x_n\}_{n=1}^{\infty}$, then it also oscillates around every sequence $\{z_n\}_{n=1}^{\infty}$, where for a fixed $k \in \mathbb{N}$, $z_n = x_n$, $n \geq k$.

To show the importance of the above definitions and remarks, we need the following result:

Theorem 1.2.1. [167] For the linear homogeneous difference equation

$$y_{n+k} = \sum_{i=0}^{k-1} a_i y_{n+i}, \qquad n \in \mathbb{N} \tag{E_1}$$

where k is a positive integer and $a_0, a_1, \cdots, a_{k-1} \in \mathbb{R}$, the following statements are equivalent:

(a) Every solution of (E_1) is oscillatory.

(b) The characteristic equation of (E_1), i.e. $\lambda^k = \sum_{i=0}^{k-1} a_i \lambda^i$ has no positive roots.

Corollary 1.2.2. The equation

$$\Delta^k y_n = a y_n, \qquad n \in \mathbb{N} \tag{E_2}$$

oscillates, if and only if,

$$\begin{cases} a < 0 & \text{in the case } k \text{ is even,} \\ a < -1 & \text{in the case } k \text{ is odd.} \end{cases}$$

Further, if k is odd, and $1 \le \ell < k$ the equation

$$\Delta^k y_n = a y_{n+\ell}, \qquad n \in \mathbb{N}$$

possesses nonoscillatory solution for any real number a, whereas if k is even it possesses nonoscillatory solution if a is positive.

Now suppose that the condition (b) of Theorem 1.2.1 is satisfied. Since all solutions of (E_1) are oscillatory around the null sequence θ and equation (E_1) is linear, the difference of any two solutions $\{y_n\}_{n=1}^{\infty}$ and $\{x_n\}_{n=1}^{\infty}$, i.e. the sequence $\{y_n - x_n\}_{n=1}^{\infty}$ is also the solution of (E_1) and oscillates around θ. Therefore, there exists $\{n_k\}_{k=1}^{\infty}$ such that $(y_{n_k} - x_{n_k})(y_{n_k+1} - x_{n_k+1}) \le 0$. This means that both solutions oscillate around themselves. Hence, the set of all solutions of (E_1) forms a bunch of sequences which oscillate around each other.

Next, we shall consider the nonhomogeneous equation

$$y_{n+k} = \sum_{i=0}^{k-1} a_i y_{n+i} + \phi_n, \qquad n \in \mathbb{N} \tag{E_3}$$

and suppose that for the corresponding homogeneous equation (E_1) the condition (b) of Theorem 1.2.1 is satisfied. Let $\{y_n\}_{n=1}^{\infty}$ and $\{x_n\}_{n=1}^{\infty}$ be any two solutions of (E_3). Then, the difference $\{y_n - x_n\}_{n=1}^{\infty}$ is an oscillatory solution of (E_1) around θ. Hence, there exists $\{n_k\}_{k=1}^{\infty}$ such that $((y_{n_k} - x_{n_k}) - 0)((y_{n_k+1} - x_{n_k+1}) - 0) = (y_{n_k} - x_{n_k})(y_{n_k+1} - x_{n_k+1}) \le$

0, i.e. all solutions of (E_3) oscillate around themselves, and therefore form a similar bunch as the set of solutions of (E_1).

Example 1.2.1. Consider the equation

$$y_{n+k} = \sum_{i=0}^{k-1} a_i y_{n+i} + b, \quad n \in \mathbb{N} \tag{E_4}$$

where b is any constant. Assume that the condition (b) of Theorem 1.2.1 is satisfied. Then, $1 - \sum_{i=0}^{k-1} a_i \neq 0$, otherwise the characteristic equation $\lambda^k = \sum_{i=0}^{k-1} a_i \lambda^i$ will have a positive root 1. It is clear that the constant sequence $\{y_n\}_{n=1}^{\infty}$, where

$$y_n = b\left(1 - \sum_{i=0}^{k-1} a_i\right)^{-1}, \quad n \in \mathbb{N}$$

is a solution of (E_4). Therefore, from the above remarks, every solution of (E_4) oscillates around the constant $b\left(1 - \sum_{i=0}^{k-1} a_i\right)^{-1}$.

As an another example of the theory of oscillation we consider the difference equation

$$\Delta_a y(n) + \sum_{i=1}^{m} a_n^i f^i(y(d_i(n))) = 0, \quad n \in \mathbb{N} \tag{E_5}$$

where

$$a^i : \mathbb{N} \to \mathbb{R},\ a_n^i > 0 \quad \text{for all } n \in \mathbb{N}_\mu = \{\mu, \mu+1, \cdots\}, \quad i = 1, 2, \cdots, m \tag{1.2.2}$$

$$d_i : \mathbb{N} \to \mathbb{N}, \quad \lim_{n\to\infty} d_i(n) = \infty, \quad i = 1, 2, \cdots, m \tag{1.2.3}$$

$$f^i : \mathbb{R} \to \mathbb{R},\ xf(x) > 0 \quad \text{for } x \neq 0, \quad i = 1, 2, \cdots, m. \tag{1.2.4}$$

Theorem 1.2.3. If $0 < a < 1$, then every solution $\{y_n\}_{n=1}^{\infty}$ of (E_5) is oscillatory, or $\lim_{n\to\infty} \delta^n y(n) = 0$ for some $\delta > 1$.

Proof. If $\{y_n\}_{n=1}^{\infty}$ is a nonoscillatory solution of (E_5), say positive for $n \geq n_0$, then by (1.2.3) there exists an $n_1 \geq n_0$ such that $d_i(n) \geq n_0$ for all $n \geq n_1$ and all $i \in \{1, 2, \cdots, m\}$. Hence, by (1.2.2) and (1.2.4) we have $a_n^i f^i(y(d_i(n))) \geq 0$ for all $n \geq \nu = \max\{\mu, n_1\}$, $i = 1, 2, \cdots, m$. Therefore, $\Delta_a y(n) \leq 0$, $n \geq \nu$, which is the same as

$$y(n+1) \leq ay(n) \quad \text{for } n \geq \nu. \tag{1.2.5}$$

From (1.2.5), we obtain

$$y(n+\nu) \;\le\; ay(n+\nu-1) \;\le\; a^2y(n+\nu-2) \;\le\; \cdots \;\le\; a^n y(\nu). \quad (1.2.6)$$

Now since $a \in (0,1)$, there exists a $\delta > 1$ such that $a\delta < 1$. For such δ, in view of (1.2.6), we have

$$\delta^{n+\nu} y(n+\nu) \;\le\; (a\delta)^{n+\nu}\left[y(\nu)a^{-\nu}\right],$$

which immediately gives

$$\lim_{n\to\infty} \delta^n y(n) \;=\; \lim_{n\to\infty} \delta^{n+\nu} y(n+\nu) \;\le\; \lim_{n\to\infty} (a\delta)^{n+\nu}\left[y(\nu)a^{-\nu}\right] \;=\; 0. \qquad \blacksquare$$

Now we shall study *regular oscillations,* i.e. having some fixed pattern. For this, we recall [25] that the sequence y is called *periodic* (τ–period) if there exists an integer $\tau \in \mathbb{N}$ such that $y_{n+\tau} = y_n$ for all $n \in \mathbb{N}$. Thus, for the sequence y the oscillatory behavior with some regularity is the same as its periodicity.

Suppose that $\{y_n\}_{n=1}^{\infty}$ is an oscillatory sequence. To this sequence we can attribute a new sequence of signs as follows:

$$\text{sgn }(y,n) \;=\; \begin{cases} 1 & \text{if } y(n) > 0 \\ -1 & \text{if } y(n) < 0 \\ 0 & \text{if } y(n) = 0. \end{cases}$$

Definition 1.2.4. A sequence y is said to be *periodic oscillatory* if the sequence $\{\text{sgn }(y,n)\}_{n=1}^{\infty}$ is periodic.

It is interesting to note that the regular oscillation can have different forms. For example,

$$\{\text{sgn }(y,n)\}_{n=1}^{\infty} \;=\; 1,-1,1,1,-1,-1,1,1,1,-1,-1,-1,\cdots$$

or

$$\{\text{sgn }(y,n)\}_{n=1}^{\infty} \;=\; 1,-1,1,1,-1,1,1,1,-1,\cdots.$$

The problem connecting the distance between two consecutive zeros and regular oscillation has never been studied. We shall present a very simple result in this direction. For this, let us denote

$$pos(y,m,n) \;=\; \sum_{i=m}^{n} \max(\text{sgn }(y,i),0),$$

$$neg(y,m,n) \;=\; \sum_{i=m}^{n} \max(-\text{sgn }(y,i),0).$$

These functions define the number of positive and negative terms of the sequence $y(m), y(m+1), \cdots, y(n)$.

A simple result for the linear difference equation

$$y(n+k) = \sum_{i=0}^{k-1} a_n^i y(n+i), \qquad n \in \mathbb{N} \tag{E_6}$$

where the sequences $a^i : \mathbb{N} \to \mathbb{R}$, $i = 0, 1, \cdots, k-1$, is stated in the following theorem:

Theorem 1.2.4. If the sequences $\{\text{sgn}\ (a^i, n)\}$ are τ–periodic, $\tau \leq k$ and such that

$$pos(a^{k-j}, 1, \tau) = \tau - s, \qquad neg(a^{k-j}, 1, \tau) = s$$

for $j = 2\lambda\tau + s$; $\lambda = 0, 1, \cdots$; $s = 1, 2, \cdots, \tau$; $2\lambda\tau + s \leq k$;

$$pos(a^{k-j}, 1, \tau) = s, \qquad neg(a^{k-j}, 1, \tau) = \tau - s$$

for $j = (2\lambda + 1)\tau + s$; $\lambda = 0, 1, \cdots$; $s = 1, 2, \cdots, \tau$; $(2\lambda + 1)\tau + s \leq k$;

$$\text{sgn}\ (a^{k-j}, n) = b_{p_n, j}, \qquad j = 1, 2, \cdots, \tau; \quad n = 1, 2, \cdots, \tau - 1$$

where $(p_1, \cdots, p_\tau)$ is some cyclic permutation of $(1, 2, \cdots, \tau)$,

$$[b_{s,\tau}]_{\tau\times\tau} = \begin{pmatrix} -1 & -1 & \cdots & -1 \\ 1 & -1 & \cdots & -1 \\ 1 & 1 & \cdots & -1 \\ \cdots & \cdots & \cdots & \cdots \\ 1 & 1 & \cdots & -1 \end{pmatrix},$$

$$\text{sgn}\ (a^i, n) = \text{sgn}\ (a^{i+\tau}, n), \qquad i = 0, 1, \cdots, k - \tau - 1, \quad n = 1, 2, \cdots, \tau - 1$$

then the equation (E_6) possesses 2τ–periodically oscillated solutions with τ positive and τ negative terms in each 2τ consecutive terms.

A sequence which oscillates around 0 consists of a 'string' of nonnegative terms followed by a string of negative terms, or vice versa, and so on. We call these strings positive and negative semicycles, respectively.

When we study the oscillation around α, the semicycles are defined relative to α and consist of strings greater than or equal to α followed by strings of terms less than α, and so on. More precisely we have the following definitions of the semicycles.

Definition 1.2.5. A *positive semicycle* of a sequence y consists of a string of terms $\{y(m), y(m+1), \cdots, y(n)\}$, all greater than or equal to α, and such that

$$\text{either } m = 1 \text{ or } m > 1 \text{ and } y(m-1) < \alpha$$

and

$$\text{either } n = \infty \text{ or } n < \infty \text{ and } y(n+1) < \alpha.$$

Similarly, a *negative semicycle* of a sequence y consists of a string of terms $\{y(m), y(m+1), \cdots, y(n)\}$, all less than α, and such that

$$\text{either } m = 1 \text{ or } m > 1 \text{ and } y(m-1) \geq \alpha$$

and

$$\text{either } n = \infty \text{ or } n < \infty \text{ and } y(n+1) \geq \alpha.$$

It is clear that a sequence may have a finite number of semicycles or infinitely many semicycles. Further, periodic oscillation can be described in terms of semicycles.

1.3. Oscillation of Orthogonal Polynomials

We shall need the following:

Theorem 1.3.1. If $a = (\{a_n\}_{n=1}^{\infty}) : \mathbb{N} \to \mathbb{R}_-$, and $\liminf_{n\to\infty} a_n = \alpha < 0$, then for the difference equation

$$\Delta^2 y(n) = a_n y(n), \quad n \in \mathbb{N} \ (n \in \mathbb{N} \cup \{0\}) \qquad (E_7)$$

the following hold:

(a) Every solution of (E_7) oscillates.

(b) For every nontrivial solution of (E_7) the distance between successive generalized zeros is greater than 1.

Proof. (a) Let us assume that y is a nonoscillatory (say eventually positive) solution of (E_7). Let $y(n) > 0$ for $n \geq n_1$, $n_1 \in \mathbb{N}$. Then,

$$\Delta y(n+1) - \Delta y(n) = \Delta^2 y(n) = a_n y(n) < 0, \quad n \geq n_1.$$

Therefore, $\Delta y(n+1) < \Delta y(n)$, and the sequence $\{\Delta y(n)\}_{n=1}^{\infty}$ is decreasing for $n \geq n_1$. We need now to consider the following two possible cases:

(a_1) there exists a $\nu \geq n_1$ such that $\Delta y(n) < 0$ for all $n \geq \nu$, and

(a_2) $\Delta y(n) > 0$ for all $n \geq n_1$.

In the case (a_1) we have $\Delta y(n) \leq \Delta y(\nu)$ for all $n \geq \nu$. Thus, on summing, it follows that

$$y(n+1) - y(\nu) = \sum_{j=\nu}^{n} \Delta y(j) \leq (n+1-\nu)\Delta y(\nu),$$

i.e.

$$y(n+1) \;\le\; y(\nu) + (n+1-\nu)\Delta y(\nu), \qquad n \ge \nu.$$

In the above inequality the left side is positive whereas the right side tends to $-\infty$ as $n \to \infty$. This contradiction shows that the case (a_1) is impossible.

Now assume that the case (a_2) holds. Since $\Delta y(n) > 0$ and $\Delta^2 y(n) < 0$ the sequence $\{\Delta y(n)\}_{n=1}^{\infty}$ is decreasing and bounded from below by 0. Therefore, $\lim_{n\to\infty} \Delta y(n) = g \ge 0$ exists. But, then

$$\lim_{n\to\infty} \Delta^2 y(n) \;=\; \lim_{n\to\infty} (\Delta y(n+1) - \Delta y(n)) \;=\; g - g \;=\; 0.$$

On the other hand, because $\Delta y(n) > 0$, we have $y(n) \ge y(n_1) > 0$, and consequently

$$a_n y(n) \;\le\; a_n y(n_1), \qquad n \ge n_1.$$

The above inequality yields

$$\begin{aligned} 0 \;&=\; \liminf_{n\to\infty} \Delta^2 y(n) \;=\; \liminf_{n\to\infty} a_n y(n) \;\le\; y(n_1) \liminf_{n\to\infty} a_n \\ &<\; (\alpha+\epsilon) y(n_1) \;<\; 0 \end{aligned}$$

for some small $\epsilon > 0$ such that $(\alpha + \epsilon) < 0$. Thus, the case (a_2) is also not possible. Hence, there does not exist any eventually positive solution of (E_7).

The case of eventually negative solution can be reduced, by the linearity of the equation (E_7), to the previous one. Thus, every solution of (E_7) is oscillatory.

(b) Suppose that the statement is false. Then, there exists an increasing sequence of positive integers $\{n_k\}_{k=1}^{\infty}$ such that for every $k \in \mathbb{N}$ one of the following cases hold

$$(b_1) \qquad \begin{cases} y(n_k) \;=\; y(n_k+1) \;=\; 0, \\ y(n_k) \;=\; 0,\; y(n_k+1) \;>\; 0,\; y(n_k+2) \;=\; 0, \\ y(n_k) \;=\; 0,\; y(n_k+1) \;<\; 0,\; y(n_k+2) \;=\; 0; \end{cases}$$

$$(b_2) \qquad \begin{cases} y(n_k) \;=\; 0,\; y(n_k+1) \;>\; 0,\; y(n_k+2) \;<\; 0, \\ y(n_k) \;=\; 0,\; y(n_k+1) < 0,\; y(n_k+2) > 0; \end{cases}$$

$$(b_3) \qquad \begin{cases} y(n_k) \;<\; 0,\; y(n_k+1) \;>\; 0,\; y(n_k+2) \;<\; 0, \\ y(n_k) \;>\; 0,\; y(n_k+1) \;<\; 0,\; y(n_k+2) \;>\; 0. \end{cases}$$

In the case (b_1), we get $y(n_k+2) = 0$, and consequently $y(n) = 0$ for all $n \ge n_k$, which is a contradiction to the nontriviality of the solution y.

In the case (b_2), we get $y(n_k+2) - 2y(n_k+1) = 0$, i.e. $y(n_k+2) = 2y(n_k+1)$ which is a contradiction to the fact that $y(n_k+2)$ and $y(n_k+1)$ are of opposite sign.

In the case (b_3), we get the relation $y(n_k+2) + y(n_k) - a_{n_k}y(n_k) = 2y(n_k+1)$, in which the left hand side is negative (positive) while the right hand side is positive (negative).

The above contradictions prove (b). ∎

Following [251], we note that the general solution of (E_7) can be written as

$$y(n) \;=\; 2^{n-3}(1-a_1)y(1)\sum_{d(n-4,2)} p_i + 2^{n-2}y(2)\sum_{d(n-3,1)} p_i, \qquad n \in \mathbb{N}$$

where $p_i = (-1+a_i)/4$, and

$$\sum_{d(m,k)} x_i \;=\; \sum_{\substack{d_1,\cdots,d_m = 0\\ d_i d_{i+1} = 0\\ i = 1,\cdots,m-1}}^{1} \prod_{i=1}^{m} (x_{i+k})^{d_i}, \qquad m > 1, \;\; k \geq 0,$$

$$\sum_{d(1,k)} x_i \;=\; \sum_{d=0}^{1} (x_{1+k})^d, \qquad \sum_{d(0,k)} x_i \;=\; \sum_{d(-1,k)} x_i \;=\; 1, \qquad \sum_{d(-2,k)} x_i \;=\; 0.$$

From the above representation of the solution of (E_7) the following result is immediate.

Theorem 1.3.2. If there exists an increasing sequence of positive integers $\{n_k\}_{k=1}^{\infty}$ such that one of the sequences

$$\left\{\sum_{d(n_k-4,2)} p_i\right\}_{k=1}^{\infty}, \qquad \text{or} \qquad \left\{\sum_{d(n_k-3,1)} p_i\right\}_{k=1}^{\infty}$$

oscillates, where $p_i = (-1+a_i)/4$, then the equation (E_7) possesses oscillatory solutions.

Remark 1.3.1. Theorem 1.3.1 holds if a is eventually negative, i.e. $a : \mathbb{N} \to \mathbb{R}$ and there exists a $\nu \in \mathbb{N}$ such that $a_n < 0$ for all $n \geq \nu$. Also the condition $\liminf_{n\to\infty} a_n = \alpha < 0$ can be replaced by $\liminf_{n\to\infty} a_n = -\infty$.

Chebyshev Polynomials. Chebyshev polynomials of the first kind $\{T_n(x)\}_{n=0}^{\infty}$ satisfy the difference equation

$$T_{n+2}(x) \;=\; 2xT_{n+1}(x) - T_n(x), \qquad n \in \mathbb{N}\cup\{0\}, \;\; x \in [-1,1] \qquad (E_8)$$

where $T_0(x) = 1,\ T_1(x) = x$.

We arrange the equation (E_8) as

$$T_{n+2}(x) - 2xT_{n+1}(x) + x^2T_n(x) \ = \ x^2T_n(x) - T_n(x),$$

take any $x \in (-1,1)\backslash\{0\}$, and put $x \to a,\ T_n(a) \to y(n)$, to obtain

$$y(n+2) - 2ay(n+1) + a^2y(n) \ = \ (a^2-1)y(n).$$

Now on dividing the above equation by a^{n+2} and substituting $z(n) = y(n)/a^n$, we get

$$\Delta^2 z(n) \ = \ \frac{a^2-1}{a^2}z(n), \qquad n \in \mathbb{N} \cup \{0\}. \tag{1.3.1}$$

Theorem 1.3.3. For every $x \in [-1,1)$ the sequence $\{T_n(x)\}_{n=1}^{\infty}$ is oscillatory.

Proof. We note that $(a^2-1)/a^2 < 0$ for $a \in (-1,1)\backslash\{0\}$, therefore for the equation (1.3.1) the assumptions of Theorem 1.3.1 are satisfied. Let $x \in (0,1)$. Since $z(n) = T_n(x)/x^n,\ x^n > 0$, and $\{z(n)\}_{n=1}^{\infty}$ oscillates by Theorem 1.3.1, it is clear that $\{T_n(x)\}_{n=1}^{\infty}$ also oscillates. For $x = 0$, equation (E_8) reduces to $T_{n+2}(0) = -T_n(0)$. Now the initial condition $T_0(0) = 1$ yields

$$T_2(0) = -1, \ T_4(0) = 1, \ \cdots, \ T_{4k}(0) = 1, \ T_{4k+2}(0) = -1, \ k = 0,1,\cdots.$$

Next, let $x \in (-1,0)$. By Theorem 1.3.1 part (b) the sequence $\{z(n)\}_{n=1}^{\infty}$ where $z(n) = T_n(x)/x^n$ is not quickly oscillatory, i.e. in every string of three successive, different from zero, terms

$$\frac{T_k(x)}{x^k}, \quad \frac{T_{k+1}(x)}{x^{k+1}}, \quad \frac{T_{k+2}(x)}{x^{k+2}}$$

we have

$$\frac{T_k(x)}{x^k}\frac{T_{k+1}(x)}{x^{k+1}} > 0, \quad \text{or} \quad \frac{T_{k+1}(x)}{x^{k+1}}\frac{T_{k+2}(x)}{x^{k+2}} > 0.$$

However, since for $x \in (-1,0)$ the products $x^kx^{k+1},\ x^{k+1}x^{k+2}$ are negative, it follows that $T_k(x)T_{k+1}(x) < 0$, or $T_{k+1}(x)T_{k+2}(x) < 0$. This means that the sequence $\{T_n(x)\}_{n=1}^{\infty}$ oscillates for $x \in (-1,0)$.

Finally, let $x = -1$. Then, the equation (E_8) takes the form

$$T_{n+2}(-1) \ = \ -2T_{n+1}(-1) - T_n(-1).$$

Suppose that $\{T_n(-1)\}_{n=1}^{\infty}$ is nonoscillatory, and let $T_n(-1) < 0$ for all $n \geq n_1$. Then, $T_{n_1+2}(-1) = -2T_{n_1+1}(-1) - T_{n_1}(-1) > 0$, which is a contradiction to $T_{n_1+2}(-1) < 0$. ∎

Chebyshev polynomials of the second kind $\{U_n(x)\}_{n=0}^{\infty}$ satisfy the same difference equation (E_8), however $U_0(x) = 1$, $U_1(x) = 2x$.

Theorem 1.3.4. For every $x \in [-1, 1)$ the sequence $\{U_n(x)\}_{n=1}^{\infty}$ is oscillatory.

Proof. From Theorem 1.3.3 the sequence $\{T_n(x)\}_{n=0}^{\infty}$ is oscillatory for each fixed $x \in [-1, 1)$. Therefore, for any $a \in [-1, 1)$ the equation

$$y_{n+2} \;=\; 2ay_{n+1} - y_n$$

possesses an oscillatory solution, namely $\{T_n(a)\}_{n=0}^{\infty}$. Further, this equation can be written in self–adjoint form

$$\Delta(p_n \Delta y_n) \;=\; q_n y_{n+1}, \qquad n \in \mathbb{N} \cup \{0\} \tag{1.3.2}$$

where $p_n = 1$ and $q_n = 2a - 2$ for all $n \in \mathbb{N} \cup \{0\}$. Therefore, from a well–known result [2] every solution of (1.3.2) is oscillatory. Since for each $a \in [-1, 1)$ the sequence $\{U_n(a)\}_{n=0}^{\infty}$ also satisfies (1.3.2), it has to be oscillatory. ∎

Hermite Polynomials. These polynomials $\{H_n(x)\}_{n=0}^{\infty}$ satisfy the difference equation

$$H_{n+2}(x) - xH_{n+1}(x) + (n+1)H_n(x) \;=\; 0, \qquad n \in \mathbb{N} \cup \{0\}, \;\; x \in \mathbb{R} \tag{E_9}$$

where $H_0(x) = 1$, $H_1(x) = x$.

Theorem 1.3.5. For every $x \in \mathbb{R}$ the sequence $\{H_n(x)\}_{n=0}^{\infty}$ is oscillatory.

Proof. For $x = 0$, equation (E_9) reduces to $H_{n+2}(0) = -(n+1)H_n(0)$ from which the oscillation follows. For $x = a \neq 0$, we let $z_n = H_n(a)$ so that the equation (E_9) can be written as

$$z_{n+2} - az_{n+1} + (n+1)z_n \;=\; 0.$$

On multiplying the above equation by $2^{n+2}/a^{n+2}$ and arranging the terms, we obtain

$$\frac{2^{n+2}}{a^{n+2}} z_{n+2} - 2\frac{2^{n+1}}{a^{n+1}} z_{n+1} + \frac{2^n}{a^n} z_n \;=\; \frac{2^n}{a^n} z_n - \frac{4}{a^2}(n+1)\frac{2^n}{a^n} z_n,$$

which is the same as

$$\Delta^2 y_n \;=\; \left(1 - \frac{4(n+1)}{a^2}\right) y_n, \tag{1.3.3}$$

where $y_n = (2^n/a^n)\, z_n$. This equation is of the form (E_7). Furthermore, if $a \neq 0$, then $\liminf_{n\to\infty}\left(1 - \dfrac{4(n+1)}{a^2}\right) = -\infty$, and $\left(1 - \dfrac{4(n+1)}{a^2}\right) < 0$ for $n > (a^2/4) - 1$. Hence, by Theorem 1.3.1 together with Remark 1.3.1 it follows that every solution $\{y_n\}_{n=0}^{\infty}$ of (1.3.3) is oscillatory. But, since $y_n = (2^n/a^n)\, H_n(a)$, we conclude that $\{H_n(a)\}_{n=0}^{\infty}$ is oscillatory for $a \in \mathbb{R}_+$.

For $a \in \mathbb{R}_-$, we have

$$\frac{2^n}{a^n}\frac{2^{n+1}}{a^{n+1}} < 0, \qquad n \in \mathbb{N}. \tag{1.3.4}$$

Further, from part (b) of Theorem 1.3.1 we find that for all $n \in \mathbb{N}$ in every string of three successive terms different from zero

$$\frac{2^n}{a^n}H_n(a), \quad \frac{2^{n+1}}{a^{n+1}}H_{n+1}(a), \quad \frac{2^{n+2}}{a^{n+2}}H_{n+2}(a),$$

we have

$$\frac{2^n}{a^n}H_n(a)\frac{2^{n+1}}{a^{n+1}}H_{n+1}(a) > 0, \quad \text{or} \quad \frac{2^{n+1}}{a^{n+1}}H_{n+1}(a)\frac{2^{n+2}}{a^{n+2}}H_{n+2}(a) > 0.$$

The oscillatory behavior of the sequence $\{H_n(x)\}_{n=0}^{\infty}$ for $a \in \mathbb{R}_-$ now follows from (1.3.4).

Thus, for each $x \in \mathbb{R}$ the sequence $\{H_n(x)\}_{n=0}^{\infty}$ is oscillatory. ∎

Legendre Polynomials. These polynomials $\{P_n(x)\}_{n=0}^{\infty}$ satisfy the difference equation

$$(n+2)P_{n+2}(x) = (2n+3)xP_{n+1}(x) - (n+1)P_n(x), \quad n \in \mathbb{N}\cup\{0\}, \quad x \in [-1,1] \tag{E_{10}}$$

where $P_0(x) = 1$, $P_1(x) = x$.

Theorem 1.3.6. For every $x \in (-1,1)$ the sequence $\{P_n(x)\}_{n=0}^{\infty}$ is oscillatory.

Proof. For $x = 0$ we get the equation $P_{n+2}(0) = -\left(\dfrac{n+1}{n+2}\right)P_n(0)$ from which the oscillation follows directly. For $x \neq 0$, we let $x = a$, $y_0 = P_0(a)$, $y_n = \dfrac{2^n}{a^n}\displaystyle\prod_{j=0}^{n-2}\dfrac{j+2}{2j+3}P_n(a)$ for $n \geq 1$, so that the equation (E_{10}) can be written as

$$\Delta^2 y_n = \left(1 - \frac{4(n+1)^2}{a^2(2n+1)(2n+3)}\right) y_n. \quad n \in \mathbb{N}\cup\{0\}.$$

Since for arbitrary $a \in (-1,1)\backslash\{0\}$,

$$\lim_{n\to\infty}\left(1-\frac{4(n+1)^2}{a^2(2n+1)(2n+3)}\right) = 1-\frac{1}{a^2} < 0,$$

from Theorem 1.3.1 and Remark 1.3.1 it follows that for every $a \in (-1,1)\backslash \{0\}$ the sequence $\{P_n(a)\}_{n=0}^{\infty}$ is oscillatory. ∎

1.4. Oscillation of Functions Recurrence Equations

Theorems 1.3.3 – 1.3.6 prove the oscillation of orthogonal polynomials in the point–wise sense. We define such oscillatory behavior in the following:

Definition 1.4.1. The sequence $\{f_n(x)\}_{n=1}^{\infty}$ of real–valued functions defined on the set $\mathbf{D}$ is said to be *oscillatory in the point–wise sense* on the set $\mathbf{D}$ if for each $a \in \mathbf{D}$ the sequence $\{f_n(a)\}_{n=1}^{\infty}$ is oscillatory, i.e. for every $a \in \mathbf{D}$ there exists an increasing sequence of positive integers $\eta(a) = \{n_k(a)\}_{k=1}^{\infty}$ such that $f_{n_k}(a)f_{n_k+1}(a) \le 0$.

In our next definition we shall introduce oscillatory behavior in the global sense.

Definition 1.4.2. The sequence $\{f_n(x)\}_{n=1}^{\infty}$ of real–valued functions defined on the set $\mathbf{D}$ is said to be *oscillatory in the global sense* if there exists an increasing sequence of positive integers $\eta = \{n_k\}_{k=1}^{\infty}$ such that $f_{n_k}(x)f_{n_k+1}(x) \le 0$.

It is clear that if the sequence $\{f_n(x)\}_{n=1}^{\infty}$ is global oscillatory on the set $\mathbf{D}$, then it is point–wise oscillatory on the same set. However, generally the converse does not hold.

Definition 1.4.3. The sequence $\{f_n(x)\}_{n=1}^{\infty}$ of real–valued functions defined on the set $\mathbf{D}$ is said to be *nonoscillatory* if for each $a \in \mathbf{D}$ the sequence $\{f_n(a)\}_{n=1}^{\infty}$ is nonoscillatory (eventually of constant sign).

We remark that oscillatory and nonoscillatory sequences do not exhaust all possible behavior of sequences of functions. Indeed, there exist sequences of functions defined on the set $\mathbf{D}$ such that they are point–wise oscillatory on some proper subset $\mathbf{D_1}$ of $\mathbf{D}$ and nonoscillatory on the nonempty subset $\mathbf{D_2} \subset \mathbf{D}$, where $\mathbf{D_1} \cup \mathbf{D_2} = \mathbf{D}$.

Example 1.4.1. Let $\mathbf{D} = [0,\infty)$. The sequence $\{(-1)^n x^n\}_{n=1}^{\infty}$, $x \in [0,\infty)$ which is the solution of the problem

$$f_{n+1}(x) = xf_n(x), \quad f_1(x) = -x$$

is global oscillatory on the set $\mathbf{D}$.

Example 1.4.2. The sequence of functions defined by

$$f_n(x) = (-1)^n \left(x^2 - \frac{1}{n^2}\right), \qquad x \in \mathbf{D} = [-1, 1], \quad n \in \mathbb{N}$$

is point–wise oscillatory, but not global oscillatory on the set $\mathbf{D}$.

Many results known for scalar difference equations have their analogs for difference equations on function spaces. In fact, now we shall present such analogs whose scalar forms appeared in [2,252]. For this, let $\mathbf{X}$ be the set of real–valued functions defined on $\mathbf{D}$. For $a \in \mathbf{X}$ and any sequence $\{y_n\}_{n=1}^{\infty}$ of elements of the set $\mathbf{X}$ the difference operator Δ_a is defined exactly as in the scalar case, i.e. $\Delta_a y_n = y_{n+1} - a y_n$, $n \in \mathbb{N}$. For $u, v \in \mathbf{X}$ we say $u > v$ if $u(x) > v(x)$ for all $x \in \mathbf{D}$. By θ we shall denote the null function of $\mathbf{D}$.

The difference equation we shall consider is of the form

$$\Delta_a^2 y_n = F(n, y_n, \Delta_b y_n), \qquad n \in \mathbb{N} \tag{E_{11}}$$

where a, b are some given elements of $\mathbf{X}$, and $F : \mathbb{N} \times \mathbf{X}^2 \to \mathbf{X}$. For the equation (E_{11}) we shall consider only the nontrivial solutions, i.e. solutions for which $\sup_{n \geq m}[\ |y_n|\ ] > \theta$ for every $m \in \mathbb{N}$.

Theorem 1.4.1. Let $a > \theta$ and

$$\begin{cases} F(n,u,v) = \theta \text{ for } (n,u,v) \in \mathbf{S} = \mathbb{N} \times \left\{(u,v) \in \mathbf{X}^2 : v + (b-a)u = \theta\right\} \\ F(n,u,v)\,[v + (b-a)u] + a\,[v + (b-a)u]^2 > \theta \\ \hfill \text{for } (n,u,v) \in \mathbb{N} \times \left(\mathbf{X}^2 \backslash \mathbf{S}\right). \end{cases} \tag{1.4.1}$$

Then, every solution of (E_{11}) is nonoscillatory.

Proof. Condition $v + (b-a)u = \theta$ which describes the set $\mathbf{S}$ applying to (E_{11}) is equivalent to $y_{n+1} - a y_n = \theta$. We denote by $\mathbf{W}$ the set of all solutions of (E_{11}), by $\mathbf{W}_1$ a subset of $\mathbf{W}$ such that $y \in \mathbf{W}_1$, if and only if, there exists a $k \in \mathbb{N}$ for which $y_{k+1} - a y_k = \theta$, and by $\mathbf{W}_2 = \mathbf{W} \backslash \mathbf{W}_1$ the complement set of $\mathbf{W}_1$ with respect to $\mathbf{W}$.

Now let $y \in \mathbf{W}_1$, so that for some $s \in \mathbb{N}$ we have $y_{s+1} - a y_s = \theta$. Therefore, from (E_{11}) in view of (1.4.1), we get

$$\Delta_a^2 y_s = y_{s+2} - 2a y_{s+1} + a^2 y_s = y_{s+2} - 2a y_{s+1} + a y_{s+1} = y_{s+2} - a y_{s+1} = \theta.$$

Continuing in this way, we obtain $y_{s+i} - a y_{s+i-1} = \theta$, i.e. $y_{s+i} = a^i y_s$ for all $i \geq 1$. Since $a > \theta$, we also have $a^i > \theta$. Therefore, $\operatorname{sgn} y_{s+i}(x) = \operatorname{sgn} y_s(x)$ for every $x \in \mathbf{D}$ and this solution is nonoscillatory. Let $y \in \mathbf{W}_2$, and suppose that there exists a $\xi \in \mathbf{D}$ such that the sequence

$\{y_n(\xi)\}_{n=1}^{\infty}$ is oscillatory. Let $m \in \mathbb{N}$ be such that $y_m(\xi) > 0$, $y_{m+1}(\xi) \leq 0$. Then, it is clear that

$$\Delta_{a(\xi)} y_m(\xi) \;<\; 0. \tag{1.4.2}$$

Now multiplying (E_{11}) by $\Delta_a y_m$, using the relations

$$\Delta_a^2 y_m \;=\; \Delta_a y_{m+1} - a\Delta_a y_m, \qquad \Delta_a y_m \;=\; \Delta_b y_m + (b-a)y_m,$$

and (1.4.1), to obtain (at the point ξ)

$$\begin{aligned}\Delta_{a(\xi)} y_{m+1}(\xi)\Delta_{a(\xi)} y_m(\xi) \;&=\; F\left(m, y_m(\xi), \Delta_{b(\xi)} y_m(\xi)\right) \\ &\quad\times \left[\Delta_{b(\xi)} y_m(\xi) + (b(\xi) - a(\xi))y_m(\xi)\right] \\ &\quad+ a(\xi)\left[\Delta_{b(\xi)} y_m(\xi) + (b(\xi) - a(\xi))y_m(\xi)\right]^2 \\ &> \; 0.\end{aligned}$$

Therefore, by (1.4.2) we find $\Delta_{a(\xi)} y_{m+1}(\xi) < 0$. On repeating this reasoning, we get $\Delta_{a(\xi)} y_n(\xi) < 0$ for all $n \geq m$. This means that $y_n(\xi) < 0$ for all $n > m+1$, but this contradicts our assumption. The proof for the case $y_m(\xi) \geq 0$, $y_{m+1}(\xi) < 0$ is similar. ∎

We note that for the solutions of (E_{11}) which belong to the set $\mathbf{W}_2$ we have proved more than we have stated in the theorem. Indeed, we have proved that these solutions are nonoscillatory at each point $x \in \mathbf{D}$.

A dual result of Theorem 1.4.1 is the following:

Theorem 1.4.2. Let $a < \theta$ and

$$\begin{cases} F(n,u,v) = \theta \text{ for } (n,u,v) \in \mathbf{S} = \mathbb{N} \times \left\{(u,v) \in \mathbf{X}^2 : v + (b-a)u = \theta\right\} \\ F(n,u,v)\left[v + (b-a)u\right] + a\left[v + (b-a)u\right]^2 < \theta \\ \qquad\qquad\qquad\qquad \text{for } (n,u,v) \in \mathbb{N} \times \left(\mathbf{X}^2 \backslash \mathbf{S}\right). \end{cases} \tag{1.4.3}$$

Then, every solution of (E_{11}) is point-wise oscillatory.

Proof. Let y be a solution of (E_{11}) which is not point-wise oscillatory. Then, there should be at least one point $\xi \in \mathbf{D}$ such that the sequence $\{y_n(\xi)\}_{n=1}^{\infty}$ is nonoscillatory, say, eventually positive for all $n \geq m$. Let m be even. The difference $\Delta_{a(\xi)} y_m(\xi)$ cannot be equal to zero, if not, then the equality $y_{m+1}(\xi) = a(\xi)y_m(\xi)$ leads to the contradiction that $y_{m+1}(\xi) < 0$. Now since $\Delta_{a(\xi)} y_m(\xi) \neq 0$, $(y_m, \Delta_b y_m) \notin \left\{(u,v) \in \mathbf{X}^2 : v + (b-a)u = \theta\right\}$. On multiplying (E_{11}) (taken at the point ξ) by $\Delta_{a(\xi)} y_m(\xi)$, and using (1.4.3), we get

$$\begin{aligned}\Delta_{a(\xi)} y_{m+1}(\xi)\Delta_{a(\xi)} y_m(\xi) = \Delta_{a(\xi)} y_m(\xi) F\left(m, y_m(\xi), \Delta_{b(\xi)} y_m(\xi)\right) \\ + a(\xi)\left(\Delta_{a(\xi)} y_m(\xi)\right)^2 \;<\; 0.\end{aligned}$$

Next, since $\Delta_a y_m = a^{m+1}\Delta\left(y_m/a^m\right)$, it follows that

$$(a(\xi))^{m+2}\Delta\left(\frac{y_{m+1}(\xi)}{(a(\xi))^{m+1}}\right)(a(\xi))^{m+1}\Delta\left(\frac{y_m(\xi)}{(a(\xi))^m}\right) \ < \ 0$$

and hence

$$\Delta\left(\frac{y_{m+1}(\xi)}{(a(\xi))^{m+1}}\right)\Delta\left(\frac{y_m(\xi)}{(a(\xi))^m}\right) \ > \ 0.$$

If $\Delta\left(y_m(\xi)/(a(\xi))^m\right) > 0$, then $\left(y_{m+1}(\xi)/(a(\xi))^{m+1}\right) > \left(y_m(\xi)/(a(\xi))^m\right) > 0$. Therefore, $y_{m+1}(\xi) < 0$, and we obtain a contradiction. Hence it is necessary that $\Delta\left(y_m(\xi)/(a(\xi))^m\right) < 0$. But, then $\Delta\left(y_{m+1}(\xi)/(a(\xi))^{m+1}\right) < 0$, i.e. $\left(y_{m+2}(\xi)/(a(\xi))^{m+2}\right) < \left(y_{m+1}(\xi)/(a(\xi))^{m+1}\right) < 0$, and hence $y_{m+2}(\xi) < 0$. This contradiction completes the proof. ■

Theorem 1.4.3. Let $a > \theta$ and for $(n,u,v) \in \left(\mathbb{N}\times\mathbf{X}^2\right)\setminus\left(\mathbb{N}\times\{(u,v)\in \mathbf{X}^2 : v+bu=\theta\}\right)$

$$F(n,u,v)(v+bu) + a(v+bu)[v+(b-a)u] \ > \ \theta. \tag{1.4.4}$$

Then, every solution of (E_{11}) is nonoscillatory.

Proof. Let y be an arbitrary solution of (E_{11}). Condition $v+bu=\theta$ applying to (E_{11}) is equivalent to $y_{n+1} = \theta$. Since we are considering nontrivial solutions only, there exists a $m \in \mathbb{N}$ such that $y_{m+1} \neq \theta$. For this m we have $\Delta_b y_m + by_m = y_{m+1} \neq \theta$. Hence, by (1.4.4) it follows that

$$F\left(m,y_m,\Delta_b y_m\right)\left(\Delta_b y_m + by_m\right) + a\left(\Delta_b y_m + by_m\right)\left[\Delta_b y_m + (b-a)y_m\right] \ > \ \theta. \tag{1.4.5}$$

However, since

$$\begin{aligned} F\left(m,y_m,\Delta_b y_m\right)\left(\Delta_b y_m + by_m\right) &+ a\left(\Delta_b y_m + by_m\right)\left[\Delta_b y_m + (b-a)y_m\right] \\ &= F\left(m,y_m,\Delta_b y_m\right)y_{m+1} + ay_{m+1}\Delta_a y_m \end{aligned}$$

from (1.4.5) it follows that

$$F\left(m,y_m,\Delta_b y_m\right)y_{m+1} + ay_{m+1}\Delta_a y_m \ > \ \theta.$$

Now from (E_{11}), we have

$$F\left(m,y_m,\Delta_b y_m\right)y_{m+1} + ay_{m+1}\Delta_a y_m \ = \ y_{m+1}\Delta_a y_{m+1}$$

and hence

$$y_{m+1}\Delta_a y_{m+1} \ > \ \theta. \tag{1.4.6}$$

From (1.4.6) it follows that $y_{m+1}(\xi) \neq 0$ for every $\xi \in \mathbf{D}$. Now let for some $\xi \in \mathbf{D}$, $y_{m+1}(\xi) > 0$, then by (1.4.6) we have $\Delta_{a(\xi)}y_{m+1}(\xi) > 0$, which implies that $y_{m+2}(\xi) > a(\xi)y_{m+1}(\xi) > 0$. On repeating the above reasoning with y_{m+2} instead of y_{m+1} we get $y_{m+3}(\xi) > 0$. Inductively, we obtain $y_n(\xi) > 0$ for all $n > m$. Similarly, if $y_{m+1}(\eta) < 0$ for some $\eta \in \mathbf{D}$ then $y_n(\eta) < 0$ for all $n > m$. Therefore, for every $\xi \in \mathbf{D}$ the sequence $\{y_n(\xi)\}_{n=1}^{\infty}$ is nonoscillatory. ∎

In the following result we shall change slightly the definition of ordering in the space $\mathbf{X}$. For $u,\ v \in \mathbf{X}$ we shall say $u \succ v$ if $u(x) \geq v(x)$ for all $x \in \mathbf{D}$ and there exists at least one point $\xi \in \mathbf{D}$ such that $u(\xi) > v(\xi)$.

Theorem 1.4.4. Let $a(\xi) < 0$ for all $\xi \in \mathbf{D}$ and for $(n, u, v) \in (\mathbb{N} \times \mathbf{X}^2) \setminus (\mathbb{N} \times \{(u,v) \in \mathbf{X}^2 : v + bu = \theta\})$

$$F(n,u,v)(v+bu) + a(v+bu)[v+(b-a)u] \prec \theta. \tag{1.4.7}$$

Then, every solution of (E_{11}) is point-wise oscillatory over the set it is equal to zero.

Proof. Let y be an arbitrary solution of (E_{11}). As in Theorem 1.4.3 condition $v + bu = \theta$ is equivalent to $y_{n+1} = \theta$. For the considered nontrivial solution, there exists a $m \in \mathbb{N}$ such that $y_{m+1} \neq \theta$. For this m we have $\Delta_b y_m + b y_m = y_{m+1} \neq \theta$, i.e. $v + bu \neq \theta$, and hence by (1.4.7), we have

$$F(m, y_m, \Delta_b y_m)(\Delta_b y_m + b y_m) + a(\Delta_b y_m + b y_m)[\Delta_b y_m + (b-a) y_m] \prec \theta,$$

which implies that

$$\begin{aligned} & y_{m+1} \Delta_a y_{m+1} \\ &= F(m, y_m, \Delta_b y_m)\, y_{m+1} + a y_{m+1} \Delta_a y_m \\ &= F(m, y_m, \Delta_b y_m)(\Delta_b y_m + b y_m) + a(\Delta_b y_m + b y_m) \\ &\quad \times [\Delta_b y_m + (b-a) y_m] \prec \theta. \end{aligned} \tag{1.4.8}$$

The above inequality implies that $y_{m+2} \neq \theta$, because $y_{m+2} = \theta$ leads to a contradiction that $-a y_{m+1}^2 \prec \theta$. Hence, $y_n \neq \theta$ for all $n \geq m+1$.

Next, we note that the inequality (1.4.8) is the same as

$$y_{m+1} y_{m+2} - a y_{m+1}^2 \prec \theta. \tag{1.4.9}$$

We denote by $\mathbf{D}_+(n)$, $\mathbf{D}_-(n)$, $\mathbf{D}_0(n)$ the disjoint subsets of $\mathbf{D}$ such that $y_n(\xi) > 0$ for $\xi \in \mathbf{D}_+(n)$, $y_n(\xi) < 0$ for $\xi \in \mathbf{D}_-(n)$, and $y_n(\xi) = 0$ for $\xi \in \mathbf{D}_0(n)$, respectively, furthermore by $H_{y,m}(\xi) = y_m y_{m+1} - a y_m^2$, and $D_+^{H,y}(m) = \{\xi \in \mathbf{D} : H_{y,m}(\xi) > 0\}$, $D_-^{H,y}(m) = \{\xi \in \mathbf{D} : H_{y,m}(\xi) < 0\}$, $D_0^{H,y}(m) = \{\xi \in \mathbf{D} : H_{y,m}(\xi) = 0\}$. It is clear that $D_+^{H,y}(m)$ is empty. Let $\xi \in \mathbf{D}_+(m+1) \cap D_-^{H,y}(m+1)$, then by (1.4.9) we obtain $y_{m+2}(\xi) < a(\xi) y_{m+1}(\xi) < 0$. Similarly, for $\xi \in \mathbf{D}_+(m+1) \cap D_0^{H,y}(m+1)$, we get $y_{m+2}(\xi) = a(\xi) y_{m+1}(\xi) < 0$. These two inequalities imply that $\mathbf{D}_+(m+1) \subset \mathbf{D}_-(m+2)$. Similarly, we have $\mathbf{D}_-(m+1) \subset \mathbf{D}_+(m+2)$. Therefore, the sequence $\{y_n(\xi)\}_{n=1}^{\infty}$ is eventually quickly oscillatory for $\xi \in \bigcup_{j=1}^{\infty}(\mathbf{D}_+(j) \cup \mathbf{D}_-(j))$ and is a null sequence for $\xi \in \bigcap_{j=1}^{\infty}(\mathbf{D}\setminus(\mathbf{D}_+(j) \cup \mathbf{D}_-(j)))$. This completes the proof. ∎

From the Definition 1.2.1 it is clear that the oscillation of the sequence y around a implies that there are infinitely many terms y_n which are greater than a and there are infinitely many less than equal to a. This means that the sequence of differences, i.e. $\{\Delta y_n\}_{n=1}^{\infty}$ is oscillatory around 0. In fact, if $y_n < a$ and $y_{n+1} > a$ then $y_{n+1} - y_n > 0$, while if $y_n > a$ and $y_{n+1} < a$ then $y_{n+1} - y_n < 0$. Thus, oscillatory behavior of the sequence $\{\Delta y_n\}_{n=1}^{\infty}$ is a necessary condition for the oscillation of $\{y_n\}_{n=1}^{\infty}$, but not a sufficient condition. For example, for the nonoscillatory sequence $\{y_n\}_{n=1}^{\infty} = \{1, 3, 2, 4, 3, 5, \cdots\}$, we have $\{\Delta y_n\}_{n=1}^{\infty} = \{2, -1, 2, -1, \cdots\}$ which is oscillatory. This condition is also necessary for the global oscillation of the sequences of real-valued functions. From this we can immediately deduce that no solutions of the equations $y_{n+1}(x) - y_n(x) = x^2 + 1$ and $y_{n+1}(x) - y_n(x) = -1/n$ can be oscillatory. As a consequence of this we have the following necessary condition:

Theorem 1.4.5. If the sequence $\{y_n\}_{n=1}^{\infty}$ is oscillatory, then the sequences of differences $\{\Delta^k y_n\}_{n=1}^{\infty}$ are oscillatory for all $k \in \mathbb{N}$.

1.5. Oscillation in Ordered Sets

Consider the sequence $\{\mathcal{K}(n)\}_{n=1}^{\infty}$ of subsets of the plane Oxy defined as follows:

$$\mathcal{K}(n) = \{(x, y) : x^2 + y^2 = (1 - (-1)^n/n)^2\}, \quad n \in \mathbb{N}$$

i.e. $\mathcal{K}(1) = x^2 + y^2 = 2^2$, $\mathcal{K}(2) = x^2 + y^2 = (1 - 1/2)^2$, $\mathcal{K}(3) = x^2 + y^2 = (1 + 1/3)^2$, $\mathcal{K}(4) = x^2 + y^2 = (1 - 1/4)^2, \cdots$. We can say that this sequence of circles oscillates around the set of points which form the circle described by the equation $x^2 + y^2 = 1$ $(\mathcal{K}(\infty))$.

In general, the system oscillates around some element a if it is sometimes 'greater' and sometimes 'less' than a. Thus, in the system we consider, there should be a possibility to compare at least some elements.

Let $\mathbf{X}$ be any nonempty set. We recall that a relation $x \prec y$, defined for some pairs (x, y) of elements of the set $\mathbf{X}$ is called an order relation (partial order relation) in $\mathbf{X}$ if the following conditions are satisfied: (i) $x \prec x$ for any $x \in \mathbf{X}$ (reflexivity), (ii) $x \prec y$ and $y \prec z$ implies $x \prec z$ (transitivity), and (iii) $x \prec y$ and $y \prec x$ implies $x = y$ (antisymmetry). A set $\mathbf{X}$ with an order relation is called an ordered set. If $\mathbf{X}$ is an ordered set and if for given x, $y \in \mathbf{X}$ the relations $x \prec y$ and $y \prec x$ do not hold then these elements are called incomparable. An ordered set $\mathbf{X}$ is called a totally ordered set if it has no incomparable elements. Any totally ordered subset $\mathbf{Y}$ of the ordered set $\mathbf{X}$ is called a chain in $\mathbf{X}$.

If in the set $\mathbf{K}$ of circles on the plane described by the equation $x^2 + y^2 = r^2$, $r \in \mathbb{R}$, for any $K(r) = \{(x,y) : x^2 + y^2 = r^2\}$, $K(s) = \{(x,y) : x^2 + y^2 = s^2\}$, we say $K(r) \prec K(s)$ if $r \leq s$, then we have endowed the set $\mathbf{K}$ with the order relation. It is clear that in this set, for our sequence $\{\mathcal{K}(n)\}_{n=1}^{\infty}$, defined earlier, we have $\mathcal{K}(2k) \prec \mathcal{K}(\infty)$ and $\mathcal{K}(\infty) \prec \mathcal{K}(2k-1)$ for all $k \in \mathbb{N}$.

Definition 1.5.1. Let $(\mathbf{X}, \prec)$ be an ordered set. A function $y : \mathbb{N} \to \mathbf{X}$ is said to *oscillate around* $a \in \mathbf{X}$ if there exist infinite increasing sequences $\{n_k\}_{k=1}^{\infty}$ and $\{m_k\}_{k=1}^{\infty}$ of positive integers such that $y_{n_k} \prec a$ and $a \prec y_{m_k}$ for every $k \in \mathbb{N}$.

If a, b are two elements of $(\mathbf{X}, \prec)$, such that $a \prec b$, then the set $[a,b] = \{x \in \mathbf{X} : a \prec x \prec b\}$ is called an interval, and a and b its extremities. Definition 1.5.1 implies that the element a belongs to infinite number of intervals with extremities are the elements of the sequence y. We also note that these definitions allow the sequence y to contain some incomparable elements.

If all elements of the sequence y belong to the same chain and the cardinal of the set of all values of the sequence y is finite, then this sequence oscillates (or is constant starting from some term) around some element of the sequence.

Two ordered sets $(\mathbf{X}, \prec)$ and $(\mathbf{Y}, <)$ are said to be isomorphic if there exists a one–to–one mapping f of $\mathbf{X}$ onto $\mathbf{Y}$ such that $x \prec y$ implies $f(x) < f(y)$. Let $(\mathbb{R}, \leq)$ be the set of reals with the natural ordering, and let $(\mathbf{Y}, \prec)$ be any chain in the ordered set $(\mathbf{X}, \prec)$, isomorphic with $\mathbb{R}$. Let $f : \mathbb{R} \to \mathbf{Y}$ be an isomorphism such that $f(0) = e \in \mathbf{Y}$.

Definition 1.5.2. A sequence $\{y_n\}_{n=1}^{\infty}$ of elements of the chain $\mathbf{Y}$ is said to $(f; \mathbb{R}, \leq)$*-oscillate* if the sequence of reals $\{f^{-1}(y_n)\}_{n=1}^{\infty}$ oscillates.

Example 1.5.1. Let $\mathbf{X}$ be the set of lines on the plane Oxy not parallel to the Oy axis, and $\ell = \{(x,y) : y = mx + k$ for some fixed m, $k \in \mathbb{R}$, $x, y \in \mathbb{R}\}$. Let $\ell_1 = \{(x,y) : y = m_1 x + k_1\}$ and $\ell_2 = \{(x,y) : y = m_2 x + k_2\}$. We shall write $\ell_1 \prec \ell_2$ if $(m_1 + k_1) \leq (m_2 + k_2)$. Since this relation does not satisfy antisymmetry property it is not an order relation on $\mathbf{X}$. Let $\mathbf{Y} \subset \mathbf{X}$ be such that $\ell \in \mathbf{Y}$, if and only if, there exists $a \in \mathbb{R}$ such that $\ell = \{(x,y) : y = a(x - 0.5),\ x \in \mathbb{R}\}$. It is clear that $(\mathbf{Y}, \prec)$ is a totally ordered subset of $(\mathbf{X}, \prec)$. Next, an isomorphism f between $(\mathbb{R}, \leq)$ and $(\mathbf{Y}, \prec)$ can be defined as follows:

$f(r) = \{(x,y) : y = r(x-0.5)\}$, $r \in \mathbb{R}$. Now let the sequence $\{r_n\}_{n=1}^{\infty}$ be oscillatory in $\mathbb{R}$, then the sequence $\{f(r_n)\}_{n=1}^{\infty}$ is oscillatory in $(\mathbf{Y}, \prec)$, i.e. $(f; \mathbb{R}, \leq)$–oscillatory.

Example 1.5.2. In the space $\mathbf{X}$ of real, linear, and continuous functions on $[0,1]$, for the initial value problem

$$\begin{aligned} y(n+1,t) &= -y(n,t) + [1+(-1)^n](t-0.5) \\ y(1,t) &= t, \quad t \in [0,1] \end{aligned} \tag{1.5.1}$$

we shall consider the following problem: Is it possible to define on $\mathbf{X}$ (on some subset $\mathbf{Y}$ ($\mathbf{Y} \subseteq \mathbf{X}$), containing the solution $\{y(n,t)\}_{n=1}^{\infty}$ of (1.5.1)), an order relation $\prec$, and find an isomorphism such that $\{y(n,t)\}_{n=1}^{\infty}$ is $(f; \mathbb{R}, \leq)$–oscillatory ?

The set $\mathbf{X}$ can be totally ordered in the following manner: Let $\ell_1 = \{(t,y) : y = m_1 t + k_1\}$, $\ell_2 = \{(t,y) : y = m_2 t + k_2\}$. We say $\ell_1 \prec \ell_2$ if $m_1 < m_2$, and in the case $m_1 = m_2$ if $k_1 \leq k_2$. We define a subset $\mathbf{Y} \subseteq \mathbf{X}$ as follows:

$$\ell \in \mathbf{Y} \text{ if and only if } \ell = \left\{(t,y) : y = at - \frac{a}{2} + \frac{\operatorname{sgn}(a)}{2}\right\} \text{ for some } a \in \mathbb{R}.$$

Since every subset of totally ordered set is totally ordered under the same order relation, $(\mathbf{Y}, \prec)$ is totally ordered. The solution of the initial value problem (1.5.1) can be written as

$$y(n,t) = (-1)^{n+1}\left[n - \frac{1.+(-1)^n}{2}\right] t - \frac{1}{2}(-1)^{n+1}\left[n - \frac{1+(-1)^n}{2} - 1\right]$$

and every element of this sequence belongs to the set $\mathbf{Y}$ with $a = (-1)^{n+1}\left[n - \dfrac{1+(-1)^n}{2}\right]$. The isomorphism from $\mathbb{R}$ to $\mathbf{Y}$ can be defined as follows:

$$f(r) = \left\{(t,y) : y = rt - \frac{r}{2} + \frac{\operatorname{sgn}(r)}{2}\right\}.$$

Finally, it suffices to observe that the sequence

$$\{f^{-1}(y(n,t))\}_{n=1}^{\infty} = \left\{(-1)^{n+1}\left[n - \frac{1+(-1)^n}{2}\right]\right\}_{n=1}^{\infty}$$

oscillates around 0 (in fact around every real number), therefore the sequence $\{y(n,t)\}_{n=1}^{\infty}$ is $(f; \mathbb{R}, \leq)$–oscillates around $f(0)$.

We remark that the definition of $(f; \mathbb{R}, \leq)$–oscillations in the ordered set essentially depends on the isomorphism f. It is possible that the same sequence y is $(f;\mathbb{R},\leq)$–oscillatory but not $(g;\mathbb{R},\leq)$–oscillatory, where g is some other isomorphism between $(\mathbb{R},\leq)$ and $(\mathbf{X},\prec)$. To provide an example for this, we recall that every strictly increasing bijection h on $\mathbb{R}$ preserves natural ordering of the set, if f is an isomorphism between $(\mathbb{R},\leq)$ and $(\mathbf{X},\prec)$, then $f\circ h$ is an isomorphism between $(\mathbb{R},\leq)$ and $(\mathbf{X},\prec)$ also. Now let the sequence $\{y_n\}_{n=1}^{\infty}$ be $(f;\mathbb{R},\leq)$–oscillatory, and the sequence $\{f^{-1}(y_n)\}_{n=1}^{\infty}$ be bounded, i.e. $-m < f^{-1}(y_n) < M$ for all $n\in\mathbb{N}$ and some positive numbers m, M. Then, the function $g(r+m) = f(r)$ is an isomorphism between $\mathbb{R}$ and $\mathbf{X}$, but $\{g^{-1}(y_n)\}_{n=1}^{\infty}$ is not oscillatory (around 0) sequence of reals.

Finally, we shall illustrate here one more example which shows that the above phenomenon depends on the order relation in the set $\mathbb{R}$.

Example 1.5.3. Let $\mathbf{X}=\mathbb{R}$. The following function h is one–to–one mapping from $\mathbb{R}$ to $\mathbb{R}$

$$h(x) = \begin{cases} x \text{ for } x\in(-1,1)\cup\bigcup\limits_{k\in\mathbb{N}}\{[2k,2k+1)\cup(-2k-1,-2k)\} \\ -x \text{ for } x\in\bigcup\limits_{k\in\mathbb{N}}\{(-2k,-2k+1]\cup[2k-1,2k)\}. \end{cases}$$

For r_1, $r_2\in\mathbf{X}$ we shall write $r_1\prec r_2$ if $h(r_1)\leq h(r_2)$. The isomorphism between $(\mathbb{R},\leq)$ and $(\mathbf{X},\prec)$ is given by the function h^{-1}. We can check that for all real numbers r_1, r_2 if $r_1\leq r_2$ then $h^{-1}(r_1)\prec h^{-1}(r_2)$. Let $\{y_n\}_{n=1}^{\infty}=\left\{n+\dfrac{1}{n}\right\}_{n=1}^{\infty}$. Under this order relation for the sequence $\left\{n+\dfrac{1}{n}\right\}_{n=1}^{\infty}$, $n\in\mathbb{N}$ we have the ordering

$$\cdots 5\frac{1}{6}\prec 3\frac{1}{4}\prec 1\frac{1}{2}\prec 2\frac{1}{3}\prec 4\frac{1}{5}\cdots.$$

This nonoscillatory sequence $\left\{n+\dfrac{1}{n}\right\}_{n=1}^{\infty}$ is $(h^{-1};\mathbb{R},\leq)$–oscillatory because the sequence $\left\{\left(h^{-1}\right)^{-1}(y_n)\right\}_{n=1}^{\infty}=\{h(y_n)\}_{n=1}^{\infty}$ is oscillatory.

1.6. Oscillation in Linear Spaces

In the previous section we have used the concept of ordering. In the following we shall omit this assumption, but suppose that $\mathbf{X}$ is a real vector space. We recall that if a, $b\in\mathbf{X}$ and $a\neq b$, then the set of elements $x=\lambda a+(1-\lambda)b$, $\lambda\in\mathbb{R}$ is called the line passing through a

and b. The set $\{x \in \mathbf{X} : x = \lambda a + (1-\lambda)b,\ 0 \le \lambda \le 1\}$ is called the segment (interval) determined by a and b, and a and b are called the extremal points of the segment.

Definition 1.6.1. A sequence $\{y_n\}_{n=1}^{\infty}$ of elements of the real vector space $\mathbf{X}$ is said to *oscillate around* a if there exists an increasing sequence of indices $\{n_k\}_{k=1}^{\infty}$ such that a belongs to the segments determined by y_{n_k} and y_{n_k+1}.

We shall now present two very simple results on the oscillatory behavior of solutions of difference equations in real vector space. Our first result is for the difference equation

$$\Delta y_n = \lambda_n(a - y_n), \qquad n \in \mathbb{N}. \tag{E_{12}}$$

Theorem 1.6.1. Let $\{\lambda_n\}_{n=1}^{\infty}$ be a sequence of real numbers. If there exists an increasing sequence of positive integers $\{n_k\}_{k=1}^{\infty}$ such that $\lambda_{n_k} > 1$, then every solution of the equation (E_{12}) oscillates around a.

Proof. Let $\{y_n\}_{n=1}^{\infty}$ be any solution of (E_{12}), and $j \in \{n_k\}_{k=1}^{\infty}$, i.e. $\lambda_j > 1$. Then, from (E_{12}) it follows that

$$y_{j+1} = \lambda_j a + (1-\lambda_j)y_j,$$

which implies that

$$a = \frac{1}{\lambda_j}y_{j+1} + \frac{\lambda_j - 1}{\lambda_j}y_j.$$

However, since

$$\frac{1}{\lambda_j} > 0, \qquad \frac{\lambda_j - 1}{\lambda_j} > 0, \qquad \text{and} \qquad \frac{1}{\lambda_j} + \frac{\lambda_j - 1}{\lambda_j} = 1$$

a is in the segment determined by y_j and y_{j+1}.

It is clear that if all $\lambda_n \ge 1$ then every solution of (E_{12}) is quickly oscillatory around a (a is in all segments determined by y_n and y_{n+1} for all $n \in \mathbb{N}$). Furthermore, every solution is in the straight line passing through a and y_1. ∎

The second result is concerned with the following difference equation

$$y_{n+2} + by_{n+1} + cy_n = \theta, \qquad n \in \mathbb{N} \tag{E_{13}}$$

where b, c are real constants, and θ is the null element of the linear space $\mathbf{X}$.

Theorem 1.6.2. Let $b \in \mathbb{R}_0$ and $c \in \mathbb{R}_-$. Then, the equation (E_{13}) possesses a family of oscillatory solutions around θ.

Proof. Let y_1 be any element of the linear space $\mathbf{X}$ and let $\{\delta_n\}_{n=1}^{\infty}$ be any solution of the scalar difference equation

$$\delta_{n+1}\delta_n + b\delta_n + c = 0, \qquad n \in \mathbb{N}. \tag{1.6.1}$$

We note that every solution of (1.6.1) never vanishes on $\mathbb{N}$. Further, equation (1.6.1) can be written as

$$\delta_{n+1} = -b - \frac{c}{\delta_n}, \qquad n \in \mathbb{N}.$$

Now consider the function $f(x) = -b - (c/x)$, where $b \geq 0$ and $c < 0$. It is clear that $f(x) < 0$ for $x < 0$. Therefore, if we take $\delta_1 < 0$, then the suitable solution of (1.6.1) remains negative for all $n \in \mathbb{N}$.

By direct substitution we can check that the sequence $\{y_n\}_{n=1}^{\infty}$ of elements of the space $\mathbf{X}$ defined by the formula

$$y_n = \left(\prod_{j=1}^{n-1} \delta_j\right) y_1, \qquad n \in \mathbb{N}$$

is the solution of the equation (E_{13}). Furthermore,

$$y_{n+1} = \delta_n y_n, \quad \text{i.e.} \quad y_{n+1} - \delta_n y_n = \theta. \tag{1.6.2}$$

Now let $\lambda_n = 1/(1-\delta_n)$, $n \in \mathbb{N}$. Since $\delta_n < 0$, we find that $\lambda_n \in (0,1)$. Hence, from (1.6.2) we obtain $y_{n+1} + ((1-\lambda_n)/\lambda_n)y_n = \theta$, and consequently $\lambda_n y_{n+1} + (1-\lambda_n)y_n = \theta$ for all $n \in \mathbb{N}$. This means that θ belongs to every interval determined by y_{n+1} and y_n, i.e. this solution $\{y_n\}_{n=1}^{\infty}$ quickly oscillates around θ. ∎

The following generalization of Theorem 1.6.2 can be proved similarly.

Theorem 1.6.3. Let $k \geq 2$ and $(-1)^j a_{k-j} < 0$ for $j = 2, 3, \cdots, k$, $a_{k-1} \geq 0$. Then, the equation

$$y_{n+k} + \sum_{i=0}^{k-1} a_i y_{n+i} = \theta, \qquad n \in \mathbb{N}$$

possesses a family of quickly oscillatory solutions around θ.

1.7. Oscillation in Archimedean Spaces

A real vector space $\mathbf{X}$ is said to be an ordered vector space if an order relation has been endowed such that the following conditions are satisfied

(a) if $x_1, x_2 \in \mathbf{X}$ and $x_1 \prec x_2$, then $x_1 + x \prec x_2 + x$ for all $x \in \mathbf{X}$,

(b) if $x_1, x_2 \in \mathbf{X}$ and $x_1 \prec x_2$, then $ax_1 \prec ax_2$ for any $a \in \mathbb{R}_0$.

Let $x, x_1, x_2, x_3, x_4 \in \mathbf{X}$, $a, b \in \mathbb{R}$, and θ be the zero element of the vector space $\mathbf{X}$. The following properties of the ordered vector spaces are fundamental:

(i) if $x_1 \prec x_2$ and $x_3 \prec x_4$, then $x_1 + x_3 \prec x_2 + x_4$,

(ii) if $\theta \prec x_1$ and $\theta \prec x_2$, then $\theta \prec x_1 + x_2$,

(iii) if $x_1 \prec x_2$ and $a \leq 0$, then $ax_2 \prec ax_1$,

(iv) if $\theta \prec x$ and $a \leq b$, then $ax \prec bx$.

An element $x \in \mathbf{X}$ is said to be positive if $\theta \prec x$, and negative if $x \prec \theta$. The set of all positive (negative) elements we shall denote by $\mathbf{X}_+$ ($\mathbf{X}_-$). Obviously, $\mathbf{X}_+ \cap \mathbf{X}_- = \theta$. An ordered vector space $\mathbf{X}$ we shall call an 'Archimedean space' if for any element x, which is not negative, the set $\{ax : a \in \mathbb{R}_+\}$ is not bounded from above [47]. In the Archimedean space the following properties can be proved rather easily.

(P1) For any element x, which is not positive, the set $\{ax : a \in \mathbb{R}_+\}$ is not bounded from below.

(P2) If the set $\{u_j : j \in J\}$ is not bounded from above, then the set $\{u_j + x : j \in J\}$ is also not bounded from above for any element $x \in \mathbf{X}$.

(P3) For any element x which is not negative, and any sequence $\{a_n\}_{n=1}^{\infty}$ of positive real numbers such that $\lim_{n\to\infty} a_n = \infty$ the set $\{a_n x : n \in \mathbb{N}\}$ is not bounded from above.

Definition 1.7.1. A sequence $\{y_n\}_{n=1}^{\infty}$ of elements in the ordered real vector space $\mathbf{X}$ is said to be ***nonoscillatory around*** θ if eventually it consists of positive or negative elements. Furthermore, we suppose that $card\{n \in \mathbb{N} : y_n = \theta\} < \aleph_0$.

Here we shall study the equation

$$\Delta^3 y_n = a_n y_n, \quad n \in \mathbb{N} \tag{E_{14}}$$

in the Archimedean space.

Theorem 1.7.1. Let $a : \mathbb{N} \to (-\infty, -1)$. Then, every solution of the equation (E_{14}) is not nonoscillatory.

Proof. Suppose that there exists an eventually nonoscillatory solution $y = \{y_n\}_{n=1}^{\infty}$ of the equation (E_{14}). Since (E_{14}) is linear we can assume that $\theta \prec y_n$ for $n \geq \nu$. Now because $a < -1$ by (iv) we have $a_n y_n \prec -y_n$. Hence, by the transitivity of the order relation we find from (E_{14}) that

$$\Delta^3 y_n \prec -y_n \prec \theta \quad \text{for} \quad n \geq \nu. \tag{1.7.1}$$

Next, from (1.7.1) and (a) and (b), we get

$$\frac{1}{3}y_{n+3} \prec \Delta y_{n+1} \quad \text{for} \quad n \geq \nu, \tag{1.7.2}$$

which implies that

$$\theta \prec \frac{1}{3}y_{n+2} \prec \Delta y_n \quad \text{for} \quad n \geq \nu + 1. \tag{1.7.3}$$

This leads to $y_{\nu+1} \prec y_{\nu+2} \prec y_{\nu+3} \prec \cdots \prec y_n \prec y_{n+1} \prec \cdots$, i.e. the sequence $\{y_n\}_{n=\nu+1}^{\infty}$ is positive and increasing (in the order sense). Hence from (1.7.1), we obtain $\Delta^3 y_n \prec -y_k$ for $n \geq k \geq \nu+1$. Taking $k = \nu+2$ and summing the inequality $\Delta^3 y_j \prec -y_{\nu+2}$, (i.e. using (i) suitable number of times) from $j = p$ to $j = n-1$, we obtain

$$\Delta^2 y_n - \Delta^2 y_p \prec -\sum_{j=p}^{n-1} y_{\nu+2}, \tag{1.7.4}$$

which is in view of (a) for $p = \nu + 2$ is the same as

$$\Delta^2 y_n \prec \Delta^2 y_{\nu+2} - (n - \nu - 2)y_{\nu+2} \quad \text{for} \quad n \geq \nu + 2. \tag{1.7.5}$$

Thus, $\Delta^2 y_n$ is eventually negative (in the scalar case from (1.7.5) it is clear that $\Delta^2 y_n \to -\infty$ which leads to the contradiction to the positivity of y_n.) Now from (1.7.1) – (1.7.3) on using the suitable properties of the order relation, we get

$$\begin{aligned} \Delta^2 y_{\nu+2} &\prec \Delta^2 y_{\nu+1} = \Delta y_{\nu+2} - \Delta y_{\nu+1} \prec \Delta y_{\nu+2} = y_{\nu+3} - y_{\nu+2} \\ &\prec 3\Delta y_{\nu+1} - y_{\nu+2} = 2y_{\nu+2} - 3y_{\nu+1}. \end{aligned}$$

Therefore, from (1.7.5), we have

$$\Delta^2 y_n \prec \Delta^2 y_{\nu+2} - (n - \nu - 2)y_{\nu+2} \prec -3y_{\nu+1} - (n - \nu - 4)y_{\nu+2},$$

which yields

$$\Delta^2 y_n \prec \theta \quad \text{for} \quad n \geq \nu + 4.$$

Next, on taking $p \geq \nu + 4$ in (1.7.4), we obtain

$$\Delta^2 y_n \prec \Delta^2 y_p - (n - p)y_{\nu+2} \prec -(n - p)y_{\nu+2} \quad \text{for} \quad n \geq p,$$

which implies that

$$\Delta y_n - \Delta y_{p+1} \prec -\sum_{j=p+1}^{n-1} (j - p)y_{\nu+2}$$

and hence

$$\left(\sum_{j=p+1}^{n-1}(j-p)\right) y_{\nu+2} - \Delta y_{p+1} \prec -\Delta y_n \quad \text{for} \quad n \geq p+1. \qquad (1.7.6)$$

Thus, from the definition of the Archimedean space and the properties listed in the beginning of this section, we find that the set

$$\left\{\left(\sum_{j=p+1}^{n-1}(j-p)\right) y_{\nu+2} - \Delta y_{p+1}\right\}_{n=p+1}^{\infty}$$

is not bounded from above. Therefore, from (1.7.6) it follows that the set $\{-\Delta y_n\}_{n=p+1}^{\infty}$ is not bounded from above. On the other hand by (1.7.3) we have $-\Delta y_n \prec \theta$ for all $n \geq \nu+1$, therefore the set $\{-\Delta y_n\}_{n=p+1}^{\infty}$ is bounded from above by θ. This contradiction completes the proof. ∎

It is interesting to note that in view of Definition 1.7.1, the sequence $y = \{y_n\}_{n=1}^{\infty}$ which is not nonoscillatory can possess infinite subsequences from the sets $\mathbf{X}_+$ and $\mathbf{X}_-$, also subsequences which are non–comparable with θ (even every element of y can be of this type). So, if we call this type of sequences oscillatory, then Theorem 1.7.1 states that every solution of (E_{14}) is oscillatory. Further, in Theorem 1.7.1 we did not assume that the set $\mathbf{X}$ is directed (to the right), if it happens then our theorem obviously holds. As a consequence of this, Theorem 1.7.1 remains valid for (E_{14}) in the space $\mathbf{X}(\mathbf{D})$ (real–valued functions on some set $\mathbf{D}$) in which the order relation is defined in the usual way, i.e. for $x,\ y \in \mathbf{X}(\mathbf{D})$, $x \prec y$ if and only if $x(t) \leq y(t)$, $t \in \mathbf{D}$.

1.8. Oscillation of Partial Recurrence Equations

Here we shall consider the real–valued sequences in two independent variables, i.e. functions $y : \mathbb{N}^2 \to \mathbb{R}$.

Definition 1.8.1. A sequence $y = \{y(m,n)\}_{m=1}^{\infty}{}_{n=1}^{\infty}$ is said to be *nonoscillatory around* 0 if there exist positive integers μ, ν such that

$$y(m,n) > 0 \quad \text{(positive sequence)} \quad \text{for all } m \geq \mu, \ n \geq \nu$$

or

$$y(m,n) < 0 \quad \text{(negative sequence)} \quad \text{for all } m \geq \mu, \ n \geq \nu.$$

Otherwise the sequence y is called *oscillatory*.

We note that according to the above definition eventually zero sequences are oscillatory, we shall exclude this type of sequences from our consideration. Further, for the sequence $y : \{1, 2, \cdots, \mu\} \times \mathbb{N} \to \mathbb{R}$ the above definition can be changed to the following:

Definition 1.8.2. A sequence $y = \{y(m,n)\}_{m=1}^{\mu}{}_{n=1}^{\infty}$ is said to be *nonoscillatory around* 0 if there exists a positive integer ν such that

$$y(m,n) > 0 \quad \text{(positive sequence)} \quad \text{for all } m \le \mu, \ n \ge \nu$$

or

$$y(m,n) < 0 \quad \text{(negative sequence)} \quad \text{for all } m \le \mu, \ n \ge \nu.$$

Otherwise the sequence y is called *oscillatory.*

Hereafter, we shall consider sequences y which are defined on $\mathbb{N}^2$. It is clear that if the sequence y is nonoscillatory then it is nonoscillatory for each 'section' along each arbitrary, but fixed, m, $m > \mu$ as well as arbitrary, but fixed, n, $n > \nu$. Further, all these sections are of the same fixed sign (positive or negative). The sufficient condition for oscillation can be formulated as follows: There exists an increasing to infinity sequence $\mathbf{m} = \{m_k\}_{k=1}^{\infty}$ such that the sections $\{y(m_k, n)\}_{n=1}^{\infty}$ are oscillatory sequences for all $m_k \in \mathbf{m}$. Analogously, if there exists an increasing to infinity sequence $\mathbf{n} = \{n_k\}_{k=1}^{\infty}$ such that the sections $\{y(m, n_k)\}_{m=1}^{\infty}$ are oscillatory for all $n_k \in \mathbf{n}$.

This sufficient condition requires essentially more than the Definition 1.8.1. For this, we note that the sequence defined as

$$y(m,n) = \begin{cases} -1 & \text{for } m, n \in \mathbb{N} \quad \text{such that } m = n \\ 1 & \text{otherwise} \end{cases}$$

is oscillatory in the sense of Definition 1.8.1, however, every section for a fixed m as well as for a fixed n is a nonoscillatory sequence.

If all sequences $\{y(p,n)\}_{n=1}^{\infty}$ and $\{y(m,q)\}_{m=1}^{\infty}$ for each fixed p, $q \in \mathbb{N}$ are oscillatory, then we say that the sequence $y = \{y(m,n)\}_{m=1}^{\infty}{}_{n=1}^{\infty}$ possesses a stronger oscillatory property. Nonoscillatory sequences can be oscillatory along some (but finite number) sections. For example the nonoscillatory sequence defined as

$$y(m,n) = \begin{cases} (-1)^n & \text{for } m = 1, \ n \in \mathbb{N} \\ (-1)^m & \text{for } n = 1, \ m \in \mathbb{N} \\ 1 & \text{for the other } (m,n) \in \mathbb{N}^2 \end{cases}$$

has quickly oscillatory sections $\{y(1,n)\}_{n=1}^{\infty}$, $\{y(m,1)\}_{m=1}^{\infty}$.

Now we shall define the oscillation of the sequence y in a different way. For this, we need the following:

Definition 1.8.3. Any sequence $\{m_k, n_k)\}_{k=1}^{\sigma}$ where $m_k,\ n_k \in \mathbb{N}$ for $k = 1, 2, \cdots, \sigma$ such that $m_{k+1} + n_{k+1} = m_k + n_k \pm 1$ for all $k \in \mathbb{N}$ if $\sigma = \infty$, and $k = 1, 2, \cdots, \sigma - 1$, otherwise, is called a *path argument.*

Definition 1.8.4. The sequence $y = \{y(m,n)\}_{m=1}^{\infty}{}_{n=1}^{\infty}$ is called *oscillatory across the family* $\gamma = \{\gamma_k : k \in \mathbb{N}\}$ *of each other disjoint path arguments* if $\left(y\Big|_{P\in\gamma_k}\right)\left(y\Big|_{Q\in\gamma_{k+1}}\right) < 0$ and $\left(y\Big|_{R\in\gamma_k}\right)\left(y\Big|_{S\in\gamma_k}\right) > 0$ for all $P \in \gamma_k$, $Q \in \gamma_{k+1}$, $R,\ S \in \gamma_k$, and for all $k \in \mathbb{N}$.

Example 1.8.1. The sequence $\{y(m,n)\}_{m=1}^{\infty}{}_{n=1}^{\infty}$, where $y(m,n) = (-1)^{\max\{m,n\}}$ is oscillatory across the family of path arguments γ_k defined by

$$\gamma_k \ = \ \{(m,k),\ m = 1,2,\cdots,k\} \cup \{(k,n),\ n = 1,2,\cdots,k\}, \quad k \in \mathbb{N}.$$

For the sequence $y = \{y(m,n)\}_{m=1}^{\infty}{}_{n=1}^{\infty}$, we define the partial difference operators

$$\Delta_1 y(m,n) \ = \ y(m+1,n) - y(m,n) \ \text{ and } \ \Delta_2 y(m,n) = y(m,n+1) - y(m,n), \quad m,n \in \mathbb{N}.$$

We shall consider the following first order partial difference equation

$$\Delta_1 y(m,n) + a_m b_n \Delta_2 y(m,n) \ = \ 0, \qquad m,n \in \mathbb{N}. \qquad (E_{15})$$

Theorem 1.8.1. Let one of the following holds

(a) the sequence $\{a_n\}_{n=1}^{\infty}$ is eventually negative and there exists a negative constant a such that $a_m \le a$ for all $m \ge \mu \in \mathbb{N}$,

(b) the sequence $\{b_n\}_{n=1}^{\infty}$ is eventually negative and there exists a negative constant b such that $b_n \ge b$ for all $n \ge \nu \in \mathbb{N}$.

Then, the equation (E_{15}) possesses a family of oscillatory solutions around the zero solution.

Proof. To prove this result we shall use the method of separation of variables. Suppose that $\{y(m,n)\}_{m=1}^{\infty}{}_{n=1}^{\infty}$ is never vanishing solution of (E_{15}) which can be represented as the product, i.e.

$$y(m,n) \ = \ u(m)w(n) \quad \text{for all } m,n \in \mathbb{N}. \qquad (1.8.1)$$

We substitute (1.8.1) in the difference equation (E_{15}) and divide by $u(m)$ $w(n)a_m$, to obtain

$$\frac{\Delta_1 u(m)}{a_m u(m)} \ = \ -\frac{b_n \Delta_2 w(n)}{w(n)}, \qquad m,n \in \mathbb{N}. \qquad (1.8.2)$$

The left side of (1.8.2) depends on m only, whereas the right side is independent of m. Thus, $y(m,n) = u(m)w(n)$ is a solution of (E_{15}) if and only if u and w satisfy two ordinary difference equations

$$\frac{\Delta u(m)}{a_m u(m)} = C, \quad m \in \mathbb{N} \quad \text{and} \quad \frac{b_n \Delta w(n)}{w(n)} = -C, \quad n \in \mathbb{N}$$

or

$$\Delta u(m) = C a_m u(m), \quad m \in \mathbb{N} \quad \text{and} \quad \Delta w(n) = -\frac{C}{b_n} w(n), \quad n \in \mathbb{N} \tag{1.8.3}$$

for some constant C.

Solving the system (1.8.3), we get

$$u(m) = u(1) \prod_{i=1}^{m-1} (1 + Ca_i), \quad m \in \mathbb{N} \quad \text{and} \quad w(n) = w(1) \prod_{j=1}^{n-1} \left(1 - \frac{C}{b_j}\right), \quad n \in \mathbb{N}.$$

Therefore, the sequence

$$y(m,n) = y(1,1) \left(\prod_{i=1}^{m-1} (1 + Ca_i)\right) \left(\prod_{j=1}^{n-1} \left(1 - \frac{C}{b_j}\right)\right),$$

where $y(1,1) = u(1)w(1)$ should be a solution of (E_{15}). In fact, by direct substitution we can check that the sequence of the form

$$\left\{ A \left(\prod_{i=1}^{m-1} (1 + Ca_i)\right) \left(\prod_{j=1}^{n-1} \left(1 - \frac{C}{b_j}\right)\right) \right\}_{m=1,n=1}^{\infty\ ,\ \infty} \tag{1.8.4}$$

satisfy (E_{15}) for arbitrary $A, C \in \mathbb{R}$.

Now suppose that the condition (a) is satisfied. Then, on taking $C > -1/a$, we obtain $-1/a_m < -1/a < C$, and hence $0 > 1 + Ca_m$ for all $m \geq \mu$. Thus, the sequence

$$\left\{ \left(\prod_{i=1}^{m-1} (1 + Ca_i)\right) \right\}_{m=1}^{\infty}$$

quickly oscillates for $m > \mu$. Therefore, the sequences

$$\left\{ A \left(\prod_{i=1}^{m-1} (1 + Ca_i)\right) \left(\prod_{j=1}^{s-1} \left(1 - \frac{C}{b_j}\right)\right) \right\}_{m=1}^{\infty}$$

oscillate for any fixed $A \in \mathbb{R}\backslash\{0\}$ and all fixed $s \in \mathbb{N}$. This means that the sufficient conditions for the oscillation of the sequence (1.8.4) are satisfied. Similarly, if we take $C < b$ then the condition (b) ensures the oscillation of the sequence (1.8.4). ∎

Method similar to that used in Theorem 1.8.1 can be employed for the equation

$$\Delta_1^2 y(m,n) + a_m b_n \Delta_2 y(m,n) = 0, \quad m,n \in \mathbb{N} \qquad (E_{16})$$

where $\Delta_1^2 y(m,n) = y(m+2,n) - 2y(m+1,n) + y(m,n)$, $m,n \in \mathbb{N}$.

Theorem 1.8.2. Let one of the following holds

(a) the sequence $\{a_n\}_{n=1}^{\infty}$ is eventually negative and there exists a negative constant a such that $\liminf_{m\to\infty} a_m = a$,

(b) the sequence $\{a_n\}_{n=1}^{\infty}$ is eventually positive and there exists a positive constant a such that $\limsup_{m\to\infty} a_m = a$,

(c) the sequence $\{b_n\}_{n=1}^{\infty}$ is eventually negative and there exists a negative constant b such that $b_n \geq b$ for all $n \geq \nu \in \mathbb{N}$.

Then, the equation (E_{16}) possesses a family of oscillatory solutions around the zero solution.

Proof. Applying the same method as in the proof of Theorem 1.8.1, we obtain the system of ordinary difference equations

$$\Delta^2 u(m) = C a_m u(m), \quad m \in \mathbb{N} \quad \text{and} \quad \Delta w(n) = -\frac{C}{b_n} w(n), \quad n \in \mathbb{N} \qquad (1.8.5)$$

where C is an arbitrary constant. The equation (E_{16}) possesses a family of solutions of the form

$$\left\{ Au(m) \left(\prod_{j=1}^{n-1} \left(1 - \frac{C}{b_j}\right) \right) \right\}_{m=1,n=1}^{\infty\ ,\ \infty},$$

where $u(m)$ is a solution of the first equation in (1.8.5). In the case (a) ((b)) the assumptions of Theorem 1.3.1 are satisfied with any positive (negative) constant C. Thus, every solution $\{u(m)\}_{m=1}^{\infty}$ of the equation $\Delta^2 u(m) = C a_m u(m)$, $m \in \mathbb{N}$ is oscillatory. Hence for all $s \in \mathbb{N}$ the sequence

$$\left\{ Au(m) \left(\prod_{j=1}^{s-1} \left(1 - \frac{C}{b_j}\right) \right) \right\}_{m=1}^{\infty}$$

oscillates, and thus the sufficient conditions for the oscillatory behavior of $y(m,n)$ are satisfied. The case (c) reduces to the case (b) of Theorem 1.8.1. ∎

Finally, we remark that the Definition 1.8.1 can be generalized to sequences of arbitrary variables k as follows: Let $n = (n_1, n_2, \cdots, n_k) \in \mathbb{N}^k$ denote a multi–index. A real–valued sequence $y = \{y(n)\}_{n=(1,\cdots,1)}^{(\infty,\cdots,\infty)}$ is said to be nonoscillatory around 0 if there exists a positive multi–index $\nu = (\nu_1, \nu_2, \cdots, \nu_k)$, $\nu_i > 0$ such that

$$y(n) > 0 \quad \text{(positive sequence)} \ \text{ for all } \ n_i \geq \nu_i, \ i = 1, \cdots, k$$

or

$$y(n) < 0 \quad \text{(negative sequence)} \ \text{ for all } \ n_i \geq \nu_i, \ i = 1, \cdots, k.$$

Otherwise, the sequence y is called oscillatory.

1.9. Oscillation of System of Equations

In the system case a solution is a sequence of vectors. Thus, here oscillation can be defined for some or all of its components. We begin with the system of two linear difference equations with constant coefficients

$$y_{n+1}^i = a_{i1} y_n^1 + a_{i2} y_n^2, \quad i = 1, 2, \ n \in \mathbb{N}. \qquad (E_{17})$$

Theorem 1.9.1. Let $a_{ij} > 0$ for $i, j = 1, 2$. Then,

(a) every solution of (E_{17}) is nonoscillatory, i.e. both $y^1 = \{y_n^1\}_{n=1}^\infty$ and $y^2 = \{y_n^2\}_{n=1}^\infty$ are nonoscillatory sequences,

(b) both y^1, y^2 are together eventually positive or eventually negative, except the case when

$$y_1^2 = \frac{a_{22} - a_{11} - \sqrt{\delta}}{2a_{12}} y_1^1, \quad \text{where} \quad \delta = (a_{11} - a_{22})^2 + 4a_{12}a_{21}. \quad (1.9.1)$$

Further, when (1.9.1) holds then $y_n^1 y_n^2 < 0$ for all $n \in \mathbb{N}$.

Proof. From our consideration we exclude the trivial solution of (E_{17}), which is obtained when $y_1^1 = y_1^2 = 0$.

(i) It is clear that if $y_k^1 > 0$ and $y_k^2 > 0$ for some $k \in \mathbb{N}$ then $y_n^1 > 0$ and $y_n^2 > 0$ for all $n \geq k$. Further, if $y_k^1 < 0$ and $y_k^2 < 0$ for some $k \in \mathbb{N}$ then $y_n^1 < 0$ and $y_n^2 < 0$ for all $n \geq k$.

(ii) If $y_1^1 = 0$ or $y_1^2 = 0$ then we obtain the case (i) with $k = 2$.

(iii) In view of (i) and (ii) we need to consider only the cases $y_1^1 < 0$, $y_1^2 > 0$ and $y_1^1 > 0$, $y_1^2 < 0$. We shall consider only the case $y_1^1 < 0$, $y_1^2 > 0$, as the arguments are similar for the other case. For the system (E_{17}) there corresponds the square matrix $A = \begin{pmatrix} a_{11} & a_{12} \\ a_{21} & a_{22} \end{pmatrix}$, which possesses two distinct real eigenvalues $\left(a_{11}+a_{22}+\sqrt{\delta}\right)/2$ and $\left(a_{11}+a_{22}-\sqrt{\delta}\right)/2$, and two family of eigenvectors $w_1 = \left[\sigma, \sigma\left(\sqrt{\delta}-a_{11}+a_{22}\right)/(2a_{12})\right]$ and $w_2 = \left[\tau, -\tau\left(\sqrt{\delta}+a_{11}-a_{22}\right)/(2a_{12})\right]$, where σ, τ are parameters. We observe that both components of the vector w_1 are of the same sign, whereas of w_2 are of the opposite sign. For this reason our interest is in the vector $w_2(1) = \left[1, \left(a_{22}-a_{11}-\sqrt{\delta}\right)/(2a_{12})\right]$. We shall denote by $m_w = \left(a_{22}-a_{11}-\sqrt{\delta}\right)/(2a_{12})$ the direction of the vector $w_2(1)$ in the plane Oy^1y^2.

Now let $\det A > 0$. Then, $m_w \in (-a_{11}/a_{12}, -a_{21}/a_{22})$. (In the case $\det A < 0$, we have $m_w \in (-a_{21}/a_{22}, -a_{11}/a_{12})$, while if $\det A = 0$ then $m_w = -a_{21}/a_{22} = -a_{11}/a_{12}$).

From the equation (E_{17}) it follows that if the point (y_n^1, y_n^2) belongs to the line $y^2 = my^1$ with $m \neq -a_{11}/a_{12}$, then (y_{n+1}^1, y_{n+1}^2) belongs to the line

$$y^2 = \frac{a_{21}+a_{22}m}{a_{11}+a_{12}m}y^1. \tag{1.9.2}$$

Let $\mathbf{D}_1$ be the region of the second quadrant of the plane Oy^1y^2 bounded by the lines $y^2 = 0$ and $y^2 = -(a_{21}/a_{22})y^1$, and $\mathbf{D}_2$ bounded by the lines $y^2 = -(a_{11}/a_{12})y^1$ and $y^1 = 0$. Take any point $(y_1^1, y_1^2) \in \mathbf{D}_2$, (i.e. $y_1^2 = my_1^1$, $m \in (-\infty, -a_{11}/a_{12})$, $y_1^1 < 0$). From the first equation of (E_{17}), i.e. $y_{n+1}^1 = a_{11}y_n^1 + a_{12}y_n^2$ it follows that $y_2^1 = a_{11}y_1^1 + a_{12}my_1^1 > 0$. Furthermore, the function

$$f(m) = \frac{a_{21}+a_{22}m}{a_{11}+a_{12}m} \tag{1.9.3}$$

is positive for $m \in (-\infty, -a_{11}/a_{12})$ and hence, by (1.9.2), $y_2^2 > 0$. So the point (y_2^1, y_2^2) belongs to the first quadrant and we are back to the case (i).

Similarly, if $(y_1^1, y_1^2) \in \mathbf{D}_1$, (i.e. $y_1^2 = my_1^1$, $m \in (-a_{21}/a_{22}, 0)$, $y_1^1 < 0$), then we find that (y_2^1, y_2^2) belongs to the third quadrant, and so here also we return to the case (i).

The image of the line $y^2 = -(a_{11}/a_{12})y^1$ is the line $y^1 = 0$ (y^2–axis).

The image of the line $y^2 = -(a_{21}/a_{22})y^1$ is the line $y^2 = 0$ (y^1–axis).

For the rest of the proof we should show that every solution with the initial point (y_1^1, y_1^2) which belongs to the open region of the second quadrant bounded by the lines $y^2 = -(a_{11}/a_{12})y^1$ and $y^2 = -(a_{21}/a_{22})y^1$, and such that $y_1^2 \neq m_w y_1^1$, after finite number of steps comes in the first or third quadrant (in fact by the previous remarks, to the set $\mathbf{D}_2 \cup$ {first quadrant}, or $\mathbf{D}_1 \cup$ {third quadrant}). Since

$$f'(m) = \frac{\det A}{(a_{11} + a_{12}m)^2} > 0, \tag{1.9.4}$$

$$f''(m) = \frac{-2a_{12}\det A}{(a_{11} + a_{12}m)^3} < 0 \quad \text{for} \quad m \in (-a_{11}/a_{12}, \infty), \tag{1.9.5}$$

and

$$f(m_w) = m_w \quad \text{and} \quad f\left(-\frac{a_{21}}{a_{22}}\right) = 0,$$

we have

$$f(m) > m + a \quad \text{for every} \quad m \in (m_w, -a_{21}/a_{22}), \tag{1.9.6}$$

where

$$a = (m_w - m)\frac{a_{21}}{a_{21} + a_{22}m_w} > 0.$$

Now we observe that if $m > -a_{11}/a_{12}$ and (y_n^1, y_n^2) belongs to the second quadrant, then (y_{n+1}^1, y_{n+1}^2) also belongs to the second quadrant as far as $f(m) < 0$.

Let $(y_1^1, y_1^2) \in \mathbf{D}_3 = \{(y^1, y^2) \in Oy^1y^2 : y^1 < 0,\ y^2 = my^1, \text{ and } m \in (m_w, -a_{21}/a_{22})\}$, i.e. $y_1^1 < 0$, $y_1^2 = m_1 y_1^1$, and $m_1 \in (m_w, -a_{21}/a_{22})$.

For our purpose, suppose that all elements of the corresponding solution $\{y_n^1, y_n^2\}_{n=1}^{\infty}$ $(y_n^2 = m_n y_n^1)$ belongs to the second quadrant. In fact, all the points should be in the region $\mathbf{D}_3$, because of $m_1 \in (m_w, -a_{21}/a_{22})$, all $m_n < 0$, and (1.9.4). The sequence $\{m_n\}_{n=1}^{\infty}$ is increasing and bounded from above by $-a_{21}/a_{22}$. However, this is impossible, because from (1.9.6), we have

$$m_{n+1} = f(m_n) > m_n + a = f(m_{n-1}) + a > m_{n-1} + 2a > \cdots > m_1 + na \to \infty$$

as $n \to \infty$, where $a = (m_w - m_1)a_{21}/(a_{21} + a_{22}m_w) > 0$.

Let $\mathbf{D}_4 = \{(y^1, y^2) \in Oy^1y^2 : y^1 < 0,\ y^2 = my^1, \text{ and } m \in (-a_{11}/a_{12}, m_w)\}$. We note that for $m \in (-a_{11}/a_{12}, m_w)$, by (1.9.4) and (1.9.5), we obtain

$$f(m) < \frac{\det A}{(a_{11} + a_{12}m_w)^2}(m - m_w) + m_w < m. \tag{1.9.7}$$

Let $(y_1^1, y_1^2) \in \mathbf{D}_4$ and $\{y_n^1, y_n^2\}_{n=1}^{\infty}$ be the corresponding solution of (E_{17}). If for some $k > 1$, we have $y_k^1 < 0$, $y_k^2 = m_k y_k^1$, $m_k \in (-a_{11}/a_{12}, m_w)$, but $m_{k+1} = f(m_k) < -a_{11}/a_{12}$, then $(y_{k+1}^1, y_{k+1}^2) \in$ {third quadrant} and we are back to the case (i). Therefore, suppose that $(y_n^1, y_n^2) \in \mathbf{D}_4$ for all $n \in \mathbb{N}$, which yields $y_n^2 = m_n y_n^1$ with $m_n \in (-a_{11}/a_{12}, m_w)$, $n \in \mathbb{N}$. Now from (1.9.7) it is clear that the sequence $\{m_n\}_{n=1}^{\infty}$ is decreasing and bounded from below by $-a_{11}/a_{12}$. On the other hand, again from (1.9.7), we obtain

$$m_{n+1} = f(m_n) < m_n - a = f(m_{n-1}) - a < m_{n-1} - 2a < \cdots < m_1 - na \to -\infty$$

as $n \to \infty$, where $a = (m_1 - m_w)\left[1 - \dfrac{\det A}{(a_{11} + a_{12}m_w)^2}\right] > 0$, which is a contradiction.

Similarly, we can verify that if $y_n^1 < 0$, $y_n^2 = m_w y_n^1$ then $y_{n+1}^1 < 0$, $y_{n+1}^2 = m_w y_{n+1}^1$. Thus, we have covered all the possible cases in the second quadrant. A similar reasoning can be applied for the points of the fourth quadrant. The case $\det A = 0$ immediately leads to (i) or (ii). Finally, in the case $\det A < 0$ we note that the proof of (iii) needs some modifications because now the trajectory can have jumps between second and fourth quadrants. ∎

As a consequence of Theorem 1.9.1 we have the following:

Theorem 1.9.2. Let $a_{ij} < 0$ for $i, j = 1, 2$. Then,

(a) every solution of (E_{17}) is quickly oscillatory, i.e. both $y^1 = \{y_n^1\}_{n=1}^{\infty}$ and $y^2 = \{y_n^2\}_{n=1}^{\infty}$ are quickly oscillatory sequences,

(b) both y^1, y^2 are eventually of the same sign, i.e. $y_n^1 y_n^2 > 0$, except the case

$$y_1^2 = \frac{a_{22} - a_{11} + \sqrt{\delta}}{2a_{12}} y_1^1, \quad \text{where} \quad \delta = (a_{11} - a_{22})^2 + 4a_{12}a_{21}$$

when $y_n^1 y_n^2 < 0$ for all $n \in \mathbb{N}$.

Proof. In (E_{17}) let $x_n^i = (-1)^n y_n^i$, $i = 1, 2$, $n \in \mathbb{N}$, to obtain the system $x_{n+1}^i = -a_{i1}x_n^1 - a_{i2}x_n^2$, $i = 1, 2$ for which the Theorem 1.9.1 can be applied. Theorem 1.9.2 now follows from the substitution employed. ∎

Finally, we shall state here a result which provides sufficient conditions for the one component of the following system to be quickly oscillatory

$$\Delta y_n^i = \sum_{j=1}^{m} a_n^{ij} y_n^j, \quad i = 1, 2, \cdots, m, \quad n \in \mathbb{N}. \tag{E_{18}}$$

Theorem 1.9.3. If

$$a^{ij} : \mathbb{N} \to \begin{cases} [-1,\infty) & \text{for } i = j, i \neq k \\ (-\infty,-1] & \text{for } i = j = k \\ [0,\infty) & \text{for } i \neq j, i \neq k, j \neq k \end{cases}$$

and a^{ik}, a^{kj} for $i \neq k$, $j \neq k$ are quickly oscillatory on $\mathbb{N}$ with $a_1^{ik} < 0$, $a_1^{kj} > 0$. Then, every solution of the system (E_{18}) with the initial conditions $y_1^i > 0$, $y_1^k < 0$ has positive components y^i for $i \neq k$ and quickly oscillatory component y^k on the whole $\mathbb{N}$, while every solution with the initial conditions $y_1^i < 0$, $y_1^k > 0$ has negative components y^i for $i \neq k$ and quickly oscillatory component y^k on the whole $\mathbb{N}$.

1.10. Oscillation Between Sets

We shall now present another generalization of the concept of oscillation. We recall that the oscillation around zero of the real valued sequence $\{y_n\}_{n=1}^{\infty}$ means that there exist two disjoint sets $\mathbb{R}_0$ and $\mathbb{R}_-$, and two increasing sequences of indices $\{n_k\}_{k=1}^{\infty}$, $\{m_j\}_{j=1}^{\infty}$ such that $y_{n_k} \in \mathbb{R}_0$, $y_{m_j} \in \mathbb{R}_-$ for all $k, j \in \mathbb{N}$.

Definition 1.10.1. Let $\{y_n\}_{n=1}^{\infty}$ be a sequence of elements of some set $\mathbf{X}$ and $\{\mathbf{X}_\gamma : \gamma \in \Gamma\}$ be any family of disjoint subsets of $\mathbf{X}$, (i.e. $\mathbf{X}_\tau \cap \mathbf{X}_\sigma = \emptyset$ for all τ, $\sigma \in \Gamma$, $\tau \neq \sigma$). We say that this sequence *oscillates in relation to the family* $\mathbf{X}_\gamma$ if for all $\gamma \in \Gamma$ there exists $\{n_{\gamma,k}\}_{k=1}^{\infty}$ such that for all $k \in \mathbb{N}$, $y_{n_{\gamma,k}} \in \mathbf{X}_\gamma$.

In the above definition we can assume that $\mathbf{X} = \cup_{\gamma\in\Gamma}\mathbf{X}_\gamma$.

Example 1.10.1. Consider the initial value problem

$$y_{n+1} = \cos\left(\frac{n\pi}{2}\right)(1 + y_n) + \frac{1}{n}, \qquad n \in \mathbb{N}$$
$$y_1 = 1.$$

The approximate solution of this problem we present in the following:

n	1	2	3	4	5	6	7	8	9
y_n	1	1	−1.5	0.3333	1.5833	0.2	−1.033	0.1428	1.2678
n	10	11	⋯	26	27	28	29	30	31
y_n	0.1111	−1.011	⋯	0.04	−1.015	0.0370	1.0728	0.0345	−1.001

We can say that this solution oscillates in relation to the family of sets $\{(-\infty,-1], (-1,1), [1,\infty)\}$. The terms y_{4k+3} belong to $(-\infty,-1]$; y_{4k+1} and y_2 belong to $[1,\infty)$, and the rest are in $(-1,1)$.

In what follows we shall consider the following system of two equations

$$\begin{cases} x_{n+1} = \dfrac{a}{x_n} + a y_n \\ y_{n+1} = \dfrac{b x_n}{1 + x_n y_n}, \quad n \in \mathbb{N} \end{cases} \tag{E_{19}}$$

We shall denote by

$$\begin{aligned}
\mathbf{S}_1 &= \left\{(x,y) \in \mathbb{R}^2 : x = 0 \vee y = 0\right\}, \\
\mathbf{S}_2 &= \left\{(x,y) \in \mathbb{R}^2 : y = -1/x \wedge x \neq 0\right\}, \\
\mathbf{X} &= \mathbb{R}^2 \setminus \{\mathbf{S}_1 \cup \mathbf{S}_2\}, \\
\mathbf{D}_1 &= \left\{(x,y) \in \mathbf{X} : y = \frac{ab}{x} \wedge x \in (0, \sqrt{a(ab+1)}]\right\}, \\
\mathbf{D}_2 &= \left\{(x,y) \in \mathbf{X} : y = \frac{ab}{x} \wedge x \in (\sqrt{a(ab+1)}, \infty)\right\}, \\
\mathbf{D}_3 &= \left\{(x,y) \in \mathbf{X} : y = \frac{ab}{x} \wedge x \in (-\infty, -\sqrt{a(ab+1)})\right\}, \\
\mathbf{D}_4 &= \left\{(x,y) \in \mathbf{X} : y = \frac{ab}{x} \wedge x \in [-\sqrt{a(ab+1)}, 0)\right\}, \\
\mathbf{D} &= \{\mathbf{D}_1 \cup \mathbf{D}_2 \cup \mathbf{D}_3 \cup \mathbf{D}_4\}.
\end{aligned}$$

Theorem 1.10.1. Let $a,\ b > 0$. Then, every solution $\{(x_n, y_n)\}_{n=1}^{\infty}$ of the system (E_{19}) with the initial condition in the space $\mathbf{X}$ is oscillatory (at least for $n \geq 2$) between the sets $\mathbf{D}_1$ and $\mathbf{D}_2$, or between the sets $\mathbf{D}_3$ and $\mathbf{D}_4$.

Proof. Let $P = (x_1, y_1)$ be an arbitrary point of the phase space $\mathbf{X}$. We denote by $f(x,y) = \left(\dfrac{a}{x} + ay, \dfrac{bx}{1+xy}\right)$, which defines the dynamical system described by (E_{19}). It can be easily checked that $f(P) \in \mathbf{D}$. Therefore, it suffices to show that for any point $(x,y) \in \mathbf{D}_1$ there is $f(x,y) \in \mathbf{D}_2$ and also if $(x,y) \in \mathbf{D}_2$ then $f(x,y) \in \mathbf{D}_1$. Similarly, if $(x,y) \in \mathbf{D}_3$ then $f(x,y) \in \mathbf{D}_4$ and if $(x,y) \in \mathbf{D}_4$ then $f(x,y) \in \mathbf{D}_3$. Let $P_k = (x_k, y_k) \in \mathbf{D}_1$ then from (E_{19}), we obtain

$$x_{k+1} > \frac{a}{\sqrt{a(ab+1)}} + \frac{ba^2}{\sqrt{a(ab+1)}} = \frac{a + ba^2}{\sqrt{a(ab+1)}} = \sqrt{a(ab+1)}.$$

Further, since

$$y_{n+1} = \frac{ab}{x_{n+1}} \quad \text{for all} \quad n \in \mathbb{N}$$

it is clear that $P_{k+1} = (x_{k+1}, y_{k+1}) \in \mathbf{D}_2$. The rest of the relations can be proved analogously. ∎

We note that in the considered system every solution (at least from the point (x_2, y_2)) is 2–periodic. For a system (equation) which has periodic or asymptotically periodic solutions we can always find a family of sets for which these solutions are oscillatory (at least eventually) between them. Further, in the space with a cone of positive elements and (or) a cone of negative elements we can define oscillations between these cones. Moreover, the concept of oscillation can also be useful in considerations of difference fuzzy equations and leads to a concept of fuzzy oscillations.

Finally, to show how different forms oscillation between sets can have, we present here two examples.

Example 1.10.2. Let $\mathbf{X}$ denote the set of $m \times m$ matrices. By $\mathbf{X}_L$ $(\mathbf{X}_P)$ we denote subset of $\mathbf{X}$ consisting of all lower triangular (upper triangle) matrices. Consider the matrix difference equation

$$\mathbf{Y}_{n+1} = \mathbf{A}_n \mathbf{Y}_n^T, \quad n \in \mathbb{N} \tag{E_{20}}$$

where $\{\mathbf{A}_n\}_{n=1}^{\infty}$ is a sequence of $m \times m$ diagonal matrices, and $\mathbf{Y}^T$ denotes the transpose of $\mathbf{Y}$. We can check that every solution of (E_{20}) with the initial condition $\mathbf{Y}_1 \in \mathbf{X}_L \cup \mathbf{X}_P$ oscillates between sets $\mathbf{X}_L$ and $\mathbf{X}_P$. We note that $\mathbf{X}_L \cap \mathbf{X}_P = \mathbf{X}_D$ where $\mathbf{X}_D$ is the set of $m \times m$ diagonal matrices. This set, in our example, plays the role similar to zero element in the ordinary oscillation definition. If for some $k \in \mathbb{N}$ we find that $\mathbf{Y}_k \in \mathbf{X}_D$ then $\mathbf{Y}_n \in \mathbf{X}_D$ for all $n \geq k$.

For the next example, we need a different version of Definition 1.10.1. For this, suppose that we have a set $\mathbf{X}$, and the set of properties $\{p_1, p_2, \cdots, p_k\}$ which are possessed by some elements of $\mathbf{X}$. By $\mathbf{X}_{p_i}$ we shall denote the subset of $\mathbf{X}$ which possess property p_i.

Definition 1.10.2. Let $\{x_n\}_{n=1}^{\infty}$ be a sequence of elements of $\mathbf{X}$. We say that this sequence is *oscillatory (strictly) in relation to the set of properties* $(p_1, p_2, \cdots, p_k)$ if there exist (disjoint) sequences of indices $n^i = \{n_j^i\}_{j=1}^{\infty}$, $i = 1, \cdots, k$ such that for all $i \in \{1, 2, \cdots, k\}$ and $j \in \mathbb{N}$, $x_{n_j^i} \in \mathbf{X}_{p_i}$.

Example 1.10.3. Consider a scalar difference equation. We say that this equation possesses property p_c if every solution of this equation converges, and property p_{nc} if every (may be over a finite set) solution is nonconvergent. We now consider the sequence $\{x_m\}_{m=1}^{\infty}$ of difference equations of the form

$$x_m : y_{n+1} = \left(1 + \frac{(-1)^m}{m}\right) y_n, \quad n \in \mathbb{N}, \quad m \in \mathbb{N}. \tag{E_{21}}$$

(It is clear that (E_{21}) is a particular case of the equation $y_{n+1} = f(n, y_n, m)$, $n \in \mathbb{N}$ with the natural valued parameter m, and oscillation is in the sense of variation of parameters).

It is easy to check that for m odd the equation x_m has the property p_c, whereas for m even property p_{nc}. We can say that this sequence of equations oscillate in relation to the set of properties $\{p_c, p_{nc}\}$.

1.11. Oscillation of Continuous–Discrete Recurrence Equations

Let J be an interval (bounded or unbounded) on the real line, and let y be a real-valued function defined on J. Usually, the function y is called oscillatory on J if it possesses an infinite number of zeros contained in J, where by a zero of the function y we mean a point x such that $y(x) = 0$. For continuous functions this definition is quite satisfactory. However, the function y defined on $[0, 1]$, where

$$y(x) = \begin{cases} 1 & \text{for} \quad x \text{ rational} \\ -1 & \text{for} \quad x \text{ irrational} \end{cases}$$

can be called oscillatory, but it does not vanish on $[0, 1]$. Thus, in this case we should talk about generalized zeros.

Let $(\mathbf{X}, d)$ be a metric space and $y : \mathbf{X} \to \mathbb{R}$ be a given function. We say that at the point x the function y possesses a generalized zero, if in every sufficiently small neighborhood $o_\epsilon(x)$ of the point x there exist two points $p, q \in o_\epsilon(x)$ such that

$$y(p) > 0 \wedge y(q) = 0 \vee y(p) < 0 \wedge y(q) = 0 \vee y(p) > 0 \wedge y(q) < 0. \quad (1.11.1)$$

Now let $\mathbf{J}$ be a subset of $(\mathbf{X}, d)$, by $o_\epsilon(x)\Big|_{\mathbf{J}}$ we shall denote the neighborhood of the point x in the set $\mathbf{J}$, i.e. $o_\epsilon(x)\Big|_{\mathbf{J}} = o_\epsilon(x) \cap \mathbf{J}$. The set of all generalized zeros of the function y contained in the set $\mathbf{J}$ we shall denote by $gzer(y, \mathbf{J})$.

Definition 1.11.1. Let $\mathbf{J}$ be a subset of the metric space $(\mathbf{X}, d)$ and $y : \mathbf{J} \to \mathbb{R}$. We say that the function y *oscillate on the set* $\mathbf{J}$ if $card(gzer(y, \mathbf{J})) \geq \aleph_0$, and y does not oscillate if $card(gzer(y, \mathbf{J})) < \aleph_0$.

The space $\mathbf{X}$ for example can be m–dimensional Euclidean space $\mathbb{R}^m$. We also note that in the above definition the set $\mathbb{R}$ can be replaced by any ordered set $(\mathbf{S}, \prec)$, zero by $a \in (\mathbf{S}, p)$, $y : \mathbf{J} \to \mathbf{S}$ and the condition (1.11.1) by

$$y(p) \prec a \wedge a \prec y(q) \wedge (y(p) \neq a \neq y(q))$$

or

$$y(p) = a \wedge a \prec y(q) \wedge a \neq y(q)$$

or

$$y(p) = a \wedge y(q) \prec a \wedge a \neq y(q).$$

The above remarks are enough to study equations of functions in more than one variable, one of which takes discrete values and the second continuous.

Definition 1.11.2. Let $\mathbf{X} = N \times \mathbf{J}$, where $\mathbf{J} \subset \mathbb{R}^m$, and $y : \mathbf{X} \to \mathbb{R}$. We say that the function y is *oscillatory* if there exists a sequence $\{x_k\}_{k=1}^{\infty} = \{(n_k, t_k)\}_{k=1}^{\infty}$ with $\{n_k\}_{k=1}^{\infty}$ strictly increasing sequence of positive integers such that the sequence $\{y(n_k, t_k)\}_{k=1}^{\infty}$ is oscillatory.

In the above definition all t_k, $k \in \mathbb{N}$ can be the same, so all sections $y(n,t)$ can be nonoscillatory as the functions of the variable t for each fixed $n \in \mathbb{N}$.

A more restrictive definition is the following:

Definition 1.11.3. Let $\mathbf{X} = \mathbb{N} \times \mathbf{J}$, where $\mathbf{J} \subset \mathbb{R}^m$, and $y : \mathbf{X} \to \mathbb{R}$. We say that the function y is *oscillatory in relation to the set of discrete sections,* if there exists a sequence $\{n_k\}_{k=1}^{\infty}$ such that for each n_k the suitable section $y(n_k, t)$ is an oscillatory function on $\mathbf{J}$.

It is clear that if the function y is oscillatory in relation to the set of discrete sections, then it is oscillatory in the sense of Definition 1.11.2, however, the converse is not true.

In our first result here, we shall consider the equation

$$y(n+1, Tt) = a(t)y(n,t), \quad n \in \mathbb{N},\ t \in \mathbb{R}^m \qquad (E_{22})$$

where T is some one–to–one mapping from the space $\mathbb{R}^m$ onto $\mathbb{R}^m$. Let T^{-1} denote the inverse of T and $T^{-j} = T^{-1} \circ T^{-1} \circ \cdots \circ T^{-1}$, i.e. T^{-1} composed j–times.

Theorem 1.11.1. Every solution of (E_{22}) with the initial function $y(1,t)$ such that

$$y\left(1, T^{-n_k} t_k\right) y\left(1, T^{-n_k+1} t_k\right) a\left(T^{-n_k} t_k\right) < 0, \quad k \in \mathbb{N} \qquad (1.11.2)$$

where $\{n_k\}_{k=1}^{\infty}$ is some strictly increasing sequence of positive integers, and $\{t_k\}_{k=1}^{\infty}$ is a sequence of elements of the space $\mathbb{R}^m$, is oscillatory in the sense of Definition 1.11.2.

Proof. The solution of the equation (E_{22}) can be written as

$$y(n,t) = \left[\prod_{i=1}^{n-1} a\left(T^{-i} t\right)\right] y\left(1, T^{-n+1} t\right).$$

Hence, for any $\tau \in \mathbb{R}^m$ we have

$$y(n+1,\tau)y(n,\tau) = \left[\prod_{i=1}^{n-1} a\left(T^{-i}\tau\right)\right]^2 \left\{a\left(T^{-n}\tau\right) y\left(1,T^{-n}\tau\right) y\left(1,T^{-n+1}\tau\right)\right\}$$

and now the result follows from (1.11.2). ∎

The operator T can be for example $Tt = t+1$, $T : \mathbb{R} \to \mathbb{R}$, in fact in this case we have the equation $y(n+1,t+1) = a(t)y(n,t)$, $n \in \mathbb{N}$, $t \in \mathbb{R}$, or $Tt = (c_1t_1, c_2t_2)$, $c_1, c_2 \in \mathbb{R}$, $T : \mathbb{R}^2 \to \mathbb{R}^2$ which yields the equation $y(n+1, c_1t_1, c_2t_2) = a(t_1,t_2)y(n,t_1,t_2)$, $n \in \mathbb{N}$, $t_1, t_2 \in \mathbb{R}$.

Finally, we shall consider the following difference–differential equation

$$\Delta_1 y(n,t) + a_n \frac{\partial^2 y(n,t)}{\partial t^2} = 0, \quad n \in \mathbb{N},\ t \in R_0. \qquad (E_{23})$$

Theorem 1.11.2. Let $a : \mathbb{N} \to \mathbb{R}$ be such that $a_n < \alpha$ for all $n \in \mathbb{N}$ and some $\alpha \in \mathbb{R}_+$. Then, (E_{23}) possesses both oscillatory and nonoscillatory solutions in the sense of Definition 1.11.2.

Proof. We shall seek the solution of (E_{23}) in the form $y(n,t) = u(n)v(t)$. For this, let y to be of the form

$$y(n,t) = \left[\prod_{j=1}^{n-1} (1 + Ca_j)\right] v(t), \quad n \in \mathbb{N},\ t \in \mathbb{R}_0 \qquad (1.11.3)$$

where C is an arbitrary constant. We take C such that $1 + Ca_n \neq 0$ for all $n \in \mathbb{N}$. On substituting (1.11.3) in (E_{23}), we obtain

$$a_n \left[\prod_{j=1}^{n-1} (1 + Ca_j)\right] [v''(t) + Cv(t)] = 0. \qquad (1.11.4)$$

From (1.11.4) it follows that for (1.11.3) to be a solution of (E_{23}) the function v must satisfy the differential equation

$$v''(t) + Cv(t) = 0. \qquad (1.11.5)$$

Let $C = -d^2$ be such that $1 - d^2a_n > 0$ for all $n \geq \eta \in \mathbb{N}$. This choice is always possible because for $0 < d < 1/\sqrt{\alpha}$, we have $1 - d^2a_n > 1 - d^2\alpha > 0$. For such a C the sequence $\left\{\prod_{j=1}^{n-1}(1 - d^2a_j)\right\}_{n=1}^{\infty}$ is nonoscillatory for $n \geq \eta$, and

$$\operatorname{sgn} \prod_{j=1}^{n-1}(1 - d^2a_j) = \operatorname{sgn} \prod_{j=1}^{\eta-1}(1 - d^2a_j)$$

for all $n \geq \eta$. Furthermore, every solution of (1.11.5) is in this case nonoscillatory, so the same is for $y(n,t) = u(n)v(t)$ with $u(n) = \left[\prod_{j=1}^{n-1}(1 + Ca_j)\right]$. Similarly, if we take $C = d^2$ then every solution of (1.11.5) is oscillatory on $[0, \infty)$. Hence, for each $k \in \mathbb{N}$ the section

$$y(k,t) = \left[\prod_{j=1}^{k-1}(1 + d^2 a_j)\right] v(t)$$

is oscillatory on $[0, \infty)$. Therefore, the solution $y(n,t)$ is oscillatory in the sense of Definition 1.11.3 and hence in the sense of Definition 1.11.2. ∎

1.12. Second Order Quasilinear Difference Equations

Here we shall obtain sufficient conditions for the oscillation of all solutions, as well as necessary conditions for the existence of nonoscillatory solutions of the second order quasilinear difference equations of the type

$$\Delta\left(a_n|\Delta y_n|^{\alpha-1}\Delta y_n\right) + q_{n+1}f(y_{n+1}) = 0, \quad n \in \mathbb{N}(n_0) \qquad (E_{24})$$

where $n_0 \in \mathbb{N} \cup \{0\}$, $\alpha > 0$. In what follows with respect to the equation (E_{24}) we shall assume that

(i) $a_n > 0$ for all $n \in \mathbb{N}(n_0)$, and $R(n) = \sum_{s=n_0}^{n-1} 1/a_s^{1/\alpha} \to \infty$ as $n \to \infty$,

(ii) $\{q_n\}$ is a given real sequence and is allowed to change sign infinitely often on $\mathbb{N}(n_0)$,

(iii) $f : \mathbb{R} \to \mathbb{R}$ is continuous and $uf(u) > 0$ for all $u \neq 0$,

(iv) $f(u) - f(v) = g(u,v)\ (u - v)$ for all $u, v \neq 0$, where g is a nonnegative function,

(v) $\{q_n\}$ is (conditionally) summable on $\mathbb{N}(n_0)$, i.e.

$$\sum_{n=n_0}^{\infty} q_{n+1} = \lim_{j\to\infty} \sum_{n=n_0}^{j} q_{n+1}$$

exists and is finite, in which case the function $Q(n) = \sum_{s=n}^{\infty} q_{s+1}$ is well-defined on $\mathbb{N}(n_0)$, and

(vi) $\lim_{|u|\to\infty} f(u) = \infty$.

We shall need the following two lemmas:

Lemma 1.12.1. [2] Let the function $K(n,s,u) : \mathbb{N}(n_0) \times \mathbb{N}(n_0) \times \mathbb{R}_0 \to \mathbb{R}$ be nondecreasing in u for fixed n and s. Let $\{p_n\} \subset \mathbb{R}$ be a given sequence, and $\{y_n\}$, $\{x_n\}$ be sequences defined on $\mathbb{N}(n_0)$, satisfying

$$y_n \geq p_n + \sum_{s=n_0}^{n-1} K(n,s,y_s)$$

and

$$x_n = p_n + \sum_{s=n_0}^{n-1} K(n,s,x_s), \quad n \in \mathbb{N}(n_0).$$

Then, $x_n \leq y_n$ for all $n \in \mathbb{N}(n_0)$.

Lemma 1.12.2. Let $\{y_n\}$ be a real sequence such that $y_n \neq 0$ on $\mathbb{N}(n_0)$. Then, for any $\alpha > 0$ and $N \in \mathbb{N}(n_0)$, we have

$$\limsup_{n\to\infty} \left\{ \frac{a_n|\Delta y_n|^{\alpha-1}\Delta y_n}{f(y_n)} + \sum_{s=N}^{n-1} \frac{a_s|\Delta y_s|^{\alpha+1} g(y_s,y_{s+1})}{f(y_s)f(y_{s+1})} \right\} \geq 0, \quad (1.12.1)$$

where $\{a_n\}$ and f satisfy conditions (i), (iii) and (iv).

Proof. Suppose that (1.12.1) is false for some α and $N \in \mathbb{N}(n_0)$. Then, there exists a constant $k > 0$ and an integer $N_1 > N$ such that

$$\frac{a_n|\Delta y_n|^{\alpha-1}\Delta y_n}{f(y_n)} + \sum_{s=N_1}^{n-1} \frac{a_s|\Delta y_s|^{\alpha+1} g(y_s,y_{s+1})}{f(y_s)f(y_{s+1})} \leq -k, \quad n \geq N_1$$

or

$$k + \sum_{s=N_1}^{n-1} \frac{a_s|\Delta y_s|^{\alpha+1} g(y_s,y_{s+1})}{f(y_s)f(y_{s+1})} \leq -\frac{a_n|\Delta y_n|^{\alpha-1}\Delta y_n}{f(y_n)}, \quad n \geq N_1. \quad (1.12.2)$$

This implies that $y_n\Delta y_n < 0$ for $n \geq N_1$. If $\{y_n\}$ is positive, then (1.12.2) becomes

$$w_n \geq kf(y_n) + \sum_{s=N_1}^{n-1} \frac{f(y_n)g(y_s,y_{s+1})(-\Delta y_s)}{f(y_s)f(y_{s+1})} w_s,$$

where $w_n = -a_n|\Delta y_n|^{\alpha-1}\Delta y_n$. Define

$$K(n,s,x) = \frac{f(y_n)g(y_s,y_{s+1})(-\Delta y_s)}{f(y_s)f(y_{s+1})} x, \quad n \geq N_1, \quad x \in \mathbb{R}_0.$$

Notice that, for each fixed n and s, the function $K(n,s,x)$ is nondecreasing in x. So we can use Lemma 1.12.1 to obtain $w_n \geq v_n$, where v_n satisfies

$$v_n = kf(y_n) + \sum_{s=N_1}^{n-1} K(n,s,v_s)$$

provided $v_s \in \mathbb{R}_0$ for all $s \geq N_1$. Multiplying the last equality by $1/f(y_n)$, and then applying the operator Δ, we obtain $\dfrac{\Delta v_n}{f(y_n)} = 0$ so that $v_n = v_{N_1} = kf(y_{N_1})$ for all $n \geq N_1$. Thus,

$$-a_n|\Delta y_n|^{\alpha-1}\Delta y_n \;\geq\; kf(y_{N_1})$$

or

$$-\Delta y_n \;\geq\; \frac{k^{1/\alpha}f^{1/\alpha}(y_{N_1})}{a_n^{1/\alpha}}, \quad n \geq N_1. \tag{1.12.3}$$

Summing (1.12.3), we obtain $\lim_{n\to\infty} y_n = -\infty$, a contradiction. The proof for the case $\{y_n\}$ is negative follows from a similar argument by taking $w_n = a_n|\Delta y_n|^{\alpha-1}\Delta y_n$. ■

Theorem 1.12.3. If $\{y_n\}$ is a nonoscillatory solution of (E_{24}) such that $\liminf_{n\to\infty} |y_n| > 0$, then

$$\frac{a_n|\Delta y_n|^{\alpha-1}\Delta y_n}{f(y_n)} \;=\; Q(n) + \sum_{s=n}^{\infty} \frac{a_s|\Delta y_s|^{\alpha+1}g(y_s, y_{s+1})}{f(y_s)f(y_{s+1})}, \quad n \geq n_0. \tag{1.12.4}$$

Proof. There exist constants m_1, $m_2 > 0$ and an integer $n_1 > n_0$ such that $|y_n| \geq m_1$, and $|f(y_n)| \geq m_2$ for $n \in \mathbb{N}(n_1)$. Dividing the equation (E_{24}) by $f(y_{n+1})$ and summing it from n to $j-1$, we have

$$\frac{a_j|\Delta y_j|^{\alpha-1}\Delta y_j}{f(y_j)} - \frac{a_n|\Delta y_n|^{\alpha-1}\Delta y_n}{f(y_n)} + \sum_{s=n}^{j-1} \frac{a_s|\Delta y_s|^{\alpha+1}g(y_s, y_{s+1})}{f(y_s)f(y_{s+1})} + \sum_{s=n}^{j-1} q_{s+1} \;=\; 0, \tag{1.12.5}$$

for $j > n \geq n_1$. We claim that

$$\sum_{s=n}^{\infty} \frac{a_s|\Delta y_s|^{\alpha+1}g(y_s, y_{s+1})}{f(y_s)f(y_{s+1})} \;<\; \infty, \quad n \geq n_1. \tag{1.12.6}$$

If (1.12.6) does not hold, then there is an integer $N > n$ such that

$$-\frac{a_n|\Delta y_n|^{\alpha-1}\Delta y_n}{f(y_n)} + \sum_{s=n}^{N-1} \frac{a_s|\Delta y_s|^{\alpha+1}g(y_s, y_{s+1})}{f(y_s)f(y_{s+1})} + \sum_{s=n}^{j-1} q_{s+1} \;\geq\; 1, \quad j \geq N.$$

From (1.12.5), it follows that

$$\frac{a_j|\Delta y_j|^{\alpha-1}\Delta y_j}{f(y_j)} + \sum_{s=N}^{j-1} \frac{a_s|\Delta y_s|^{\alpha+1}g(y_s, y_{s+1})}{f(y_s)f(y_{s+1})}$$

$$= \frac{a_n|\Delta y_n|^{\alpha-1}\Delta y_n}{f(y_n)} - \sum_{s=n}^{N-1} \frac{a_s|\Delta y_s|^{\alpha+1} g(y_s, y_{s+1})}{f(y_s)f(y_{s+1})} - \sum_{s=n}^{j-1} q_{s+1} \le -1, \quad j \ge N$$

which contradicts Lemma 1.12.2. Hence our claim (1.12.6) holds. Now letting $j \to \infty$ in (1.12.5), we see that the finite limit

$$\beta = \lim_{j\to\infty} \left[\frac{a_j|\Delta y_j|^{\alpha-1}\Delta y_j}{f(y_j)}\right]$$

exists, and

$$\frac{a_n|\Delta y_n|^{\alpha-1}\Delta y_n}{f(y_n)} = \beta + Q(n) + \sum_{s=n}^{\infty} \frac{a_s|\Delta y_s|^{\alpha+1} g(y_s, y_{s+1})}{f(y_s)f(y_{s+1})}, \quad n \ge n_1. \tag{1.12.7}$$

To prove (1.12.4) it suffices to show that $\beta = 0$ in (1.12.7). Suppose that $y_n > 0$ for all $n \in \mathbb{N}(n_1)$. The proof for the case $y_n < 0$ for all large n is similar. Let $\beta > 0$. From (1.12.7), there exists an integer $N_1 > n_1$ such that

$$y_n \Delta y_n > 0 \quad \text{and} \quad \frac{a_n(\Delta y_n)^\alpha}{f(y_n)} \ge \frac{\beta}{2} \quad \text{for all} \quad n \in \mathbb{N}(N_1). \tag{1.12.8}$$

From (1.12.8), we have the inequality

$$\Delta y_n \ge \left(\frac{\beta}{2}\frac{f(y_{N_1})}{a_n}\right)^{1/\alpha}, \quad n \ge N_1$$

which on summation, leads to

$$\lim_{n\to\infty} y_n = \infty. \tag{1.12.9}$$

Since

$$\frac{a_n|\Delta y_n|^{\alpha+1} g(y_n, y_{n+1})}{f(y_n)f(y_{n+1})} \ge \frac{\beta}{2}\frac{g(y_n, y_{n+1})\Delta y_n}{f(y_{n+1})}$$

we have

$$\sum_{s=N_1}^{n-1} \frac{a_s|\Delta y_s|^{\alpha+1} g(y_s, y_{s+1})}{f(y_s)f(y_{s+1})} \ge \frac{\beta}{2}\sum_{s=N_1}^{n-1} \frac{g(y_s, y_{s+1})\Delta y_s}{f(y_{s+1})}.$$

Let $h(t) = f(y_{n+1}) + g(y_n, y_{n+1})(t - y_n)$ for $y_n \le t \le y_{n+1}$, $n \ge N_1$. Then, $h'(t) = g(y_n, y_{n+1}) > 0$ and $h(y_n) = f(y_{n+1})$, $n \ge N_1$. Hence,

$$\sum_{s=N_1}^{n-1} \frac{a_s|\Delta y_s|^{\alpha+1} g(y_s, y_{s+1})}{f(y_s)f(y_{s+1})} \ge \frac{\beta}{2}\sum_{s=N_1}^{n-1} \int_{y_s}^{y_{s+1}} \frac{h'(t)}{h(t)}dt = \frac{\beta}{2}\ln\frac{h(y_{n+1})}{h(y_{N_1})},$$

which, in view of (1.12.9), tends to infinity as $n \to \infty$. This however contradicts (1.12.6). Next, assume that $\beta < 0$. Choose an integer $N_2 > n_0$ such that

$$-\frac{a_n|\Delta y_n|^{\alpha-1}\Delta y_n}{f(y_n)} + \sum_{s=n}^{j-1} q_{s+1} \geq -\frac{\beta}{2}, \quad j \geq n \geq N_2.$$

In view of (1.12.5), we then have

$$\frac{a_j|\Delta y_j|^{\alpha-1}\Delta y_j}{f(y_j)} + \sum_{s=n}^{j-1} \frac{a_s|\Delta y_s|^{\alpha+1} g(y_s, y_{s+1})}{f(y_s)f(y_{s+1})} \leq \frac{\beta}{2} < 0, \quad j \geq n \geq N_2$$

which contradicts Lemma 1.12.2. Thus, we must have $\beta = 0$. ∎

Theorem 1.12.4. Assume that

$$ug(u,v) \geq f(u) \quad \text{for} \quad u, v \neq 0. \tag{1.12.10}$$

If $Q(n) \geq 0$ for $n \in \mathbb{N}(n_0)$, then every nonoscillatory solution $\{y_n\}$ of (E_{24}) satisfies eventually 'a priori' estimate

$$c_1 \leq |y_n| \leq c_2 R(n) \tag{1.12.11}$$

for some positive constants c_1 and c_2 (depending on y_n).

Proof. We may assume that $\{y_n\}$ is eventually positive, i.e. $y_n > 0$ for $n \in \mathbb{N}(n_0)$. Since $Q(n) \geq 0$, we have from Theorem 1.12.3 that $\Delta y_n \geq 0$, and hence $y_n \geq y_{n_0}$ for $n \geq n_0$. Thus, the first inequality $y_n \geq c_1$, $n \geq n_0$ in (1.12.11) holds with $c_1 = y_{n_0}$. From (1.12.4), we have

$$(\Delta y_n)^\alpha \geq \frac{Q(n)f(y_n)}{a_n}, \quad n \geq n_0. \tag{1.12.12}$$

Summing equation (E_{24}) from n to $j-1$, we obtain

$$a_n(\Delta y_n)^\alpha = a_j(\Delta y_j)^\alpha - Q(j)f(y_j) + Q(n)f(y_n) + \sum_{s=n}^{j-1} Q(s)g(y_s, y_{s+1})\Delta y_s \tag{1.12.13}$$

for $j > n \geq n_0$. Note that $\sum_{s=n}^{\infty} Q(s)g(y_s, y_{s+1})\Delta y_s < \infty$, otherwise, it would follow from (1.12.13) that $a_j(\Delta y_j)^\alpha - Q(j)f(y_j) \to -\infty$ as $j \to \infty$, which contradicts (1.12.12). Therefore, letting $j \to \infty$ in (1.12.13), we find

$$a_n(\Delta y_n)^\alpha = \beta + Q(n)f(y_n) + \sum_{s=n}^{\infty} Q(s)g(y_s, y_{s+1})\Delta y_s, \quad n \geq n_0 \tag{1.12.14}$$

where β denotes the finite limit

$$\beta = \lim_{j\to\infty} (a_j(\Delta y_j)^\alpha - Q(j)f(y_j)) \geq 0.$$

Define

$$K_1(n) = \sum_{s=n}^{\infty} Q(s)g(y_s, y_{s+1})\Delta y_s, \quad n \geq n_0 \tag{1.12.15}$$

and

$$K_2(n) = \sum_{s=n}^{\infty} \frac{Q^{(\alpha+1)/\alpha}(s)g(y_s, y_{s+1})f^{1/\alpha}(y_s)}{a_s^{1/\alpha}}, \quad n \geq n_0. \tag{1.12.16}$$

From (1.12.12) and (1.12.15), we see that $K_1(n) \geq K_2(n)$, and hence $K_2(n)$ is well-defined for $n \geq n_0$. From (1.12.14), we have

$$a_n^{1/\alpha}\Delta y_n \leq 3^{1/\alpha}\left\{\beta^{1/\alpha} + Q^{1/\alpha}(n)f^{1/\alpha}(y_n) + K_1^{1/\alpha}(n)\right\}, \quad n \geq n_0.$$

Summing the last inequality from n_0 to $n-1$, we obtain

$$y_n \leq y_{n_0} + 3^{1/\alpha}\left\{\beta^{1/\alpha}R(n) + \sum_{s=n_0}^{n-1} \frac{Q^{1/\alpha}(s)f^{1/\alpha}(y_s)}{a_s^{1/\alpha}} + \sum_{s=n_0}^{n-1} \frac{K_1^{1/\alpha}(s)}{a_s^{1/\alpha}}\right\}, \tag{1.12.17}$$

for $n \geq n_0$. From the decreasing property of $K_1(n)$, we obtain

$$\sum_{s=n_0}^{n-1} \frac{K_1^{1/\alpha}(s)}{a_s^{1/\alpha}} \leq K_1^{1/\alpha}(n_0)R(n), \quad n \geq n_0. \tag{1.12.18}$$

Further, from Hölder's inequality and (1.12.10), we have

$$\begin{aligned}
&\sum_{s=n_0}^{n-1} \frac{Q^{1/\alpha}(s)f^{1/\alpha}(y_s)}{a_s^{1/\alpha}} \\
&\quad = \sum_{s=n_0}^{n-1} \frac{Q^{1/\alpha}(s)(g(y_s, y_{s+1}))^{1/(\alpha+1)}f^{1/\alpha}(y_s)}{a_s^{1/\alpha}(g(y_s, y_{s+1}))^{1/(\alpha+1)}} \\
&\quad \leq \left[\sum_{s=n_0}^{n-1} \frac{Q^{(\alpha+1)/\alpha}(s)(g(y_s, y_{s+1}))f^{1/\alpha}(y_s)}{a_s^{1/\alpha}}\right]^{1/(\alpha+1)} \\
&\qquad \times \left[\sum_{s=n_0}^{n-1} \frac{f^{1/\alpha}(y_s)}{a_s^{1/\alpha}(g(y_s, y_{s+1}))^{1/\alpha}}\right]^{\alpha/(\alpha+1)} \\
&\quad \leq K_2^{1/(\alpha+1)}(n_0)y_n^{1/(\alpha+1)}R^{\alpha/(\alpha+1)}(n), \quad n \geq n_0.
\end{aligned} \tag{1.12.19}$$

Using (1.12.18) and (1.12.19) in (1.12.17), we obtain

$$\begin{aligned} y_n &\le y_{n_0} + 3^{1/\alpha}\left\{\left[\beta^{1/\alpha} + K_1^{1/\alpha}(n_0)\right] R(n)\right. \\ &\quad \left. + K_2^{1/(\alpha+1)}(n_0) y_n^{1/(\alpha+1)} R^{\alpha/(\alpha+1)}(n)\right\}, \end{aligned}$$

for $n \ge n_0$. It then follows that

$$\frac{y_n}{R(n)} \le \frac{y_{n_0}}{R(n_0)} + 3^{1/\alpha}\left\{\beta^{1/\alpha} + K_1^{1/\alpha}(n_0) + K_2^{1/(\alpha+1)}(n_0)\left(\frac{y_n}{R(n)}\right)^{1/(\alpha+1)}\right\}$$

or

$$z_n \le A + B z_n^{1/(\alpha+1)},$$

where $z_n = y_n/R(n)$, $A = y_{n_0}/R(n_0) + 3^{1/\alpha}\left(\beta^{1/\alpha} + K_1^{1/\alpha}(n_0)\right)$, and $B = 3^{1/\alpha} K_2^{1/(\alpha+1)}(n_0)$. From the above inequality it is easy to see that z_n is bounded above by a constant depending on A and B. Thus, $y_n \le c_2 R(n)$, $n \ge n_0$ for some positive constant c_2. ∎

In view of Theorem 1.12.4 solutions which are asymptotically equivalent to non-zero constant c_1 will be called the solutions of *minimal type*, while the solutions which are asymptotic to $c_2 R(n)$ will be called of *maximal type*.

Theorem 1.12.5. Suppose that $Q(n) \ge 0$ for all $n \in \mathbb{N}(n_0)$. A necessary condition for the equation (E_{24}) to have a nonoscillatory solution which tends to a nonzero constant as $n \to \infty$ is that

$$\sum_{n=n_0}^{\infty}\left(\frac{Q(n)}{a_n}\right)^{1/\alpha} < \infty \quad \text{and} \quad \sum_{n=n_0}^{\infty}\left(\frac{1}{a_n}\sum_{s=n}^{\infty}\frac{Q^{(\alpha+1)/\alpha}(s)}{a_s^{1/\alpha}}\right)^{1/\alpha} < \infty.$$

Proof. Let $\{y_n\}$ be a nonoscillatory solution of (E_{24}). We may suppose that $y_n > 0$ for $n \in \mathbb{N}(n_0)$. Then, by Theorem 1.12.3, $\Delta y_n \ge 0$ for $n \ge n_0$, and

$$\frac{a_n(\Delta y_n)^\alpha}{f(y_n)} = Q(n) + \sum_{s=n}^{\infty}\frac{a_s(\Delta y_s)^{\alpha+1} g(y_s, y_{s+1})}{f(y_s) f(y_{s+1})}, \quad n \ge n_0. \tag{1.12.20}$$

Hence,

$$\frac{(\Delta y_n)^\alpha}{f(y_n)} \ge \frac{Q(n)}{a_n} \quad \text{or} \quad \frac{\Delta y_n}{f^{1/\alpha}(y_n)} \ge \left(\frac{Q(n)}{a_n}\right)^{1/\alpha}, \quad n \ge n_0. \tag{1.12.21}$$

Summing the second inequality in (1.12.21), we obtain

$$\sum_{s=n_0}^{n-1}\left(\frac{Q(s)}{a_s}\right)^{1/\alpha} \le \sum_{s=n_0}^{n-1}\frac{\Delta y_s}{f^{1/\alpha}(y_s)} \le \frac{y_n - y_{n_0}}{f^{1/\alpha}(y_{n_0})}, \quad n \ge n_0$$

which, in view of the boundedness of $\{y_n\}$ implies that

$$\sum_{n=n_0}^{\infty}\left(\frac{Q(n)}{a_n}\right)^{1/\alpha} < \infty.$$

Next, from (1.12.20) and (1.12.21), we have

$$\begin{aligned}\frac{a_n(\Delta y_n)^\alpha}{f(y_n)} &\geq \sum_{s=n}^{\infty}\frac{a_s(\Delta y_s)^{\alpha+1}g(y_s,y_{s+1})}{f(y_s)f(y_{s+1})}\\ &\geq \sum_{s=n}^{\infty}\frac{Q^{(\alpha+1)/\alpha}(s)}{a_s^{1/\alpha}}\frac{g(y_s,y_{s+1})f^{1/\alpha}(y_s)}{f(y_{s+1})}, \quad n\geq n_0.\end{aligned}$$

Since $\{y_n\}$ is bounded, there exists an integer $n_1 > n_0$ such that

$$\frac{g(y_n,y_{n+1})f^{1/\alpha}(y_n)}{f(y_{n+1})} \geq c, \quad n\geq n_1.$$

Thus, for $n\geq n_1$, we have

$$\frac{a_n(\Delta y_n)^\alpha}{f(y_n)} \geq c\sum_{s=n}^{\infty}\frac{Q^{(\alpha+1)/\alpha}(s)}{a_s^{1/\alpha}}$$

or

$$c^{1/\alpha}\left(\frac{1}{a_n}\sum_{s=n}^{\infty}\frac{Q^{(\alpha+1)/\alpha}(s)}{a_s^{1/\alpha}}\right)^{1/\alpha} \leq \frac{\Delta y_n}{f^{1/\alpha}(y_n)}, \quad n\geq n_1.$$

Summing the last inequality from n_1 to $n-1$ and using the boundedness of $\{y_n\}$, we obtain

$$\sum_{n=n_0}^{\infty}\left(\frac{1}{a_n}\sum_{s=n}^{\infty}\frac{Q^{(\alpha+1)/\alpha}(s)}{a_s^{1/\alpha}}\right)^{1/\alpha} < \infty. \quad ■$$

Theorem 1.12.6. Suppose that $Q(n)\geq 0$ for all $n\in \mathbb{N}(n_0)$ and $\{a_n\}$ is nondecreasing. A sufficient condition for (E_{24}) to have a nonoscillatory solution which tends to a nonzero constant as $n\to\infty$ is that

$$\sum_{n=n_0}^{\infty}A^{1/\alpha}(n) < \infty \quad\text{and}\quad \sum_{n=n_0}^{\infty}\left(\sum_{s=n}^{\infty}A^{(\alpha+1)/\alpha}(s)\right)^{1/\alpha} < \infty, \tag{1.12.22}$$

where $A(n) = Q(n)/a_n$.

Proof. Define

$$\rho(n) = \sum_{s=n}^{\infty}A^{(\alpha+1)/\alpha}(s), \quad n\geq n_0$$

then by Hölder's inequality, we have

$$\begin{aligned}\sum_{s=n}^{\infty} A(s)\rho^{1/\alpha}(s) &\le \left(\sum_{s=n}^{\infty} A^{(\alpha+1)/\alpha}(s)\right)^{\alpha/(\alpha+1)} \left(\sum_{s=n}^{\infty} \rho^{(\alpha+1)/\alpha}(s)\right)^{1/(\alpha+1)} \\ &\le \rho^{\alpha/(\alpha+1)}(n)\left(\rho(n)\sum_{s=n}^{\infty}\rho^{1/\alpha}(s)\right)^{1/(\alpha+1)} \\ &= \rho(n)\left(\sum_{s=n}^{\infty}\rho^{1/\alpha}(s)\right)^{1/(\alpha+1)}, \quad n \ge n_0. \end{aligned} \tag{1.12.23}$$

Let $\lambda > 0$ be a fixed constant, and put $\mu = f(3\lambda/2)$,

$$\overline{\lambda} = \begin{cases} 3\lambda/2, & f \text{ is linear or superlinear} \\ \lambda/2, & f \text{ is sublinear.} \end{cases}$$

Let $M = \max\{g(u,v) : |u| \le \overline{\lambda},\ |v| \le \overline{\lambda}\}$, then $M > 0$. Choose $\delta > 0$ so that

$$M(2\mu)^{1/\alpha} \le \frac{\delta}{2} \tag{1.12.24}$$

and let $N \in \mathbb{N}(n_0)$ be large enough, so that

$$(2\mu)^{1/\alpha}\sum_{s=N}^{\infty} A^{1/\alpha}(s) \le \frac{\lambda}{4}, \tag{1.12.25}$$

$$(2\delta)^{1/\alpha}\sum_{s=N}^{\infty} \rho^{1/\alpha}(s) \le \frac{\lambda}{4} \tag{1.12.26}$$

and

$$M(2\delta)^{1/\alpha}\left(\sum_{s=N}^{\infty}\rho^{1/\alpha}(s)\right)^{1/(\alpha+1)} \le \frac{\delta}{2}. \tag{1.12.27}$$

Further, let

$$B = \{y = \{y_n\} : y_n \text{ is defined for } n \ge N\},$$

$$Y = \left\{ y \in B : |y_n - \lambda| \le \frac{\lambda}{2},\ n \ge N \text{ and } |y_{n_1} - y_{n_2}| \le (\mu\overline{A}(N) + \delta\rho(N))^{1/\alpha} \times |n_1 - n_2|, \quad n_1, n_2 \ge N \right\},$$

where $\overline{A}(N) = \sup\{A(n) : n \ge N\}$, and

$$X = \{x = \{x_n\} \in B : |x_n| \le \mu A(n) + \delta\rho(n),\ n \ge N\}.$$

Let T_1 and T_2 denote the mappings from $Y \times X \to B$ defined by

$$T_1(y,x)_n = \lambda - \sum_{s=n}^{\infty} |x_s|^{1/\alpha - 1} x_s, \quad n \geq N \tag{1.12.28}$$

and

$$T_2(y,x)_n = A(n)f(y_n) + \frac{1}{a_n}\sum_{s=n}^{\infty} Q(s)g(y_s, y_{s+1})|x_s|^{1/\alpha-1}x_s, \quad n \geq N \tag{1.12.29}$$

respectively. Finally, we define the mapping $T : Y \times X \to B \times B$ by

$$T(y,x) = (T_1(y,x), T_2(y,x)), \quad (y,x) \in Y \times X. \tag{1.12.30}$$

Let $(y,x) \in Y \times X$. Then, from $|x_n| \leq \mu A(n) + \delta\rho(n)$, we get

$$|x_n|^{1/\alpha} \leq (2\mu)^{1/\alpha}A^{1/\alpha}(n) + (2\delta)^{1/\alpha}\rho^{1/\alpha}(n), \quad n \geq N$$

and on using (1.12.25) and (1.12.26), we find

$$\sum_{s=n}^{\infty} |x_s|^{1/\alpha-1}x_s \leq (2\mu)^{1/\alpha}\sum_{s=n}^{\infty} A^{1/\alpha}(s) + (2\delta)^{1/\alpha}\sum_{s=n}^{\infty}\rho^{1/\alpha}(s) \leq \frac{\lambda}{2}, \quad n \geq N.$$

Thus, in view of (1.12.28), it follows that $|T_1(y,x)_n - \lambda| \leq \lambda/2$ for $n \geq N$. Also,

$$\begin{aligned}
|T_1(y,x)_{n_1} - T_1(y,x)_{n_2}| &= \left|\sum_{s=n_1}^{n_2-1} |x_s|^{1/\alpha-1}x_s\right| \\
&\leq \left|\sum_{s=n_1}^{n_2-1} (\mu A(s) + \delta\rho(s))^{1/\alpha}\right| \\
&\leq (\mu\overline{A}(N) + \delta\rho(N))^{1/\alpha}|n_1 - n_2|
\end{aligned}$$

for $n_1, n_2 \geq N$. This implies that $T_1(y,x) \in Y$. From (1.12.29), we have

$$\begin{aligned}
|T_2(y,x)_n| &\leq f(3\lambda/2)A(n) + M\sum_{s=n}^{\infty} A(s)(\mu A(s) + \delta\rho(s))^{1/\alpha} \\
&\leq \mu A(n) + M(2\mu)^{1/\alpha}\sum_{s=n}^{\infty} A^{(\alpha+1)/\alpha}(s) \\
&\quad + M(2\delta)^{1/\alpha}\sum_{s=n}^{\infty} A(s)\rho^{1/\alpha}(s) \\
&\leq \mu A(n) + \frac{\delta}{2}\rho(n) + M(2\delta)^{1/\alpha}\rho(n)\left(\sum_{s=n}^{\infty}\rho^{1/\alpha}(s)\right)^{1/(\alpha+1)} \\
&\leq \mu A(n) + \frac{\delta}{2}\rho(n) + \frac{\delta}{2}\rho(n) = \mu A(n) + \delta\rho(n), \quad n \geq N
\end{aligned}$$

where (1.12.23), (1.12.24), (1.12.27) and the nondecreasing nature of $\{a_n\}$ have been used. This shows that $T_2(y,x) \in X$. Therefore, by (1.12.30), we conclude that $T(Y \times X) \subseteq Y \times X$. It is clear that $Y \times X$ is closed and convex subset of $B \times B$ and $T(Y \times X)$ is relatively compact in $B \times B$. Therefore, by Schauder's fixed point theorem there exists an element $(y,x) \in Y \times X$ such that $T(y,x) = (y,x)$. Thus, by (1.12.30), (1.12.28) and (1.12.29), y_n and x_n satisfy the following equations for $n \geq N$,

$$\begin{aligned} y_n &= \lambda - \sum_{s=n}^{\infty} |x_s|^{1/\alpha-1} x_s, \\ x_n &= A(n)f(y_n) + \frac{1}{a_n}\sum_{s=n}^{\infty} Q(s)g(y_s,y_{s+1})|x_s|^{1/\alpha-1}x_s. \end{aligned} \tag{1.12.31}$$

From the above equations, we find that $\Delta y_n = |x_n|^{1/\alpha-1}x_n$, or $|\Delta y_n|^{\alpha-1} \times \Delta y_n = x_n$, and

$$\begin{aligned} \Delta(a_n x_n) &= -q_{n+1}f(y_{n+1}) + Q(n)g(y_n,y_{n+1})\Delta y_n - Q(n)g(y_n,y_{n+1})\Delta y_n \\ &= -q_{n+1}f(y_{n+1}), \quad n \geq N \end{aligned}$$

which shows that $\{y_n\}$ is a positive solution of the equation (E_{24}) for $n \geq N$. Since (1.12.31) implies that $\lim_{n\to\infty} y_n = \lambda > 0$, the proof of our theorem is complete. ∎

If $a_n \equiv 1$, then from Theorems 1.12.5 and 1.12.6 we have the following immediate corollary.

Corollary 1.12.7. Suppose $Q(n) \geq 0$ for all $n \in \mathbb{N}(n_0)$. A necessary and sufficient condition for the equation (E_{24}) to have a nonoscillatory solution which tends to a nonzero constant as $n \to \infty$ is that

$$\sum_{n=n_0}^{\infty} Q^{1/\alpha}(n) < \infty \quad \text{and} \quad \sum_{n=n_0}^{\infty}\left(\sum_{s=n}^{\infty} Q^{(\alpha+1)/\alpha}(s)\right)^{1/\alpha} < \infty.$$

Theorem 1.12.8. Assume that $Q(n) \geq 0$ for $n \in \mathbb{N}(n_0)$ and $g(u,v) \geq \lambda > 0$ for $u,v \neq 0$. A necessary condition for the equation (E_{24}) to have a nonoscillatory solution $\{y_n\}$ such that

$$c_1 R(n) \leq |y_n| \leq c_2 R(n), \quad n \in \mathbb{N}(n_0) \tag{1.12.32}$$

for some positive constants c_1 and c_2 is that

$$\sum_{n=n_0}^{\infty} \frac{Q^{(\alpha+1)/\alpha}(n) f^{1/\alpha}(c_1 R(n))}{a_n^{1/\alpha}} < \infty. \tag{1.12.33}$$

Proof. Let $\{y_n\}$ be a nonoscillatory solution of the equation (E_{24}) satisfying (1.12.32). We may suppose that $\{y_n\}$ is eventually positive and so $c_1 R(n) \le y_n \le c_2 R(n)$ for $n \ge n_1 > n_0$.

From the equations

$$a_n(\Delta y_n)^\alpha = Q(n)f(y_n) + f(y_n)\sum_{s=n}^{\infty} \frac{a_s(\Delta y_s)^{\alpha+1} g(y_s, y_{s+1})}{f(y_s)f(y_{s+1})}, \quad n \ge n_1 \tag{1.12.34}$$

and

$$a_n(\Delta y_n)^\alpha = a_{n_1}(\Delta y_{n_1})^\alpha - Q(n_1)f(y_{n_1}) + Q(n)f(y_n) - \sum_{s=n_1}^{n-1} Q(s)g(y_s, y_{s+1})\Delta y_s$$

we obtain, for $n \ge n_1$,

$$f(y_n)\sum_{s=n}^{\infty} \frac{a_s(\Delta y_s)^{\alpha+1} g(y_s, y_{s+1})}{f(y_s)f(y_{s+1})}$$

$$= a_{n_1}(\Delta y_{n_1})^\alpha - Q(n_1)f(y_{n_1}) - \sum_{s=n_1}^{n-1} Q(s)g(y_s, y_{s+1})\Delta y_s. \tag{1.12.35}$$

From (1.12.34) and (1.12.35) it follows that

$$\Delta y_n \ge \cdot \frac{Q^{1/\alpha}(n) f^{1/\alpha}(y_n)}{a_n^{1/\alpha}}, \quad n \ge n_1 \tag{1.12.36}$$

and

$$\sum_{s=n_1}^{\infty} Q(s)g(y_s, y_{s+1})\Delta y_s < \infty. \tag{1.12.37}$$

On combining (1.12.36) and (1.12.37), we find

$$\lambda \sum_{s=n_1}^{\infty} \frac{Q^{(\alpha+1)/\alpha}(s) f^{1/\alpha}(y_s)}{a_s^{1/\alpha}} < \infty. \tag{1.12.38}$$

The result now follows from (1.12.38) and (1.12.32). ∎

Theorem 1.12.9. If

$$\sum_{n=n_0}^{\infty} q_{n+1} = \lim_{N\to\infty} \sum_{n=n_0}^{N-1} q_{n+1} = \infty, \tag{1.12.39}$$

then all solutions of the equation (E_{24}) are oscillatory.

Proof. Assume to the contrary that (E_{24}) has a nonoscillatory solution $\{y_n\}$. Suppose that $y_n \neq 0$ for $n \geq n_1 \in \mathbb{N}(n_0)$. Dividing (E_{24}) by $f(y_{n+1})$ and summing from n_1 to $n-1$, we obtain

$$\frac{a_n|\Delta y_n|^{\alpha-1}\Delta y_n}{f(y_n)} + \sum_{s=n_1}^{n-1} \frac{a_s|\Delta y_s|^{\alpha+1}g(y_s, y_{s+1})}{f(y_s)f(y_{s+1})}$$

$$= \frac{a_{n_1}|\Delta y_{n_1}|^{\alpha-1}\Delta y_{n_1}}{f(y_{n_1})} - \sum_{s=n_1}^{n-1} q_{s+1}$$

for $n \geq n_1$, which because of (1.12.39) implies that

$$\lim_{n\to\infty}\left\{\frac{a_n|\Delta y_n|^{\alpha-1}\Delta y_n}{f(y_n)} + \sum_{s=n_1}^{n-1} \frac{a_s|\Delta y_s|^{\alpha+1}g(y_s, y_{s+1})}{f(y_s)f(y_{s+1})}\right\} = -\infty.$$

This, however, contradicts Lemma 1.12.2, and the proof is complete. ∎

Theorem 1.12.10. Suppose that $Q(n) \geq 0$ for all $n \in \mathbb{N}(n_0)$ and $g(u,v) \geq \lambda > 0$ for $u, v \neq 0$. If

$$\int_0^c \frac{dt}{f^{1/\alpha}(t^{1/\alpha})} < \infty \quad \text{and} \quad \int_{-c}^0 \frac{dt}{f^{1/\alpha}(t^{1/\alpha})} > -\infty \quad \text{for } c > 0, \tag{1.12.40}$$

$$-f(-uv) \geq f(uv) \geq f(u)f(v) \quad \text{for} \quad u, v \neq 0 \tag{1.12.41}$$

and

$$\sum_{n=n_0}^{\infty} \frac{Q^{(\alpha+1)/\alpha}(n) f^{1/\alpha}(R(n))}{a_n^{1/\alpha}} = \infty, \tag{1.12.42}$$

then all solutions of the equation (E_{24}) are oscillatory.

Proof. Assume that (E_{24}) has a nonoscillatory solution $\{y_n\}$ and we may suppose that $y_n > 0$ for $n \in \mathbb{N}(n_0)$. From the proof of Theorem 1.12.4, we see that the function $K_2(n)$ defined in (1.12.16) is convergent, i.e. $K_2(n) < \infty$, $n \geq n_0$. From (1.12.14), (1.12.15), (1.12.16) and the fact that $\beta \geq 0$, $K_1(n) \geq K_2(n)$, it follows that

$$a_n(\Delta y_n)^\alpha \geq K_2(n) \quad \text{or} \quad \Delta y_n \geq \frac{K_2^{1/\alpha}(n)}{a_n^{1/\alpha}} \quad \text{for} \quad n \geq n_0.$$

Summing the last inequality from n_0 to $n-1$, we get

$$y_n \geq \sum_{s=n_0}^{n-1} \frac{K_2^{1/\alpha}(s)}{a_s^{1/\alpha}} \geq K_2^{1/\alpha}(n)R(n), \quad n \geq n_0$$

so that for $n \geq n_0$,

$$\frac{Q^{(\alpha+1)/\alpha}(n) f^{1/\alpha}(y_n) g(y_n, y_{n+1})}{a_n^{1/\alpha} f^{1/\alpha}(K_2^{1/\alpha}(n))} \geq \frac{\lambda Q^{(\alpha+1)/\alpha}(n) f^{1/\alpha}(R(n))}{a_n^{1/\alpha}}.$$

Since

$$\Delta K_2(n) = -\frac{Q^{(\alpha+1)/\alpha}(n) f^{1/\alpha}(y_n) g(y_n, y_{n+1})}{a_n^{1/\alpha}},$$

the above inequality can be written as

$$\frac{-\Delta K_2(n)}{f^{1/\alpha}(K_2^{1/\alpha}(n))} \geq \frac{\lambda Q^{(\alpha+1)/\alpha}(n) f^{1/\alpha}(R(n))}{a_n^{1/\alpha}}, \quad n \geq n_0. \tag{1.12.43}$$

Summing (1.12.43) from n_0 to $n-1$, letting $n \to \infty$, and using the fact that

$$\int_{K_2(n)}^{K_2(n_0)} \frac{dt}{f^{1/\alpha}(t^{1/\alpha})} \geq \sum_{s=n_0}^{n-1} \frac{-\Delta K_2(s)}{f^{1/\alpha}(K_2^{1/\alpha}(s))},$$

we obtain

$$\lambda \sum_{n=n_0}^{\infty} \frac{Q^{(\alpha+1)/\alpha}(n) f^{1/\alpha}(R(n))}{a_n^{1/\alpha}} \leq \int_0^{K_2(n_0)} \frac{dt}{f^{1/\alpha}(t^{1/\alpha})} < \infty,$$

which contradicts (1.12.42). ∎

Example 1.12.1. The difference equation

$$\Delta\left(2^{-n}|\Delta y_n|^{\alpha-1}\Delta y_n\right) + \frac{1}{2^{n+1}(n+1)^3} y_{n+1}^3 = 0, \quad n \geq 1 \tag{E_{25}}$$

satisfies all the conditions of Theorem 1.12.4 and hence every nonoscillatory solution $\{y_n\}$ of (E_{25}) satisfies the conclusion of Theorem 1.12.4. In fact, $\{y_n\} = \{n\}$ is a nonoscillatory solution of (E_{25}) which satisfies $c_1 \leq n \leq c_2 2^{n/\alpha}$, where $c_1 = 1$ and $c_2 = 1/(2^\alpha - 1)$.

Example 1.12.2. The difference equation

$$\Delta\left(|\Delta y_n|^{\alpha-1}\Delta y_n\right) + \frac{(2^\alpha - 1)2^{(3-\alpha)n-2\alpha+3}}{(2^{n+1}-1)^3} y_{n+1}^3 = 0, \quad n \geq 1 \tag{E_{26}}$$

satisfies all the conditions of Corollary 1.12.7, and hence (E_{26}) has a nonoscillatory solution which tends to a nonzero constant as $n \to \infty$. In fact, $\{y_n\} = \{1 - 2^{-n}\}$ is such a solution.

Example 1.12.3. For the difference equation

$$\Delta\left(n|\Delta y_n|^{\alpha-1}\Delta y_n\right) + 2^\alpha(2n+1) y_{n+1}^\beta = 0, \quad n \geq 1 \tag{E_{27}}$$

where $\beta \geq 1$ is the ratio of odd positive integers, all conditions of Theorem 1.12.9 are satisfied, and hence every solution of (E_{27}) is oscillatory. In fact, $\{y_n\} = \{(-1)^n\}$ is such a solution of (E_{27}).

Example 1.12.4. The difference equation

$$\Delta\left(n^{-\alpha}|\Delta y_n|^{\alpha-1}\Delta y_n\right) + \frac{2^\alpha\left((n+1)^\alpha + n^\alpha\right)}{n^\alpha(n+1)^\alpha} y_{n+1}^\beta = 0, \quad n \geq 1 \qquad (E_{28})$$

where $\alpha > 1$ and β is the ratio of odd positive integers such that $1 \leq \beta < \alpha^2 < 2\alpha + 3$, satisfies all the conditions of Theorem 1.12.10, and hence all of its solutions are oscillatory. In fact, $\{y_n\} = \{(-1)^n\}$ is such a solution of (E_{28}).

1.13. Oscillation of Even Order Difference Equations

Consider the difference equation

$$\Delta^m y_n + p_n \Delta^{m-1} y_n + F\left(n, y_{n-g}, \Delta y_{n-h}\right) = 0, \quad m \text{ is even} \qquad (E_{29})$$

where $\{p_n\}$ is a sequence of real numbers , $0 \leq p_n < 1$ for $n \geq n_0 \geq 0$, $F : \mathbb{N} \cup \{0\} \times \mathbb{R}^2 \to \mathbb{R}$ is continuous for each fixed n, and g and h are in $\mathbb{N} \cup \{0\}$. We shall assume that there exist an eventually positive real sequence $\{q_n\}$ and real numbers $\lambda > 0$ and $\mu \geq 0$ such that

$$F(n, y, z)\text{sgn } y \geq q_n |y|^\lambda |z|^\mu \quad \text{for } n \geq n_0 \geq 0 \text{ and } yz \neq 0. \qquad (1.13.1)$$

We shall establish the oscillation of (E_{29}) in Theorems 1.13.3 and 1.13.4 by comparing it with the oscillatory behavior of all bounded solutions of two equations of order 1 and $n-1$, Theorem 1.13.5 deals with the oscillation of a special case of (E_{29}) when condition (1.13.1) holds with $\lambda = 1$ and $\mu = 0$ and a condition on $\{p_n\}$ which holds in Theorems 1.13.3 and 1.13.4 is violated. Finally, in Theorems 1.13.7 and 1.13.8 we shall present the oscillatory behavior of the difference of two eventually positive solutions of the equation

$$\Delta^m y_n + p_n \Delta^{m-1} y_n + q_n f(y_{n-g}) = e_n, \quad m \text{ is even} \qquad (E_{30})$$

where $f : \mathbb{R} \to \mathbb{R}$ is a continuous function such that $yf(y) > 0$ and $f'(y) \geq c > 0$ for $y \neq 0$, and $\{e_n\}$ is a sequence of real numbers. For this, we shall need the following two lemmas:

Lemma 1.13.1. [2] Let y_n be defined for $n \geq n_0$, and $y_n > 0$ with $\Delta^m y_n$ of constant sign for $n \geq n_0$ and not identically zero. Then, there exists an integer p, $0 \leq p \leq m$ with $(m+p)$ odd for $\Delta^m y_n \leq 0$ and $(m+p)$ even for $\Delta^m y_n \geq 0$ such that

(i) $p \le m-1$ implies $(-1)^{p+i}\Delta^i y_n > 0$ for all $n \ge n_0$, $p \le i \le m-1$
(ii) $p \ge 1$ implies $\Delta^i y_n > 0$ for all large $n \ge n_0$, $1 \le i \le p-1$.

Lemma 1.13.2. [2] Let y_n be defined for $n \ge n_0$, and $y_n > 0$ with $\Delta^m y_n \le 0$ for $n \ge n_0$ and not identically zero. Then, there exists a large integer $n_1 \ge n_0$ such that

$$y_n \ \ge \ \frac{1}{(m-1)!}(n-n_1)^{m-1}\Delta^{m-1}y_{2^{m-p-1}n}, \quad n \ge n_1$$

where p is defined as in Lemma 1.13.1. Further, if y_n is increasing, then

$$y_n \ \ge \ \frac{1}{(m-1)!}\left(\frac{n}{2^{m-1}}\right)^{m-1}\Delta^{m-1}y_n, \quad n \ge 2^{m-1}n_1.$$

Theorem 1.13.3. Let condition (1.13.1) hold, and

$$\lim_{n\to\infty}\sum_{k=n_0\ge 0}^{n-1}\left(\prod_{i=n_0}^{k-1}(1-p_i)\right) \ = \ \infty. \tag{1.13.2}$$

If for every $\gamma_i > 0$, $i = 1,2$ the equation

$$\Delta z_n + \gamma_1\left(\frac{1}{2^{(m-1)^2}(m-1)!}\right)^{\lambda}(n-g)^{(m-1)\lambda}q_n\left|z_{n-g}\right|^{\lambda}\operatorname{sgn} z_{n-g} \ = \ 0 \tag{1.13.3}$$

is oscillatory and all bounded solutions of the equation

$$\Delta^{m-1}x_n + \gamma_2 q_n\left|x_{n-h}\right|^{\mu}\operatorname{sgn} x_{n-h} \ = \ 0 \tag{1.13.4}$$

are oscillatory, then (E_{29}) is oscillatory.

Proof. Let $\{y_n\}$ be a nonoscillatory solution of (E_{29}), say $y_n > 0$ for $n \ge n_0 \ge 1$. First, we claim that $\{\Delta^{m-1}y_n\}$ is eventually of one sign. To this end, we assume that $\{\Delta^{m-1}y_n\}$ is oscillatory. There exists $N \ge n_0 + \max\{g,h\}$ such that $\Delta^{m-1}y_N < 0$. Let $n = N$ in (E_{29}) and then multiply the resulting equation by $\Delta^{m-1}y_N$, to obtain

$$\begin{aligned}\Delta^m y_N\Delta^{m-1}y_N \ &= \ -p_N\left(\Delta^{m-1}y_N\right)^2 - F\left(N, y_{N-g}, \Delta y_{N-h}\right)\Delta^{m-1}y_N \\ &\ge \ -p_N\left(\Delta^{m-1}y_N\right)^2,\end{aligned}$$

or

$$\Delta^{m-1}y_{N+1}\Delta^{m-1}y_N \ \ge \ (1-p_N)(\Delta^{m-1}y_N)^2 \ > \ 0,$$

which implies that

$$\Delta^{m-1}y_{N+1} \ < \ 0.$$

By induction, we obtain $\Delta^{m-1} y_n < 0$ for $n \ge N$, contradicting the assumption that $\{\Delta^{m-1} y_n\}$ is oscillatory.

Next, suppose there exists $N^* \ge n_0 + \max\{g, h\}$ such that $\Delta^{m-1} y_{N^*} = 0$. Then, letting $n = N^*$ in (E_{29}) leads to

$$\Delta^m y_{N^*} = -F(N^*, y_{N^*-g}, \Delta y_{N^*-h}) \le 0,$$

which implies that

$$\Delta^{m-1} y_{N^*+1} \le \Delta^{m-1} y_{N^*} = 0.$$

As in the above case, we have seen that this contradicts the assumption that $\{\Delta^{m-1} y_n\}$ is oscillatory.

Now, we consider the following two cases:

(A) $\Delta^{m-1} y_n < 0$ eventually (B) $\Delta^{m-1} y_n > 0$ eventually.

(A) Suppose that $\Delta^{m-1} y_n < 0$ for $n \ge n_1 \ge \max\{N, N^*\}$. From (E_{29}), it follows that

$$\Delta^m y_n + p_n \Delta^{m-1} y_n \le 0 \quad \text{for} \quad n \ge n_1.$$

Set $V_n = -\Delta^{m-1} y_n$ for $n \ge n_1$. Then,

$$\Delta V_n + p_n V_n \ge 0 \quad \text{for} \quad n \ge n_1$$

or

$$V_{n+1} \ge (1 - p_n) V_n \ge \left(\prod_{i=n_1}^{n-1} (1 - p_i) \right) V_{n_1},$$

where V_{n_1} is an arbitrary constant. Thus,

$$-\Delta^{m-1} y_k \ge \left(\prod_{i=n_1}^{k-1} (1 - p_i) \right) V_{n_1} \quad \text{for} \quad n \ge n_1.$$

Summing this inequality from n_1 to $n-1$ $(> n_1)$, we get

$$\Delta^{m-2} y_{n_1} - \Delta^{m-2} y_n \ge \left(\sum_{k=n_1}^{n-1} \prod_{i=n_1}^{k-1} (1 - p_i) \right) V_{n_1} \to \infty \quad \text{as} \quad n \to \infty.$$

This contradicts the positiveness of y_n. Therefore, we must have the case (B).

(B) Suppose that $\Delta^{m-1} y_n > 0$ for $n \ge n_1 \ge \max\{N, N^*\}$. By Lemma 1.13.1, there exists $n_2 \ge n_1$ such that $\Delta y_n > 0$ for $n \ge n_2$ and either $\Delta^2 y_n > 0$ or $\Delta^2 y_n < 0$ for $n \ge n_2$.

(I) Suppose that $\Delta^2 y_n > 0$ for $n \ge n_2$. There exist $n_3 \ge n_2$ and a constant $\alpha > 0$ such that

$$\Delta y_{n-h} \ \ge \ \alpha \quad \text{for} \quad n \ge n_3. \tag{1.13.5}$$

Using conditions (1.13.1) and (1.13.5) in (E_{29}), we have

$$\Delta^m y_n + \alpha^\mu q_n y_{n-g}^\lambda \ \le \ 0 \quad \text{for} \quad n \ge n_3. \tag{1.13.6}$$

By applying Lemma 1.13.2, there exists $n_4 \ge 2^{m-1} n_3$ such that

$$y_n \ \ge \ \frac{1}{(m-1)!}\left(\frac{n}{2^{(m-1)}}\right)^{m-1} \Delta^{m-1} y_n \quad \text{for} \quad n \ge n_4.$$

There exists $n_5 \ge n_4$ such that

$$y_{n-g} \ \ge \ \frac{1}{(m-1)!}\left(\frac{1}{2^{m-1}}\right)^{m-1} (n-g)^{m-1}\Delta^{m-1} y_{n-g} \quad \text{for} \quad n \ge n_5. \tag{1.13.7}$$

Using (1.13.7) in (1.13.6) yields

$$\Delta w_n + \alpha^\mu \left(\frac{1}{2^{(m-1)^2}(m-1)!}\right)^\lambda (n-g)^{(m-1)\lambda} q_n w_{n-g}^\lambda \ \le \ 0 \quad \text{for} \quad n \ge n_5,$$

where $w_n = \Delta^{m-1} y_n$, $n \ge n_5$. Therefore, by Lemma 5 of Section 2 in [194], (1.13.3) has an eventually positive solution, which is a contradiction.

(II) Suppose that $\Delta^2 y_n < 0$ for $n \ge n_2$. There exists $N_1 \ge n_2$ and a constant $\beta > 0$ such that

$$y_{n-g} \ \ge \ \beta \quad \text{for} \quad n \ge N_1. \tag{1.13.8}$$

Using conditions (1.13.1) and (1.13.8) in (E_{29}) yields

$$\Delta^{m-1} z_n + \beta^\lambda q_n z_{n-h}^\mu \ \le \ 0 \quad \text{for} \quad n \ge N_1,$$

where $z_n = \Delta y_n$, $n \ge N_1$. Once again by Lemma 5 of Section 2 in [194], (1.13.4) has an eventually positive bounded solution, which is a contradiction. ∎

Theorem 1.13.4. Let (1.13.4) in Theorem 1.13.3 be replaced by the equation

$$\Delta^{m-1} w_n + \left(\frac{n-g}{2}\right)^\lambda q_n \left|w_{n-\tau}\right|^{\lambda+\mu} \operatorname{sgn} w_{n-\tau} \ = \ 0, \tag{1.13.9}$$

where $\tau = \max\{g, h\}$. Then, the conclusion of Theorem 1.13.3 holds.

Proof. Let $\{y_n\}$ be a nonoscillatory solution of (E_{29}), say $y_n > 0$ for $n \geq n_0 \geq 1$. The proof is similar to that of Theorem 1.13.3 except the Case II. Thus, we consider:

(II) Suppose that $\Delta^2 y_n \leq 0$ for $n \geq n_2$. Then, there exists $n_3 \geq n_2$ such that

$$y_{n-g} \geq \left(\frac{n-g}{2}\right) \Delta y_{n-g} \geq \left(\frac{n-g}{2}\right) \Delta y_{n-\tau} \quad \text{for } n \geq n_3. \tag{1.13.10}$$

Using conditions (1.13.1) and (1.13.10) in (E_{29}), we obtain

$$\Delta^{m-1} x_n + \left(\frac{n-g}{2}\right)^{\lambda} q_n x_{n-\tau}^{\lambda} x_{n-h}^{\mu} \leq 0 \quad \text{for } n \geq n_3,$$

where $x_n = \Delta y_n$, $n \geq n_3$. The rest of the proof is similar to that of Theorem 1.13.3 (II). ∎

The following criterion deals with the oscillation of a special case of (E_{29}) when $\mu = 0$ and $\lambda = 1$, namely, the linear difference equation

$$\Delta^m y_n + p_n \Delta^{m-1} y_n + q_n y_{n-g} = 0, \quad m > 2 \tag{E_{31}}$$

when condition (1.13.2) is not satisfied.

Theorem 1.13.5. Let $\Delta p_n \leq 0$ for $n \geq n_0 \geq 0$, $g > 1$ and

$$C_n = \min\{Q_n, R_n\} > p_{n-g} \quad \text{for all large } n, \tag{1.13.11}$$

where

$$Q_n = \frac{\theta}{(m-2)!} \sum_{j=n-g}^{n-1} (j-g)^{m-2} q_j, \quad \theta = \frac{1}{2^{(m-1)^2}}$$

and

$$R_n = \sum_{j=n-g}^{n-1} q_j \sum_{\ell=j-g}^{n-1} \frac{1}{(m-3)!} (\ell + m + g - j - 3)^{(m-3)}.$$

If the difference equation

$$\Delta w_n + c_n w_{n-g} = 0, \tag{1.13.12}$$

where

$$c_n = \min\left\{C_n - p_{n-g}, \frac{\theta}{(m-1)!}(n-g)^{m-1} q_n\right\} \text{ and } \theta \text{ is defined as above} \tag{1.13.13}$$

is oscillatory, then (E_{31}) is oscillatory.

Proof. Let $\{y_n\}$ be a nonoscillatory solution of (E_{31}), say $y_n > 0$ for $n \ge n_0 \ge 1$. As in Theorem 1.13.3, we see that $\{\Delta^{m-1} y_n\}$ is eventually of one sign. Next, we consider the two cases (A) and (B) in Theorem 1.13.3.

(A) Suppose that $\Delta^{m-1} y_n < 0$ for $n \ge n_1 \ge n_0$. By Lemma 1.13.1, there exists an $n_2 \ge n_1$ such that for $n \ge n_2$

$$\Delta^{m-2} y_n > 0, \quad \text{and either} \quad \text{(i)} \ \Delta y_n > 0 \quad \text{or (ii)} \ \Delta y_n < 0.$$

(i) Suppose that $\Delta y_n > 0$ for $n \ge n_2$. By Lemma 1.13.2, there exists $n_3 \ge 2^{m-1} n_2$ such that

$$y_n \ \ge \ \frac{1}{(m-2)!}\left(\frac{n}{2^{m-2}}\right)^{m-2} \Delta^{m-2} y_n \quad \text{for} \quad n \ge n_3.$$

There exists $n_4 \ge n_3$ such that

$$y_{n-g} \ \ge \ \left(\frac{1}{(m-2)!}\frac{1}{2^{(m-2)^2}}\right)(n-g)^{m-2}\Delta^{m-2} y_{n-g} \quad \text{for} \quad n \ge n_4. \tag{1.13.14}$$

Using (1.13.14) in (E_{31}) yields

$$\Delta^2 z_n + p_n \Delta z_n + \frac{\theta}{(m-2)!}(n-g)^{m-2} q_n z_{n-g} \ \le \ 0 \quad \text{for} \quad n \ge n_4, \tag{1.13.15}$$

where $z_n = \Delta^{m-2} y_n$, $n \ge n_4$ and $\theta = 1/2^{(m-2)^2}$. Summing both sides of (1.13.15) from $n-g$ to $n-1$, we have

$$\Delta z_n - \Delta z_{n-g} + \sum_{i=n-g}^{n-1} p_i \Delta z_i + \sum_{i=n-g}^{n-1} \frac{\theta}{(m-2)!}(i-g)^{m-2} q_i z_{i-g} \ \le \ 0$$

or

$$\Delta z_n + \left[p_n z_n - p_{n-g} z_{n-g} - \sum_{i=n-g}^{n-1} z_i \Delta p_i\right] + \left(\sum_{i=n-g}^{n-1} \frac{\theta}{(m-2)!}(i-g)^{m-2} q_i\right) z_{n-g} \ \le \ 0.$$

Since $\Delta p_n \le 0$, we have

$$\Delta z_n + \left(\frac{\theta}{(m-2)!}\sum_{i=n-g}^{n-1}(i-g)^{m-2} q_i - p_{n-g}\right) z_{n-g} \ \le \ 0 \tag{1.13.16}$$

for $n \ge n_5$ for some $n_5 \ge n_4$, and hence by (1.13.13), we have

$$\Delta z_n + c_n z_{n-g} \ \le \ 0 \quad \text{for} \quad n \ge n_5.$$

Therefore, by Lemma 5 of Section 2 in [194], (1.13.12) has an eventually positive solution, which is a contradiction.

(ii) Suppose that $\Delta y_n < 0$ for $n \geq n_2$. By Lemma 1.13.1, one can easily see that

$$(-1)^i \Delta^i y_n > 0 \quad \text{for} \quad i = 0, 1, \cdots, m-1 \quad \text{and} \quad n \geq n_3 \geq n_2. \tag{1.13.17}$$

By discrete Taylor's formula [2] y_k can be expressed as

$$y_k = \sum_{i=0}^{m-3} \frac{(z+i-1-k)^{(i)}}{i!}(-1)^i \Delta^i y_z + \frac{1}{(m-3)!}\sum_{\ell=k}^{z-1}(\ell+m-k-3)^{(m-3)}\Delta^{m-2} y_\ell \tag{1.13.18}$$

for all $k \in \mathbb{N}(n_3, z) = \{n_3, n_3+1, \cdots, z\}$, where $z \in \mathbb{N}(n_3) = \{n_3, n_3 + 1, \cdots\}$. Using (1.13.17) in (1.13.18) and letting $z - 1 = n - g$, we have

$$y_k \geq \frac{1}{(m-3)!}\sum_{\ell=k}^{n-g}(\ell + m - k - 3)^{(m-3)}\Delta^{m-2} y_\ell, \quad n \geq n_3.$$

Letting $k = j - g$, and using the fact that $\Delta^{m-2} y_n$ is decreasing, we have

$$y_{j-g} \geq \frac{1}{(m-3)!}\sum_{\ell=j-g}^{n-g}(\ell + m - j + g - 3)^{(m-3)}\Delta^{m-2} y_{n-g}. \tag{1.13.19}$$

Summing both sides of (E_{31}) from $n-g$ to $n-1$, one can easily see that

$$\Delta^{m-1} y_n - p_{n-g}\Delta^{m-2} y_{n-g} + \sum_{i=n-g}^{n-1} q_i y_{i-g} \leq 0, \tag{1.13.20}$$

using (1.13.19) in (1.13.20), we have

$$\Delta w_n - p_{n-g} w_{n-g} + \left(\sum_{j=n-g}^{n-1} q_j \sum_{\ell=j-g}^{n-1} \frac{1}{(m-3)!}(\ell+m+g-j-3)^{(m-3)}\right) w_{n-g} \leq 0,$$

where $w_n = \Delta^{m-2} y_n$, $n \geq n_4 \geq n_3$. Next, by (1.13.13), we see that

$$\Delta w_n + c_n w_{n-g} \leq 0 \quad \text{for} \quad n \geq n_4.$$

The rest of the proof is similar to that of the above case.

(B) Suppose that $\Delta^{m-1} y_n > 0$ for $n \geq n_1$. From (E_{31}) it follows that

$$\Delta^m y_n + q_n y_{n-g} \leq 0 \quad \text{for} \quad n \geq n_1. \tag{1.13.21}$$

As in Theorem 1.13.3 (Case B–1), there exists $N_1 \geq n_5$ such that (1.13.7) holds for $n \geq N_1$. Now, using (1.13.7) in (1.13.21), we obtain

$$\Delta x_n + \left(\frac{1}{2^{(m-1)^2}(m-1)!}\right)(n-g)^{m-1} q_n x_{n-g} \leq 0 \quad \text{for} \quad n \geq N_1,$$

where $x_n = \Delta^{m-1} y_n$, $n \geq N_1$. By (1.13.13), we see that

$$\Delta x_n + c_n x_{n-g} \leq 0 \quad \text{for} \quad n \geq N_1.$$

The rest of the proof is similar to that of the above case. ∎

From the proof of Theorem 1.13.5, one can easily extract the following oscillation criterion for (E_{31}) when $m = 2$.

Corollary 1.13.6. Let $\Delta p_n \leq 0$ for $n \geq n_0 \geq 0$, $g > 1$, and

$$\sum_{j=n-g}^{n-1} q_j > p_{n-g} \quad \text{for all large} \quad n.$$

If (1.13.12) is oscillatory, where

$$c_n = \min\left\{\sum_{j=n-g}^{n-1} q_j - p_{n-g},\ \left(\frac{n-g}{2}\right) q_n\right\}$$

then (E_{31}) when $m = 2$ is oscillatory.

Finally, we present the following results for the forced difference equations of the form (E_{30}).

Theorem 1.13.7. Let the conditions of Theorem 1.13.5 hold with q_n is replaced by cq_n. If $\{u_n\}$ and $\{v_n\}$ are eventually positive solutions of (E_{30}), then $\{u_n - v_n\}$ is oscillatory.

Proof. Let $\{u_n\}$ and $\{v_n\}$ be two positive solutions of (E_{30}) for $n \geq n_0 \geq 1$, and let $w_n = u_n - v_n$ for $n \geq n_0$. From (E_{30}) now, we can obtain

$$\Delta^m w_n + p_n \Delta^{m-1} w_n + q_n(f(u_{n-g}) - f(v_{n-g})) = 0. \tag{1.13.22}$$

To show that $\{w_n\}$ is oscillatory we will assume that $\{w_n\}$ is eventually positive. The negative case follows from the similar steps. So, let us suppose that $w_n > 0$ for $n \geq n_1 \geq n_0$. Applying the mean value theorem,

$$f(u_{n-g}) - f(v_{n-g}) \geq c w_{n-g} \quad \text{for} \quad n \geq n_1. \tag{1.13.23}$$

From (1.13.22) and (1.13.23), we get

$$\Delta^m w_n + p_n \Delta^{m-1} w_n + c q_n w_{n-g} \leq 0, \quad n \geq n_1.$$

The rest of the proof is similar to that of Theorem 1.13.5. ∎

In the case when condition (1.13.2) holds, we have the following criterion for (E_{30}).

Theorem 1.13.8. Let condition (1.13.2) hold and (1.13.3) with $\lambda = 1$ and $\gamma_1 = c$ is oscillatory. If $\{u_n\}$ and $\{v_n\}$ are two eventually positive solutions of (E_{30}), then $\{u_n - v_n\}$ is oscillatory.

Proof. The proof is similar to those of earlier results. ∎

Remark 1.13.1. We observe that the results of this section remain valid when $p_n \equiv 0$ and when $p_n \equiv p =$ constant, $0 \le p_n < 1$, we see that the series in condition (1.13.2) is a convergent geometric series and hence condition (1.13.2) is violated. In this case, we are only able to study (E_{31}) and (E_{30}).

As an example, we consider the special case of (E_{29}), namely

$$\Delta^m y_n + p_n \Delta^{m-1} y_n + q\,|y_{n-g}|^\lambda\,|\Delta y_{n-h}|^\mu \operatorname{sgn} y_{n-g} = 0, \quad m \text{ is even} \quad (E_{32})$$

where $\{p_n\}$ is a real sequence, $0 \le p_n < 1$, $n \ge n_0 \ge 0$ and condition (1.13.2) holds, q is a positive real number, g and h are positive integers, $\lambda > 0$ and $\mu \ge 0$ are real constants.

Now, by applying Theorem 1.13.3 and Corollary 2 (iv) in [194] we obtain:

Corollary 1.13.9. If for every positive constant γ, $g > 1$ and

(I) $\gamma q > \dfrac{(m-1)^{m-1} g^g}{(g+m-1)^{g+m-1}}$ if $\mu = 1$ and all $0 < \lambda \le 1$

or

(II) any $q > 0$ if $0 < \mu < 1$ and $0 < \lambda \le 1$,

then (E_{32}) is oscillatory.

Next, we consider the constant coefficients equation

$$\Delta^m y_n + p\Delta^{m-1} y_n + q y_{n-g} = 0, \quad (E_{33})$$

where $p \ge 0$ and $q > 0$ are real constants, $p < 1$ and g is a positive integer, $g > 1$. By applying Theorem 1.13.5 and Theorem 7.5.1 in [129], we obtain:

Corollary 1.13.10. If

(I) $gq - p > \dfrac{g^g}{(1+g)^{1+g}}$ if $m = 2$

and

(II) for any $p \geq 0$ and $q > 0$ if $m > 2$,

then (E_{33}) is oscillatory.

Remark 1.13.2. From Corollary 1.13.9, we see that the characteristic equation associated with (E_{33}), namely

$$(t-1)^m + p(t-1)^{m-1} + qt^{-g} = 0$$

has no positive roots provided conditions of Corollary 1.13.9 are satisfied.

Finally, by applying Theorem 1.13.7 and Corollary 1.13.10, we obtain

Corollary 1.13.11. Let the conditions of Corollary 1.13.10 be satisfied with q is replaced by cq. If $\{u_n\}$ and $\{v_n\}$ are eventually positive solutions of (E_{30}) with constant coefficients, (i.e. p_n and q_n are constants), then $\{u_n - v_n\}$ is oscillatory.

1.14. Oscillation of Odd Order Difference Equations

Consider the difference equations

$$\Delta^m y_n = p_n \Delta^{m-1} y_{n+\tau} + q_n |y_{g_n}|^\gamma \operatorname{sgn} y_{g_n} \tag{E_{34}}$$

and

$$\Delta^m y_n + p_n \Delta^{m-1} y_{n-\tau} + q_n |y_{g_n}|^\gamma \operatorname{sgn} y_{g_n} = 0, \tag{E_{35}}$$

where $m \geq 1$ is an odd integer, $\{p_n\}$ and $\{q_n\}$, $n \geq n_0 \geq 0$ are sequences of nonnegative real numbers, $\{g_n\}$, $n \geq n_0$ is a nondecreasing sequence of nonnegative integers with $\lim_{n\to\infty} g_n = \infty$, τ is a positive integer and γ is a positive real number.

Equation (E_{34}), ((E_{35})) is said to be *almost oscillatory* if for every solution $\{y_n\}$, either $\{y_n\}$ is oscillatory or $\{\Delta^{m-1} y_n\}$ is oscillatory, $n \geq n_0 \geq 0$. We shall present explicit conditions for the almost oscillation of these equations.

Theorem 1.14.1. Let $\gamma = 1$, $g_n < n$ and $\Delta p_n \leq 0$ for $n \geq n_0 \geq 0$. If

$$\liminf_{n\to\infty} \sum_{k=n-\tau}^{n-1} p_k > \left(\frac{\tau}{1+\tau}\right)^{1+\tau} \tag{1.14.1}$$

and

$$\limsup_{n\to\infty} \sum_{k=g_n}^{n-1} C_{g_n,k}\, g_k^{m-2} q_k > 2^{2(m-2)}(m-2)!, \tag{1.14.2}$$

where

$$C_{g_n,k} = \sum_{s=g_n}^{k}\left(\prod_{j=g_n+1}^{s}\frac{1}{1+p_j}\right),$$

then equation (E_{34}) is almost oscillatory.

Proof. Let $\{y_n\}$ be an eventually positive solution of (E_{34}), say $y_n > 0$ and $y_{g_n} > 0$ for $n \geq n_1 \geq n_0 \geq 0$. There are two possibilities to consider

(I) $\Delta^{m-1}y_n > 0$ eventually, and (II) $\Delta^{m-1}y_n < 0$ eventually.

(I) Suppose $\Delta^{m-1}y_n > 0$ eventually. From (E_{34}), we see that

$$\Delta z_n - p_n z_{n+\tau} = q_n y_{g_n} \geq 0 \quad \text{eventually},$$

where $z_n = \Delta^{m-1}y_n > 0$ eventually. Now, by Lemma 1.1(b) in [194], the equation

$$\Delta z_n - p_n z_{n+\tau} = 0 \tag{1.14.3}$$

has eventually positive solutions. But, in view of Theorem 3 in [245] and condition (1.14.1), equation (1.14.3) is oscillatory, which is a contradiction.

(II) Suppose $\Delta^{m-1}y_n < 0$ eventually. Since m is odd, there exists an integer $n_2 \geq n_1$ such that

$$\Delta^{m-2}y_n > 0 \quad \text{and} \quad \Delta y_n > 0 \quad \text{for} \quad n \geq n_2.$$

Applying Lemma 1.13.2, there exists an integer $n_3 \geq n_2$ such that

$$y_n \geq \frac{2^{4-2m}}{(m-2)!}n^{(m-2)}\Delta^{m-2}y_n \quad \text{for} \quad n \geq 2^{m-2}n_3.$$

Next, there exists a sufficiently large $N \geq 2^{m-2}n_3$ so that

$$y_{g_n} \geq \frac{2^{4-2m}}{(m-2)!}g_n^{m-2}\Delta^{m-2}y_{g_n}, \quad n \geq N. \tag{1.14.4}$$

Using (1.14.4) in (E_{34}) and setting $w_n = \Delta^{m-2}y_n$, we have

$$\Delta^2 w_n \geq p_n \Delta w_{n+\tau} + \frac{2^{4-2m}}{(m-2)!}g_n^{m-2}q_n w_{g_n}, \quad n \geq N. \tag{1.14.5}$$

Summing both sides of (1.14.5) from $s \geq N$ to $n-1$, we get

$$\Delta w_n - \Delta w_s \geq \sum_{k=s}^{n-1} p_k \Delta w_{k+\tau} + \frac{2^{4-2m}}{(m-2)!}\sum_{k=s}^{n-1} g_k^{m-2}q_k w_{g_k} \quad \text{for } n-1 \geq s \geq N. \tag{1.14.6}$$

Since

$$\sum_{k=s}^{n-1} p_k \Delta w_{k+\tau} = p_n w_{n+\tau} - p_s w_{s+\tau} - \sum_{k=s}^{n-1} w_{k+\tau}(\Delta p_k)$$

and $\Delta p_k \leq 0$, and w_n is nonincreasing for $n \geq N$, we obtain

$$\sum_{k=s}^{n-1} p_k \Delta w_{k+\tau} \geq -p_s w_{s+\tau}, \quad n-1 \geq s \geq N.$$

Now, (1.14.6) takes the form

$$-(\Delta w_s - p_s w_s) \geq \frac{2^{4-2m}}{(m-2)!} \sum_{k=s}^{n-1} g_k^{m-2} q_k w_{g_k}. \tag{1.14.7}$$

Define the sequence $\{R_n\}$ by

$$R_{n+1} = \frac{R_n}{1+p_n}, \quad n = 0,1,2,\cdots, \quad R_{n_0} > 0, \quad n_0 \geq 0 \tag{1.14.8}$$

and multiply (1.14.7) by R_{s+1}, to obtain

$$-\Delta(R_s w_s) \geq \left(\frac{2^{4-2m}}{(m-2)!} R_{s+1}\right) \sum_{k=s}^{n-1} g_k^{m-2} q_k w_{g_k}. \tag{1.14.9}$$

Summing both sides of (1.14.9) from $g_n \geq N$ to $n-1 \geq g_n$, we get

$$R_{g_n} w_{g_n} \geq R_{g_n} w_{g_n} - R_n w_n \geq \left(\frac{2^{4-2m}}{(m-2)!}\right) \sum_{s=g_n}^{n-1} R_{s+1} \sum_{k=s}^{n-1} g_k^{m-2} q_k w_{g_k}.$$

Now,

$$\begin{aligned} 1 &\geq \left(\frac{2^{4-2m}}{(m-2)!}\right) \sum_{s=g_n}^{n-1} \frac{R_{s+1}}{R_{g_n}} \sum_{k=s}^{n-1} g_k^{m-2} q_k \left(\frac{w_{g_k}}{w_{g_n}}\right) \\ &\geq \left(\frac{2^{4-2m}}{(m-2)!}\right) \sum_{s=g_n}^{n-1} g_k^{m-2} q_k \left(\frac{w_{g_k}}{w_{g_n}}\right) \sum_{s=g_n}^{k} \frac{R_{s+1}}{R_{g_n}}. \end{aligned}$$

Since $w_{g_k} \geq w_{g_n}$ for $g_n \leq k \leq n-1$, we find

$$1 \geq \left(\frac{2^{4-2m}}{(m-2)!}\right) \sum_{k=g_n}^{n-1} g_k^{m-2} q_k \sum_{s=g_n}^{k} \frac{R_{s+1}}{R_{g_n}} = \left(\frac{2^{4-2m}}{(m-2)!}\right) \sum_{k=g_n}^{n-1} C_{g_n,k}\, g_k^{m-2} q_k.$$

Now taking limit superior on both sides of the above inequality as $n \to \infty$, we obtain a contradiction to condition (1.14.4). ∎

Next, we shall prove the following result:

Theorem 1.14.2. Let $0 < \gamma < 1$, $\Delta p_n \leq 0$ and $g_n < n$ for $n \geq n_0 \geq 0$, and let condition (1.14.1) hold. If

$$\sum_{k\geq n_0\geq 0}^{\infty} A_{g_k,k}\, g_k^{(m-2)\gamma} q_k = \infty, \tag{1.14.10}$$

where

$$A_{g_k,k} = \sum_{s=g_k}^{k} \left(\frac{1}{1+p_s}\right)\left(\prod_{j=1}^{s-1}\frac{1}{1+p_j}\right)^{1-\gamma},$$

then (E_{34}) is almost oscillatory.

Proof. Let $\{y_n\}$ be an eventually positive solution of (E_{34}), say $y_n > 0$ and $y_{g_n} > 0$ for $n \geq n_1 \geq n_0 \geq 0$. As in the proof of Theorem 1.14.1, we consider the two cases (I) and (II) and observe that Case (I) is impossible. Next, we consider:

(II) Suppose $\Delta^{m-1} y_n < 0$ for $n \geq n_2 \geq n_1 + 1$. Define the sequence $\{R_n\}$ by (1.14.8) and proceed as in Theorem 1.14.1(II), to obtain

$$-\Delta(R_s w_s) \geq \alpha R_{s+1} \sum_{k=s}^{n-1} g_k^{(m-2)\gamma} q_k w_{g_k}^{\gamma}, \quad n-1 \geq s \geq N \geq n_2 \tag{1.14.11}$$

where $\alpha = \left(\dfrac{2^{4-2m}}{(m-2)!}\right)^{\gamma}$.

Choose $N^* \geq N$ such that $g_n \geq N$ for $s \geq N^*$ and let $N_1 \geq N^*$ be fixed, we see that

$$-\Delta(R_s w_s) \geq \alpha R_{s+1} \sum_{k=s}^{N_1} g_k^{(m-2)\gamma} g_k^{(m-2)\gamma} q_k w_{g_k}^{\gamma} \quad \text{for} \quad N_1 \geq s \geq N. \tag{1.14.12}$$

Dividing (1.14.12) by $(R_s w_s)^{\gamma}$ and summing from N to N_1, we obtain

$$\begin{aligned}
\sum_{s=N}^{N_1} \frac{-\Delta(R_s w_s)}{(R_s w_s)^{\gamma}} &\geq \alpha \sum_{s=N}^{N_1} \frac{R_{s+1}}{R_s^{\gamma}} \sum_{k=s}^{N_1} g_k^{(m-2)\gamma} q_k \left(\frac{w_{g_k}}{w_s}\right)^{\gamma} \\
&= \alpha \sum_{s=N}^{N_1} \frac{R_s^{1-\gamma}}{1+p_s} \sum_{k=s}^{N_1} g_k^{(m-2)\gamma} q_k \left(\frac{w_{g_k}}{w_s}\right)^{\gamma} \\
&\geq \sum_{k=N^*}^{N_1} g_k^{(m-2)\gamma} q_k \sum_{s=g_k}^{k} \left(\frac{R_s^{1-\gamma}}{1+p_s}\right)\left(\frac{w_{g_k}}{w_s}\right)^{\gamma}.
\end{aligned}$$

Since $w_{g_k} \geq w_s$ for $g_k \leq s \leq k$, $N_1 \geq k \geq N^*$, we have

$$\sum_{s=N}^{N_1} \frac{-\Delta(R_s w_s)}{(R_s w_s)^\gamma} \geq \alpha \sum_{k=N^*}^{N_1} g_k^{(m-2)\gamma} q_k \sum_{s=g_n}^{k} \frac{R_s^{1-\gamma}}{1+p_s}.$$

It follows from the proof of Theorem 4.3 in [133] that

$$\sum_{s=N}^{N_1} \frac{-\Delta(R_s w_s)}{(R_s w_s)^\gamma} \quad \text{is bounded from below,} \quad N_1 \geq N$$

which contradicts condition (1.14.10). ∎

Remark 1.14.1. The proofs of the above results can be easily applied to (E_{34}) when $m = 2$ to obtain the following criteria:

Theorem 1.14.3. If $m = 2$ and condition (1.14.2) in Theorem 1.14.1 is replaced by

$$\limsup_{n \to \infty} \sum_{k=g_n}^{n-1} C_{g_n,k}\, q_k > 1, \tag{1.14.13}$$

then the conclusion of Theorem 1.14.1 holds.

Theorem 1.14.4. If $m = 2$ and condition (1.14.10) in Theorem 1.14.2 is replaced by

$$\sum_{k \geq n_0 \geq 0}^{\infty} A_{g_k,k}\, q_k = \infty, \tag{1.14.14}$$

then the conclusion of Theorem 1.14.2 holds.

Remark 1.14.2. The above results are not applicable to equations of type (E_{34}) when $p_n = 0$.

The following criterion is concerned with the almost oscillation of (E_{35}) when $\gamma > 0$.

Theorem 1.14.5. Let $\Delta p_n \leq 0$ for $n \geq n_0 \geq 0$. If

$$\liminf_{n \to \infty} \sum_{k=n-\tau}^{n-1} p_k > \left(\frac{\tau}{1+\tau}\right)^{1+\tau} \tag{1.14.15}$$

and

$$\sum^{\infty} q_k = \infty, \tag{1.14.16}$$

then (E_{35}) is almost oscillatory.

Proof. Let $\{y_n\}$ be an eventually positive solution of (E_{35}), say $y_n > 0$ and $y_{g_n} > 0$ for $n \geq n_1 \geq n_0$. Next, we consider the following two cases: (I) $\Delta^{m-1}y_n > 0$ eventually, and (II) $\Delta^{m-1}y_n < 0$ eventually.

(I) Suppose $\Delta^{m-1}y_n > 0$ eventually. From (E_{35}), we have

$$\Delta z_n + p_n z_{n-\tau} = -q_n y_{g_n}^{\gamma} \leq 0 \quad \text{eventually,} \tag{1.14.17}$$

where $z_n = \Delta^{m-1}y_n > 0$ eventually. In view of Theorem 3 in [245], inequality (1.14.17) has no eventually positive solution, a contradiction.

(II) Suppose $\Delta^{m-1}y_n < 0$ eventually. Then, there exists an integer $n_2 \geq n_1$ such that

$$\Delta^{m-2}y_n > 0 \quad \text{and} \quad \Delta y_n > 0 \quad \text{for} \quad n \geq n_2.$$

Since, y_n is increasing, there exist a constant $c > 0$ and an integer $n_3 \geq n_2$ such that

$$y_{g_n} \geq c \quad \text{for} \quad n \geq n_3.$$

Thus,

$$\Delta^m y_n + p_n \Delta^{m-1}y_n + c^{\gamma} q_n \leq 0 \quad \text{for} \quad n \geq n_3. \tag{1.14.18}$$

Summing both sides of (1.14.18) from n_3 to $N-1 \geq n_3$, we have

$$\Delta^{m-1}y_N - \Delta^{m-1}y_{n_3} + \sum_{n=n_3}^{N-1} p_n \Delta^{m-1}y_n + c^{\gamma} \sum_{n=n_3}^{N-1} q_n \leq 0$$

or

$$\Delta^{m-1}y_N + \left[p_N \Delta^{m-2}y_N - p_{n_3}\Delta^{m-2}y_{n_3} - \sum_{n=n_3}^{N-1} \left(\Delta^{m-2}y_n\right)\left(\Delta p_n\right)\right]$$

$$+c^{\gamma} \sum_{n=n_3}^{N-1} q_n \leq 0, \qquad n_3 \leq n \leq N-1.$$

Using the fact that $\Delta p_n \leq 0$ for $n \geq n_0$, we find

$$\Delta^{m-1}y_N - p_{n_3}\Delta^{m-2}y_{n_3} + c^{\gamma} \sum_{n=n_3}^{N-1} q_n \leq 0. \tag{1.14.19}$$

From (1.14.16), it follows that there exist a constant $c^* > 0$ and an integer $N^* \geq n_3 + 1$ such that

$$\Delta^{m-1}y_N \leq -c^* \quad \text{for} \quad N \geq N^*$$

and consequently

$$0 < \Delta^{m-2} y_j \rightarrow -\infty \quad \text{as} \quad j \rightarrow \infty,$$

a contradiction. ∎

Finally, in this section we present the following comparison result.

Theorem 1.14.6. Let $\gamma = 1$, $\tau \leq 0$, $\Delta p_n \leq 0$ and $g_n = n - k$ for $n \geq n_0 \geq 0$, where k is a positive integer and suppose that condition (1.14.16) holds. If every bounded solution of

$$\Delta^m w_n + q_n w_{n-k} = 0 \tag{1.14.20}$$

is oscillatory and the equation

$$\Delta z_n + Q_n z_{n-k} = 0 \tag{1.14.21}$$

is oscillatory, where

$$Q_n = \frac{2^{4-2m}}{(m-2)!} \sum_{j=n-k}^{n-1} (j-k)^{m-2} q_j - p_{n-k} > 0,$$

then (E_{35}) is almost oscillatory.

Proof. Let $\{y_n\}$ be an eventually positive solution of (E_{35}), say $y_n > 0$ and $y_{n-k} > 0$ for $n \geq n_1 \geq n_0 \geq 0$. As in Theorem 1.14.5 we consider the two Cases (I) and (II).

(I) Suppose $\Delta^{m-1} y_n > 0$ eventually. In this case, we have the following two possibilities:

(i) $\Delta y_n > 0$ eventually, and (ii) $\Delta y_n < 0$ eventually.

(i) Assume that $\Delta y_n > 0$ for $n \geq n_2 \geq n_1$. There exist a constant $c > 0$ and $n_3 \geq n_2$ such that

$$y_{n-k} \geq c \quad \text{for} \quad n \geq n_3.$$

Thus,

$$\Delta^m y_n + c q_n \leq 0 \quad \text{for} \quad n \geq n_3. \tag{1.14.22}$$

Summing both sides of (1.14.22) from n_3 to $N - 1 \geq n_3$, we get

$$0 < \Delta^{m-1} y_N \leq \Delta^{m-1} y_{n_3} - c \sum_{n=n_3}^{N-1} q_n \rightarrow -\infty \quad \text{as} \quad N \rightarrow \infty,$$

a contradiction.

(ii) Assume that $\Delta y_n < 0$ eventually. From (E_{35}), we see that

$$\Delta^m y_n + q_n y_{n-k} \le 0 \quad \text{eventually.} \tag{1.14.23}$$

But, by Theorem 1 in [194], if the inequality (1.14.23) has an eventually positive solution, then (1.14.20) has an eventually positive solution, a contradiction.

(II) Suppose $\Delta^{m-1} y_n < 0$ for $n \ge n_2 \ge n_1$. Then, there exists an integer $n_3 \ge n_2$ such that

$$\Delta^{m-2} y_n > 0 \quad \text{and} \quad \Delta y_n > 0 \quad \text{for} \quad n \ge n_3.$$

Applying Lemma 1.13.2 there exists an integer $n_4 \ge 2^{m-2} n_3$ such that

$$y_{n-k} \ge \left(\frac{2^{4-2m}}{(m-2)!}\right)(n-k)^{m-2}\Delta^{m-2} y_{n-k}, \quad n \ge n_5 = n_4 + k. \tag{1.14.24}$$

Using (1.14.24) in (E_{35}), we have

$$\Delta^2 x_n + p_n \Delta x_{n-\tau} + Q_n^* x_{n-k} \le 0 \quad \text{for} \quad n \ge n_5, \tag{1.14.25}$$

where $x_n = \Delta^{m-2} y_n > 0$, $n \ge n_2$ and

$$Q_n^* = \frac{2^{4-2m}}{(m-2)!}(n-k)^{n-2} q_n, \quad n \ge n_5.$$

Summing (1.14.25) from $n-k \ge n_4$ to $n-1$, we obtain

$$\Delta x_n - \Delta x_{n-k} + \sum_{j=n-k}^{n-1} p_j \Delta x_{j-\tau} + \sum_{j=n-k}^{n-1} Q_j^* x_{j-k} \le 0, \quad n \ge n_5$$

or

$$\Delta x_n + \left[p_n x_{n-\tau} - p_{n-k} x_{n-k-\tau} - \sum_{j=n-k}^{n-1} (x_{j-\tau})\Delta p_j\right] + \left(\sum_{j=n-k}^{n-1} Q_j^*\right) x_{n-k} \le 0 \quad \text{for} \quad n \ge n_5.$$

Using the fact that $\Delta p_n \le 0$ for $n \ge n_0$, we have

$$\Delta x_n + Q_n x_{n-k} \le 0 \quad \text{for} \quad n \ge n_5.$$

Now, by Lemma 1.1(a) in [194], the equation (1.14.21) has an eventually positive solution, a contradiction. ∎

Remark 1.14.3. Once again, from the proofs of the above results we note that Theorems 1.14.5 and 1.14.6 remain valid for (E_{35}) when $m = 2$.

Remark 1.14.4. If $p_n = 0$, $n \geq 0$ in Theorem 1.14.5, then it is easy to check that all solutions of (E_{35}) are oscillatory. Therefore, we conclude that the disruption in the oscillatory property is due to the presence of p_n.

1.15. Oscillation of Neutral Difference Equations

Here we are concerned with the oscillatory behavior of higher order difference equations of the neutral type

$$\Delta^m (y_n - y_{n-\tau}) + q_n f(y_{n-g}) = 0. \qquad (E_{36})$$

With respect to (E_{36}) we shall always assume that the following hold:

(a) $\{q_n\}$ is a real sequence with $q_n \geq 0$ eventually and $n \in \mathbb{N} \cup \{0\}$,

(b) $f : \mathbb{R} \to \mathbb{R}$ is continuous with $uf(u) > 0$ and nondecreasing for $u \neq 0$, and satisfies

$$-f(-xy) \geq f(xy) \geq f(x)f(y) \quad \text{for} \quad xy > 0,$$

(c) g, τ are positive integers.

Theorem 1.15.1. Let m be odd. If

$$\sum^{\infty} q_k < \infty \qquad (1.15.1)$$

and

$$\sum^{\infty} q_n f(nQ_n) = \infty, \qquad (1.15.2)$$

where

$$Q_n = \sum_{k=n}^{\infty} q_n, \qquad (1.15.3)$$

then (E_{36}) is oscillatory.

Proof. Let $\{y_n\}$ be an eventually positive solution of (E_{36}), say $y_n > 0$, $y_{n-\tau} > 0$ and $y_{n-g} > 0$ for $n \geq n_1 \geq n_0$. Set

$$z_n = y_n - y_{n-\tau}. \qquad (1.15.4)$$

Then, (E_{36}) becomes

$$\Delta^m z_n = -q_n f(y_{n-g}) \leq 0, \quad n \geq n_1. \qquad (1.15.5)$$

Thus, $\Delta^i z_n$ are eventually of one sign, $i = 0, 1, \cdots, m$, and there are four possible cases to consider:

(A) $z_n > 0$, $\Delta z_n > 0$, (B) $z_n > 0$, $\Delta z_n < 0$,

(C) $z_n < 0$, $\Delta z_n > 0$ and (D) $z_n < 0$, $\Delta z_n < 0$ eventually.

(A) Assume $z_n > 0$ and $\Delta z_n > 0$ for $n \geq n_1$. From (1.15.4) we see that $y_n \geq z_n$ and hence there exist an integer $n_2 \geq n_1$ and a positive constant c such that

$$y_{n-g} \geq z_{n-g} \geq c \quad \text{for} \quad n \geq n_2.$$

Thus,

$$\Delta^m z_n \leq -f(c)q_n \quad \text{for} \quad n \geq n_2. \tag{1.15.6}$$

It is easy to check that $\Delta^{m-1} z_n > 0$ for $n \geq n_2$. Summing both sides of (1.15.6) from $n \geq n_2$ to $N \geq n$ and letting $N \to \infty$, we obtain

$$\Delta^{m-1} z_n \geq f(c) \sum_{k=n}^{\infty} q_k = f(c)Q_n, \quad n \geq n_2. \tag{1.15.7}$$

Next, by applying Lemma 1.13.2, there exists an integer $n_3 \geq 2^{m-1} n_2$ so large that

$$z_n \geq \alpha n^{(m-1)} \Delta^{m-1} z_n \quad \text{for} \quad n \geq n_3, \tag{1.15.8}$$

where $\alpha = \dfrac{2^{2-2m}}{(m-1)!}$. Using (1.15.8) in (1.15.7), we have

$$y_n \geq z_n \geq \alpha_1 n^{m-1} Q_n, \quad n \geq n_3$$

where $\alpha_1 = \alpha f(c)$. Now, there exists an integer $n_4 \geq n_3$ such that

$$y_{n-g} \geq \alpha_1 (n-g)^{m-1} Q_{n-g} \geq \alpha_1 (n-g)^{m-1} Q_n, \quad n \geq n_4. \tag{1.15.9}$$

Using (1.15.5) and (1.15.9) in (1.15.5), we obtain

$$\Delta^m z_n \leq -\alpha_2 q_n f\left((n-g)^{m-1} Q_n\right), \quad n \geq n_4 \tag{1.15.10}$$

where $\alpha_2 = f(\alpha_1)$.

Summing (1.15.10) from n_4 to $n-1 \geq n_4$, we get

$$\Delta^{m-1} z_n \leq \Delta^{m-1} z_{n_4} - \alpha_2 \sum_{j=n_4}^{n-1} q_j f\left((j-g)^{m-1} Q_j\right),$$

which in view of (1.15.2) leads to

$$\Delta^{m-1} z_n \to -\infty \quad \text{as} \quad n \to \infty,$$

a contradiction.

(B) Assume $z_n > 0$ and $\Delta z_n < 0$ eventually. By Lemma 1.13.1, we see that $\Delta^{m-1} z_n > 0$ eventually. In this case, we have $y_n > y_{n-\tau}$. Hence, there exist a constant $b > 0$ and $n_2 \geq n_1$ such that

$$y_{n-g} \geq b \quad \text{for all} \quad n \geq n_2.$$

Then, from (1.15.5), it follows that

$$\Delta^m z_n \leq -f(b) q_n \quad \text{for} \quad n \geq n_2$$

and hence

$$\Delta^{m-1} z_n \geq f(b) Q_n \quad \text{for} \quad n \geq n_2. \tag{1.15.11}$$

By Taylor's formula (see [2]), we see that for $s - 1 \geq k \geq j \geq n_2$

$$\begin{aligned} z_k &= \sum_{i=0}^{m-2} \frac{(s+i-1-k)^{(i)}}{i!} (-1)^i \Delta^i z_s \\ &\quad + \sum_{j=k}^{s-1} \frac{(j+m-2-k)^{(m-2)}}{(m-2)!} (-1)^{m-1} \Delta^{m-1} z_j. \end{aligned}$$

Since m is odd, and $\Delta^{m-1} z_n$ is decreasing, one can easily see that

$$z_k \geq \left(\sum_{j=k}^{s-1} \frac{(j+m-2-k)^{(m-2)}}{(m-2)!} \right) \Delta^{m-1} z_{s-1}.$$

Replacing k with $n - \tau$ and s with $n + 1$, we have

$$z_{n-\tau} \geq \left(\sum_{j=n-\tau}^{n} \frac{(j+m-2-n+\tau)^{(m-2)}}{(m-2)!} \right) \Delta^{m-1} z_n$$

or

$$z_n \geq \beta \Delta^{m-1} z_{n+\tau} \quad \text{for} \quad n \geq n_3 \geq n_2, \tag{1.15.12}$$

where

$$\beta = \sum_{j=n}^{n+\tau} \frac{(j+m-2-n)^{(m-2)}}{(m-2)!}.$$

Using (1.15.12) in (1.15.11), we obtain

$$z_n \geq \beta_1 Q_{n+\tau}, \quad n \geq n_3 \tag{1.15.13}$$

where $\beta_1 = \beta f(b)$. From (1.15.4), we have

$$y_n \geq \beta_1 Q_{n+h} + y_{n-\tau}, \quad n \geq n_3.$$

Let an integer N be such that $n_3 + (N-1)\tau \le n \le n_3 + N\tau$. Then, we have

$$y_n \ \ge\ \beta_1 \left[Q_{n+\tau} + Q_n + \cdots + Q_{n-(N-2)\tau}\right] + y_{n-N\tau} \ \ge\ \beta_1 (N-1) Q_n,$$

which together with (1.15.5) yields

$$\Delta^m z_n \ \le\ -A_n, \tag{1.15.14}$$

where

$$A_n \ =\ f\left(\frac{\beta_1(N-1)}{n}\right) q_n f(nQ_n).$$

By noting that $n/N \to \tau$ as $n \to \infty$, we find

$$\frac{A_n}{q_n f(nQ_n)} \ =\ f\left(\beta_1\left(\frac{N-1}{n}\right)\right) \to f\left(\frac{\beta_1}{\tau}\right) \quad \text{as} \quad n \to \infty$$

and by (1.15.2), we have

$$\sum^{\infty} A_n \ =\ \infty. \tag{1.15.15}$$

Thus, (1.15.14) and (1.15.15) yield

$$\Delta^{m-1} z_n \to -\infty \quad \text{as} \quad n \to \infty,$$

a contradiction.

(C) Assume $z_n < 0$ and $\Delta z_n > 0$ eventually. Then,

$$0 \ <\ v_n \ =\ -z_n \ =\ y_{n-\tau} - y_n,$$

and hence (1.15.5) becomes

$$\Delta^m v_n \ =\ q_n f(y_{n-g}) \quad \text{eventually.}$$

Since m is odd, by Lemma 1.13.1, we must have $\Delta v_n > 0$ eventually contradicting the assumption.

(D) Assume $z_n < 0$ and $\Delta z_n < 0$ for $n \ge n_1$. Since z_n is decreasing, there exist a constant $a > 0$ and a $N \ge n_1$ such that

$$z_n \ <\ -a \quad \text{for} \quad n \ge N.$$

Therefore,

$$y_N \ =\ z_N + y_{N-\tau} \ <\ -a + y_{N-\tau}$$

and it follows that

$$y_{N+j\tau} \ <\ -a(j+1) + y_{N-\tau} \to -\infty \quad \text{as} \quad j \to \infty,$$

which contradicts $y_n > 0$ eventually. ∎

Remark 1.15.1. When $\tau = 1$, (E_{36}) is reduced to an even order equation

$$\Delta^{m+1} y_n + q_n f(y_{n-g}) = 0. \qquad (E_{37})$$

Applying Theorem 1.15.1 for (E_{37}), we obtain the following new criterion for the oscillation of (E_{37}).

Corollary 1.15.2. If

$$\sum^{\infty} q_n f\left((n-g)^m Q_n\right) = \infty, \qquad (1.15.16)$$

where Q_n is defined as in (1.15.3), then (E_{37}) is oscillatory.

Proof. The proof is contained in the proof of Theorem 1.15.1(A). ∎

Now we introduce the following notation:

$$g^* = \begin{cases} g & \text{if } m \text{ is odd} \\ g - \tau > 0 & \text{if } m \text{ if even.} \end{cases} \qquad (1.15.17)$$

Next we shall prove the following comparison result.

Theorem 1.15.3. Suppose that the first order equation

$$\Delta w_n + f\left(\frac{2^{2-2m}}{(m-1)!}\right) q_n f\left((n-g)^{m-1}\right) f\left(w_{n-g}\right) = 0 \qquad (1.15.18)$$

is oscillatory, and all bounded solutions of the equation

$$\Delta^m x_n + (-1)^{m+1} q_n f\left(x_{n-g^*}\right) = 0 \qquad (1.15.19)$$

are oscillatory. Then, (E_{36}) is oscillatory.

Proof. Let $\{y_n\}$ be an eventually positive solution of (E_{36}), and let z_n be defined as in (1.15.4). As in the proof of Theorem 1.15.1, the four Cases (A) – (D) need to be considered:

(A) Assume $z_n > 0$ and $\Delta z_n > 0$ for $n \ge n_1 \ge n_0$. Proceeding as in the proof of Theorem 1.15.1(A), there exists an integer $n_2 \ge 2^{m-1} n_1$ such that $\Delta^{m-1} z_n > 0$ and (1.15.8) holds for $n \ge n_2$. From the fact that $y_n \ge z_n$ and (c) there exists an integer $n_3 \ge n_2$ such that

$$y_{n-g} \ge z_{n-g} \ge \alpha(n-g)^{m-1} \Delta^{m-1} z_{n-g} \quad \text{for} \quad n \ge n_3, \qquad (1.15.20)$$

where $\alpha = \dfrac{2^{2-2m}}{(m-1)!}$. Using (b) and (1.15.20) in (1.15.5) and letting $v_n = \Delta^{m-1} z_n$, $n \ge n_3$, we have

$$\Delta v_n + f(\alpha) f\left((n-g)^{m-1}\right) q_n f\left(v_{n-g}\right) \le 0 \quad \text{for} \quad n \ge n_3. \qquad (1.15.21)$$

Summing both sides of (1.15.21) from $n \geq n_3$ to N and letting $N \to \infty$, we obtain

$$v_n \geq f(\alpha) \sum_{k=n}^{\infty} f\left((k-g)^{m-1}\right) q_k f\left(v_{k-g}\right). \tag{1.15.22}$$

But, by the discrete analog of a result due to Philos [241] and Theorem 1 in [194], if (1.15.22) has an eventually positive solution v_n, then the corresponding equation

$$w_n = f(\alpha) \sum_{k=n}^{\infty} f\left((k-g)^{m-1}\right) q_k f\left(w_{n-g}\right), \tag{1.15.23}$$

also has an eventually positive solution w_n. It follows then that the equation (1.15.18) has the eventually positive solution w_n. This contradicts the hypothesis that (1.15.18) is oscillatory.

(B) Assume $z_n > 0$ and $\Delta z_n < 0$ for $n \geq n_1 \geq n_0$. This is the case when m is odd. By Lemma 1.13.1 and Lemma 6 in [241], there exists an integer $n_2 \geq n_1$ such that

$$(-1)^i \Delta^i z_n > 0 \quad \text{for} \quad i = 0, 1, \cdots, m-1 \quad \text{and} \quad n \geq n_2. \tag{1.15.24}$$

Summing both sides of (1.15.5) from $n \geq n_2$ to N repeatedly m–times, using (1.15.24) and the fact that $y_n \geq z_n$ for $n \geq n_2$ and letting $N \to \infty$, we have

$$z_n \geq \sum_{j_1=n}^{\infty} \sum_{j_2=j_1}^{\infty} \cdots \sum_{j_m=j_{m-1}}^{\infty} q_{j_m} f\left(z_{j_m-g}\right). \tag{1.15.25}$$

The remainder of the proof is similar to that of Case (A) given above.

(C) Assume $z_n < 0$ and $\Delta z_n > 0$ for $n \geq n_1 \geq n_0$. This is the case when m is even. Set

$$0 < x_n = -z_n = y_{n-\tau} - y_n.$$

Then, (E_{36}) becomes

$$\Delta^m x_n = q_n f\left(y_{n-g}\right) \tag{1.15.26}$$

and

$$y_{n-\tau} \geq x_n \quad \text{or} \quad y_n \geq x_{n+\tau} \quad \text{for} \quad n \geq n_1. \tag{1.15.27}$$

Since m is even and $\Delta x_n < 0$ for $n \geq n_1$, there exists $n_2 \geq n_1$ such that

$$(-1)^i \Delta^i x_n > 0 \quad \text{for} \quad i = 0, 1, \cdots, m-1 \quad \text{and} \quad n \geq n_2. \tag{1.15.28}$$

Summing both sides of (1.15.26) from $n \geq n_2$ to N repeatedly m–times and using (1.15.27) and (1.15.28) and letting $N \to \infty$, we have

$$x_n \geq \sum_{j_1=n}^{\infty} \sum_{j_2=j_1}^{\infty} \cdots \sum_{j_m=j_{m-1}}^{\infty} q_{j_m} f\left(x_{j_m-(g-\tau)}\right).$$

The rest of the proof is similar to that of Case (A) given above.

(D) Assume $z_n < 0$ and $\Delta z_n < 0$ for $n \geq n_1 \geq n_0$. The proof is exactly the same as in the proof of Theorem 1.15.1(D). ∎

As an application of Theorem 1.15.3, we consider the equation

$$\Delta^m \left(y_n - y_{n-\tau}\right) + q y_{n-g} = 0, \qquad (E_{38})$$

where q is a real number and τ and g are positive integers. We see that (E_{38}) is oscillatory if

$$q > \frac{m^m g^g}{(m+g)^{m+g}}, \quad g \geq 1. \qquad (1.15.29)$$

1.16. Oscillation of Mixed Difference Equations

Here we shall consider the difference equations of the type

$$\Delta^m \left(y_n + a y_{n-\tau} - b y_{n+\sigma}\right) + \delta \left(q y_{n-g} + p y_{n+h}\right) = 0, \qquad (E_{39};\delta)$$

$$\Delta^m \left(y_n - a y_{n-\tau} + b y_{n+\sigma}\right) + \delta \left(q y_{n-g} + p y_{n+h}\right) = 0, \qquad (E_{40};\delta)$$

$$\Delta^m \left(y_n + a y_{n-\tau} + b y_{n+\sigma}\right) + \delta \left(q y_{n-g} + p y_{n+h}\right) = 0 \qquad (E_{41};\delta)$$

and

$$\Delta^m \left(y_n - a y_{n-\tau} - b y_{n+\sigma}\right) + \delta \left(q y_{n-g} + p y_{n+h}\right) = 0, \qquad (E_{42};\delta)$$

where $\delta = \pm 1$, a, b, p and q are nonnegative real numbers, g, h, τ and σ are nonnegative real numbers and are multiple of m.

In what follows we shall establish some sufficient conditions, involving the coefficients and the arguments only, under which all solutions of $(E_{39};\delta) - (E_{42};\delta)$ oscillate. The advantage of working with these conditions rather than the usual characteristic equations [129] associated with the equations under considerations is that they are explicit and therefore, easily verifiable, while determining whether or not a positive root to the associated characteristic equation exists may be quite a problem in itself. Furthermore, our technique is given in such a way that it can be extended in a straightforward manner to the case of difference equations with variable coefficients.

We shall need the following:

Lemma 1.16.1. [87] Assume that q is a positive real number and k is a positive integer and is a multiple of m. Then, the following hold:

(i) If

$$q > \frac{m^m(k-m)^{k-m}}{k^k} \quad \text{for} \quad k > m,$$

then the difference inequality

$$\Delta^m x_n \geq q x_{n+k} \quad \text{eventually}$$

has no eventually positive solution $\{x_n\}$ which satisfies $\Delta^j x_n > 0$ eventually, $j = 0, 1, \cdots, m$.

(ii) If

$$q > \frac{m^m k^k}{(k+m)^{k+m}} \quad \text{for} \quad k \geq 1,$$

then the difference inequality

$$(-1)^m \Delta^m x_n \geq q x_{n-k} \quad \text{eventually}$$

has no eventually positive solution $\{x_n\}$ which satisfies $(-1)^j \Delta^j x_n > 0$ eventually, $j = 0, 1, \cdots, m$.

The following two criteria are concerned with the oscillation of $(E_{39}; \delta)$.

Theorem 1.16.2. Suppose that $h > m$, $g > \tau$ and $b > 0$, and

$$\frac{p}{1+a} > \frac{m^m(h-m)^{h-m}}{h^h}. \tag{1.16.1}$$

If

$$\frac{q}{b} > \frac{m^m(g+\sigma)^{g+\sigma}}{(m+g+\sigma)^{m+g+\sigma}} \quad \text{if } m \text{ is odd} \tag{1.16.2}$$

and

$$\frac{q}{1+a} > \frac{m^m(g-\tau)^{g-\tau}}{(m+g-\tau)^{m+g-\tau}} \quad \text{if } m \text{ is even,} \tag{1.16.3}$$

then $(E_{39}; -1)$ is oscillatory.

Proof. Let $\{y_n\}$ be an eventually positive solution of $(E_{39}; -1)$, say $y_n > 0$ for $n \geq n_0 \geq 0$. Set

$$z_n = y_n + a y_{n-\tau} - b y_{n+\sigma}. \tag{1.16.4}$$

Then,

$$\Delta^m z_n = qy_{n-g} + py_{n+h} \geq 0 \quad \text{eventually} \tag{1.16.5}$$

and hence we see that $\Delta^i z_n$, $i = 0, 1, \cdots, m$ are eventually of one sign. There are two possible cases to consider:

(i) $z_n < 0$ eventually, and (ii) $z_n > 0$ eventually.

(i) Assume $z_n < 0$ for $n \geq n_1 \geq n_0$. In this case, we let

$$0 < u_n = -z_n = by_{n+\sigma} - ay_{n-\tau} - y_n \leq by_{n+\sigma} \quad \text{for} \quad n \geq n_1.$$

Thus,

$$y_{n+\sigma} \geq \frac{1}{b} u_n,$$

or

$$y_n \geq \frac{1}{b} u_{n-\sigma} \quad \text{for} \quad n \geq n_2 \geq n_1. \tag{1.16.6}$$

From (1.16.5), we have

$$\Delta^m u_n + qy_{n-g} \leq 0 \quad \text{for} \quad n \geq n_2. \tag{1.16.7}$$

Using (1.16.6) in (1.16.7), we have

$$\Delta^m u_n + \frac{q}{b} u_{n-\sigma-g} \leq 0 \quad \text{for} \quad n \geq n_2. \tag{1.16.8}$$

It is easy to check that $\Delta^{m-1} u_n > 0$ for $n \geq n_3 \geq n_2$, and either (I) $\Delta u_n > 0$ for $n \geq n_3$, or (II) $\Delta u_n < 0$ for $n \geq n_3$.

(I) Suppose $\Delta u_n > 0$ for $n \geq n_3$. There exist an integer $n_4 \geq n_3$ and a positive constant α such that

$$u_{n-\sigma-g} \geq \alpha \quad \text{for} \quad n \geq n_4. \tag{1.16.9}$$

Using (1.16.9) in (1.16.8) and summing both sides of the resulting inequality from n_4 to $k \geq n_4$, we have

$$0 < \Delta^{m-1} u_k \leq \Delta^{m-1} u_{n_4} - \frac{q}{b} \alpha (k - n_4) \to -\infty \text{ as } k \to \infty,$$

a contradiction.

(II) Suppose $\Delta u_n < 0$ for $n \geq n_3$. This is the case when m is odd, and hence we see that

$$(-1)^i \Delta^i u_n > 0 \quad \text{for} \quad i = 0, 1, \cdots, m \quad \text{and} \quad n \geq N_1 \geq n_3. \tag{1.16.10}$$

But, in view of Lemma 1.16.1(ii) and condition (1.16.2), inequality (1.16.8) has no solution such that (1.16.10) holds, a contradiction.

(ii) Assume $z_n > 0$ for $n \geq n_1 \geq n_0$. Set

$$w_n = z_n + az_{n-\tau} - bz_{n+\sigma}. \tag{1.16.11}$$

Then,

$$\Delta^m w_n = qz_{n-g} + pz_{n+h} \tag{1.16.12}$$

and since w_n satisfies $(E_{39}; -1)$, we have

$$\Delta^m (w_n + aw_{n-\tau} - bw_{n+\sigma}) = qw_{n-g} + pw_{n+h}. \tag{1.16.13}$$

Using the procedure of Case (i), we observe that $w_n > 0$ for $n \geq n_2 \geq n_1$. Next, we have two cases to consider:

$(\overline{I})$ $\Delta z_n > 0$ for $n \geq n_3 \geq n_2$, and $(\overline{II})$ $\Delta z_n < 0$ for $n \geq n_3$.

$(\overline{I})$ Let $\Delta z_n > 0$ for $n \geq n_3$. From (1.16.12), we have

$$\Delta^i w_n > 0 \quad \text{for} \quad i = m, m+1 \quad \text{and} \quad n \geq n_3$$

and hence, we see that

$$\Delta^i w_n > 0, \quad i = 0, 1, \cdots, m+1 \quad \text{and} \quad n \geq n_3. \tag{1.16.14}$$

Now, using the fact that $\Delta^m w_n$ is eventually increasing in (1.16.13), we have

$$\begin{aligned}(1+a)\Delta^m w_n &\geq \Delta^m w_n + a\Delta^m w_{n-\tau} - b\Delta^m w_{n+\tau} \\ &= qw_{n-g} + pw_{n+h} \geq pw_{n+h}, \quad n \geq n_3\end{aligned}$$

and hence

$$\Delta^m w_n \geq \frac{p}{1+a} w_{n+h} \quad \text{for} \quad n \geq n_3. \tag{1.16.15}$$

But, in view of Lemma 1.16.1(i) and condition (1.16.1), inequality (1.16.15) has no solution such that (1.16.14) holds, a contradiction.

$(\overline{II})$ Let $\Delta z_n < 0$ for $n \geq n_3$. From (1.16.5), m must be even. We claim that $\Delta w_n < 0$ for $n \geq n_4 \geq n_3$. To prove it, assume that $\Delta w_n > 0$ for $n \geq n_4$. Then, from (1.16.12) we see that

$$\Delta^m w_n > 0 \quad \text{and} \quad \Delta^{m+1} w_n < 0 \quad \text{for} \quad n \geq n_3.$$

Using this fact in (1.16.13), one can easily see that

$$(1+a)\Delta^m w_{n-\tau} \geq pw_{n+h}$$

or

$$\Delta^m w_n \geq \frac{p}{1+a} w_{n+h+\tau}, \quad n \geq n_4.$$

Since $\{w_n\}$ is an increasing sequence, we have

$$\Delta^i w_n > 0, \quad i = 0, 1, \cdots, m \quad \text{and} \quad n \geq n_4$$

and

$$\Delta^m w_n \geq \frac{p}{1+a} w_{n+h}, \qquad n \geq n_4$$

and again we are led to a contradiction. Thus, $\Delta w_n < 0$ for $n \geq n_4$ and from (1.16.12), we have

$$(-1)^i \Delta^i w_n > 0, \quad i = 0, 1, \cdots, m+1 \quad \text{for} \quad n \geq n_4. \tag{1.16.16}$$

Now, using the fact that $\{\Delta^m w_n\}$ is decreasing for $n \geq n_4$ in (1.16.13), we obtain

$$\begin{aligned}(1+a)\Delta^m w_{n-\tau} &\geq \Delta^m w_n + a\Delta^m w_{n-\tau} - b\Delta^m w_{n+\sigma} \\ &= q w_{n-g} + p w_{n+h} \geq q w_{n-g} \quad \text{for} \quad n \geq n_4\end{aligned}$$

and hence

$$\Delta^m w_n \geq \frac{q}{1+a} w_{n-(g-\tau)} \quad \text{for} \quad n \geq n_4.$$

The rest of the proof is similar to that of Case (II) given above. ■

Remark 1.16.1. From the proof of Case (II) one can easily see that condition (1.16.2) can be replaced by

$$\frac{p+q}{b} > \frac{m^m(\sigma-h)^{\sigma-h}}{(m+\sigma-h)^{m+\sigma-h}}, \qquad \sigma > h. \tag{1.16.2$'$}$$

Theorem 1.16.3. Let $b > 0$, $h > \sigma + m$ and $g > \tau$, and

$$\frac{p}{b} > \frac{m^m(h-\sigma-m)^{h-\sigma-m}}{(h-\sigma)^{h-\sigma}}. \tag{1.16.17}$$

If

$$\frac{q}{1+a} > \frac{m^m(g-\tau)^{g-\tau}}{(m+g-\tau)^{m+g-\tau}} \quad \text{when} \quad m \quad \text{is odd} \tag{1.16.18}$$

and

$$\frac{q}{b} > \frac{m^m(g+\sigma)^{g+\sigma}}{(m+g+\sigma)^{m+g+\sigma}} \quad \text{when} \quad m \quad \text{is even,} \tag{1.16.19}$$

then $(E_{39}; 1)$ is oscillatory.

Proof. Let $\{y_n\}$ be an eventually positive solution of $(E_{39}; 1)$, say $y_n > 0$ for $n \geq n_0 \geq 0$. Define z_n by (1.16.4). Then,

$$\Delta^m z_n = -q y_{n-g} - p y_{n+h} \leq 0 \quad \text{eventually} \tag{1.16.20}$$

and hence, we see that $\Delta^i z_n$, $i = 0, 1, \cdots, m$ are eventually of one sign.

Now, we consider the two cases:

(i) $z_n < 0$ eventually, and (ii) $z_n > 0$ eventually.

(i) Assume $z_n < 0$ for $n \geq n_1 \geq n_0$. As in the proof of Theorem 1.16.2 Case (i), we obtain (1.16.6). Using (1.16.6) and $-z_n = u_n$ in (1.16.20), we have

$$\Delta^m u_n = qy_{n-g} + py_{n+h} \geq \frac{q}{b}u_{n-(g+\sigma)} + \frac{p}{b}u_{n+(h-\sigma)} \quad \text{for } n \geq n_2. \tag{1.16.21}$$

Next, we consider the two cases:

(I) $\Delta u_n > 0$ for $n \geq n_3 \geq n_2$, and (II) $\Delta u_n < 0$ for $n \geq n_3$.

(I) Suppose $\Delta u_n > 0$ for $n \geq n_3$. It is easy to check that

$$\Delta^i u_n > 0, \quad i = 0, 1, \cdots, m, \quad n \geq n_3.$$

Thus,

$$\Delta^m u_n \geq \frac{p}{b}u_{n+(h-\sigma)} \quad \text{for } n \geq n_3.$$

The rest of the proof is similar to that of Case (ii)-($\overline{I}$).

(II) $\Delta u_n < 0$ for $n \geq n_3$. This is the case when m is even and hence, we see that (1.16.10) holds. Now,

$$\Delta^m u_n \geq \frac{q}{b}u_{n-(g+\sigma)} \quad \text{for } n \geq n_3. \tag{1.16.22}$$

But, in view of Lemma 1.16.1(ii) and condition (1.16.19), inequality (1.16.22) has no solution such that (1.16.10) holds, a contradiction.

(ii) Assume $z_n > 0$ for $n \geq n_1$. Define w_n by (1.16.11), and obtain

$$\Delta^m w_n + qz_{n-g} + pz_{n+h} = 0 \tag{1.16.23}$$

and

$$\Delta^m (w_n + aw_{n-\tau} - bw_{n+\sigma}) + qw_{n-g} + pw_{n+h} = 0. \tag{1.16.24}$$

Using the procedure of Case (i), we see that $w_n > 0$ eventually. Next, we consider

($\overline{I}$) $\Delta z_n > 0$ for $n \geq n_3$, and ($\overline{II}$) $\Delta z_n < 0$ for $n \geq n_3$.

($\overline{I}$) Assume $\Delta z_n > 0$ for $n \geq n_3$. From (1.16.23), we see that

$$\Delta^i w_n \leq 0, \quad i = m, m+1 \quad \text{for } n \geq n_3$$

and hence, we see that $w_n < 0$ for $n \geq n_3$, a contradiction.

($\overline{II}$) Let $\Delta z_n < 0$ for $n \geq n_3$. This is the case when m is odd and from (1.16.20), we get

$$\Delta^m w_n < 0 \quad \text{and} \quad \Delta^{m+1} w_n > 0 \quad \text{for } n \geq n_3. \tag{1.16.25}$$

We claim that $\Delta w_n < 0$ for $n \geq n_4 \geq n_3$. To prove it, assume that $\Delta w_n > 0$ for $n \geq n_4$. Using (1.16.25) in (1.16.24), we have

$$(1+a)\Delta^m w_{n-\tau} + q w_{n-g} \leq 0$$

or

$$\Delta^m w_n + \frac{q}{1+a} w_{n-g+\tau} \leq 0 \quad \text{for} \quad n \geq n_4.$$

Proceeding as in the proof of Theorem 1.16.2 Case (i), we see that $\Delta^{m-1} w_n < 0$, $n \geq n_4$ a contradiction. Thus, $\Delta w_n < 0$ for $n \geq n_4$ and (1.16.16) holds. Using (1.16.25) in (1.16.24) yields

$$\Delta^m w_n + \frac{q}{1+a} w_{n-(g-\tau)} \leq 0 \quad \text{for} \quad n \geq n_4$$

and again we are led to a contradiction. ∎

Next, we present the following two theorems for the oscillation of $(E_{40}; \delta)$.

Theorem 1.16.4. Suppose that $a > 0$, $h > \sigma + m$ and $g > \tau \geq 1$, and

$$\frac{p}{1+b} > \frac{m^m (h-\sigma-m)^{h-\sigma-m}}{(h-\sigma)^{h-\sigma}}. \tag{1.16.26}$$

If

$$\frac{q}{a} > \frac{m^m (g-\tau)^{g-\tau}}{(m+g-\tau)^{m+g-\tau}} \quad \text{for } m \text{ is odd} \tag{1.16.27}$$

and

$$\frac{q}{1+b} > \frac{m^m g^g}{(m+g)^{m+g}} \quad \text{for } m \text{ is even,} \tag{1.16.28}$$

then $(E_{40}; -1)$ is oscillatory.

Proof. Let $\{y_n\}$ be an eventually positive solution of $(E_{40}; -1)$, say $y_n > 0$ for $n \geq n_0 \geq 0$. Set

$$z_n = y_n - a y_{n-\tau} + b y_{n+\sigma}. \tag{1.16.29}$$

Then,

$$\Delta^m z_n = q y_{n-g} + p y_{n+h} \geq 0 \quad \text{eventually.} \tag{1.16.30}$$

As in the proof of Theorem 1.16.2, we see that $\Delta^i z_n$, $i = 0, 1, \cdots, m$ are eventually of one sign and the two Cases (i) and (ii) are considered.

(i) Assume $z_n < 0$ for $n \geq n_1 \geq n_0$. Set

$$0 < u_n = -z_n = a y_{n-\tau} - b y_{n+\sigma} - y_n \leq a y_{n-\tau} \quad \text{for} \quad n \geq n_2 \geq n_1 \tag{1.16.31}$$

or

$$y_n \geq \frac{1}{a} u_{n+\tau}, \quad n \geq n_2. \tag{1.16.32}$$

As in Theorem 1.16.2(i), the two Cases (I) and (II) are considered and we see that the Case (I) is impossible. Therefore, we consider (II).

(II) Suppose $\Delta u_n < 0$ for $n \geq n_3 \geq n_2$. This is the case when m is odd and we see that u_n satisfies $(-1)^i \Delta^i u_n > 0$, $i = 0, 1, \cdots, m$ and $n \geq n_4 \geq n_3$. Thus,

$$\Delta^m u_n + \frac{q}{a} u_{n-(g-\tau)} \leq 0 \quad \text{for} \quad n \geq n_4.$$

The rest of the proof is similar to that of Theorem 1.16.2(ii)–$(\overline{II})$.

(ii) Suppose that $z_n > 0$ for $n \geq n_1$. We let

$$w_n = z_n - a z_{n-\tau} + b z_{n+\sigma}. \tag{1.16.33}$$

Then,

$$\Delta^m w_n = q z_{n-g} + p z_{n+h} \tag{1.16.34}$$

and

$$\Delta^m \left(w_n - a w_{n-\tau} + b w_{n+\sigma} \right) = q w_{n-g} + p w_{n+h}. \tag{1.16.35}$$

As in Theorem 1.16.2(ii), the two Cases $(\overline{I})$ and $(\overline{II})$ are considered.

$(\overline{I})$ Assume $\Delta z_n > 0$ for $n \geq n_2 \geq n_1$. From (1.16.34), it is easy to check that $\Delta^i w_n > 0$, $i = m, m+1$ and $n \geq n_2$, and hence we see that $\Delta^i w_n > 0$, $i = 0, 1, \cdots, m+1$ and $n \geq n_2$. Using this fact in (1.16.35), we obtain

$$\Delta^m w_n \geq \frac{p}{1+b} w_{n+(h-\sigma)}, \quad n \geq n_3 \geq n_2$$

and again we are led to a contradiction.

$(\overline{II})$ Assume $\Delta z_n < 0$ for $n \geq n_2$. This is the case when m is even, and from (1.16.34) one can easily see that $\Delta^m w_n > 0$ and $\Delta^{m+1} w_n < 0$ for $n \geq n_2$. We claim that $\Delta w_n < 0$ for $n \geq n_3 \geq n_2$. Otherwise, $\Delta w_n > 0$ for $n \geq n_3$. From (1.16.35) and the fact that $\{\Delta^i w_n\}$, $i = 0, 1, \cdots, m-1$ are increasing and $\{\Delta^m w_n\}$ is decreasing, we obtain

$$(1+b)\Delta^m w_n \geq p w_{n+h} \geq p w_{n+h-\sigma} \quad \text{for} \quad n \geq n_3$$

and as in the above case, we are led to a contradiction. Thus, $\Delta w_n < 0$ for $n \geq n_3$, and hence we conclude that $(-1)^i \Delta^i w_n > 0$, $i = 0, 1, \cdots, m+1$ and $n \geq n_3$. From (1.16.35), one can easily obtain

$$\Delta^m w_n \geq \frac{q}{1+b} w_{n-g}, \quad n \geq n_3.$$

The rest of the proof is similar to that of Theorem 1.16.2(ii)–$(\overline{II})$. ∎

Theorem 1.16.5. Let $a > 0$, $h + \tau > m$ and $g > \tau \geq 1$, and

$$\frac{p}{a} > \frac{m^m (h + \tau - m)^{h+\tau-m}}{(h+\tau)^{h+\tau}}. \tag{1.16.36}$$

If

$$\frac{q}{1+b} > \frac{m^m g^g}{(m+g)^{m+g}} \quad \text{if } m \text{ is odd} \tag{1.16.37}$$

and

$$\frac{q}{a} > \frac{m^m (g-\tau)^{g-\tau}}{(m+g-\tau)^{m+g-\tau}} \quad \text{if } m \text{ is even,} \tag{1.16.38}$$

then $(E_{40};1)$ is oscillatory.

Proof. Let $\{y_n\}$ be an eventually positive solution of $(E_{40};1)$, say $y_n > 0$ for $n \geq n_0 \geq 0$. Define z_n by (1.16.29), and obtain

$$\Delta^m z_n = -qy_{n-g} - py_{n+h} \leq 0 \quad \text{eventually}$$

and hence, we conclude that $\Delta^i z_n$, $i = 0, 1, \cdots, m$ are eventually of one sign. As in Theorem 1.16.2, we consider the two Cases (i) and (ii).

(i) Assume $z_n < 0$ for $n \geq n_1 \geq n_0$. Define u_n by (1.16.31) and obtain (1.16.32), and

$$\Delta^m u_n \geq \frac{q}{a} u_{n-(g-\tau)} + \frac{p}{a} u_{n+(h+\tau)}, \quad n \geq n_2 \geq n_1.$$

Now, if $\Delta u_n > 0$ for $n \geq n_2$, we see that $\Delta^i u_n > 0$, $i = 0, 1, \cdots, m$ and $n \geq n_2$. Thus,

$$\Delta^m u_n \geq \frac{p}{a} u_{n+(h+\tau)}, \quad n \geq n_2$$

and by Lemma 1.16.1(i) and condition (1.16.36) we are led to a contradiction. Next, if $\Delta u_n < 0$, $n \geq n_2$ then m must be even, and we conclude that $(-1)^i \Delta^i u_n > 0$, $i = 0, 1, \cdots, m$ and $n \geq n_2$. Thus,

$$\Delta^m u_n \geq \frac{q}{a} u_{n-(g-\tau)}, \quad n \geq n_2$$

and again by Lemma 1.16.1(ii), and condition (1.16.38) we arrive at a contradiction.

(ii) Assume $z_n > 0$, $n \geq n_1$. Define w_n by (1.16.33), and obtain

$$\Delta^m w_n + qz_{n-g} + pz_{n+h} = 0$$

and

$$\Delta^m (w_n - aw_{n-\tau} + bw_{n+\sigma}) + qw_{n-g} + pw_{n+h} = 0. \tag{1.16.39}$$

Using the procedure of the proof of Case (i) above, one can easily see that $w_n > 0$ for $n \geq n_2 \geq n_1$. As in Theorem 1.16.4, we consider the two Cases $(\bar{I})$ and $(\bar{II})$.

$(\bar{I})$ Suppose $\Delta z_n > 0$ for $n \geq n_2$. It is easy to check that $\Delta^m w_n < 0$ and $\Delta^{m+1} w_n < 0$ for $n \geq n_2$, and hence $w_n < 0$ for $n \geq n_2$, a contradiction.

$(\overline{II})$ Suppose $\Delta z_n < 0$ for $n \geq n_2$. This is the case when m is odd and we see that $\Delta^m w_n < 0$ and $\Delta^{m+1} w_n > 0$ for $n \geq n_2$. Next, we claim that $\Delta w_n < 0$ for $n \geq n_3 \geq n_2$. Otherwise, $\Delta w_n > 0$ for $n \geq n_3$. From (1.16.39), one can easily obtain

$$(1+b)\Delta^m w_n + q w_{n-g} \ \leq \ 0, \qquad n \geq n_3.$$

Using the fact that $\{w_n\}$ is increasing, we have

$$0 \ < \ \Delta^{m-1} w_n \to -\infty \qquad \text{as} \quad n \to \infty,$$

a contradiction. Thus, $\Delta w_n < 0$ for $n \geq n_3$. Now, we have $(-1)^i \Delta^i w_n > 0$, $i = 0, 1, \cdots, m, m+1$ and $n \geq n_3$. From (1.16.39), we obtain

$$\Delta^m w_n + \frac{q}{1+b} w_{n-g} \ \leq \ 0 \quad \text{ for } \quad n \geq n_3$$

and again we are led to a contradiction. ∎

The following two results deal with the oscillation of $(E_{41}; \delta)$.

Theorem 1.16.6. Let $h - \sigma > m$ and $g > \tau$, and

$$\frac{p}{1+a+b} \ > \ \frac{m^m (h-\sigma-m)^{h-\sigma-m}}{(h-\sigma)^{h-\sigma}}. \tag{1.16.40}$$

If

$$\frac{q}{1+a+b} \ > \ \frac{m^m (g-\tau)^{g-\tau}}{(m+g-\tau)^{g+m-\tau}} \quad \text{ if } m \text{ is even}, \tag{1.16.41}$$

then equation $(E_{41}; -1)$ is oscillatory.

Proof. Let $\{y_n\}$ be an eventually positive solution of $(E_{41}; -1)$, say $y_n > 0$ for $n \geq n_0 \geq 0$. Set

$$z_n \ = \ y_n + a y_{n-\tau} + b y_{n+\sigma}. \tag{1.16.42}$$

Then,

$$\Delta^m z_n \ = \ q y_{n-g} + p y_{n+h} \ \geq \ 0 \quad \text{ eventually}$$

and hence, we see that $z_n > 0$ and $\Delta^i z_n$, $i = 0, 1, \cdots, m$ are eventually of one sign, Next, we let

$$w_n \ = \ z_n + a z_{n-\tau} + b z_{n+\sigma}. \tag{1.16.43}$$

Then,

$$\Delta^m w_n \ = \ q z_{n-g} + p z_{n+h} \tag{1.16.44}$$

and

$$\Delta^m (w_n + aw_{n-\tau} + bw_{n+\sigma}) = qw_{n-g} + pw_{n+h}. \qquad (1.16.45)$$

Clearly, $w_n > 0$ eventually. Now, we consider the cases:

$(\bar{I})$ $\Delta z_n > 0$ eventually, and $(\overline{II})$ $\Delta z_n < 0$ eventually.

$(\bar{I})$ Suppose $\Delta z_n > 0$ for $n \geq n_1 \geq n_0$. Then, $\Delta^i w_n > 0$, $i = m, m+1$ and $n \geq n_1$, and hence we conclude that $\Delta^i w_n > 0$, $i = 0, 1, \cdots, m+1$ and $n \geq n_1$. From (1.16.45), we have

$$\Delta^m w_n > \frac{p}{1+a+b} w_{n+h-\sigma}, \quad n \geq n_1$$

and by Lemma 1.16.1(i) and condition (1.16.40), we arrive at a contradiction.

$(\overline{II})$ Suppose $\Delta z_n < 0$ for $n \geq n_1$. This is the case when m is even and we see that $\Delta^m w_n > 0$, and $\Delta^{m+1} w_n < 0$, $n \geq n_1$. We claim that $\Delta w_n \leq 0$ for $n \geq n_2 \geq n_1$. Otherwise, $\Delta w_n > 0$ for $n \geq n_2$, and hence we see that $\Delta^i w_n > 0$, $i = 0, 1, \cdots, m$ and $n \geq n_2$, and $\{\Delta^m w_n\}$ is a decreasing sequence. Using these facts in (1.16.45), we have

$$\begin{aligned} \Delta^m w_n &> \frac{p}{1+a+b} w_{n+h+\tau} \geq \frac{p}{1+a+b} w_{n+h} \\ &> \frac{p}{1+a+b} w_{n+h-\sigma}, \quad n \geq n_2 \end{aligned}$$

and again we are led to a contradiction. Thus, $\Delta w_n \leq 0$ for $n \geq n_2$, and hence we conclude that $(-1)^i \Delta^i w_n \geq 0$, $i = 0, 1, \cdots, m+1$ and $n \geq n_2$. From (1.16.45), we have

$$\Delta^m w_n \geq \frac{q}{1+a+b} w_{n-(g-\tau)}, \quad n \geq n_2$$

and again we arrive at a contradiction. ∎

Theorem 1.16.7. If $g + \sigma \geq 1$, and

$$\frac{q}{1+a+b} > \frac{m^m (g+\sigma)^{g+\sigma}}{(m+g+\sigma)^{m+g+\sigma}} \quad \text{for } m \text{ is odd,} \qquad (1.16.46)$$

then $(E_{41}; 1)$ is oscillatory.

Proof. Let $\{y_n\}$ be an eventually positive solution of $(E_{41}; 1)$, say $y_n > 0$ for $n \geq n_0 \geq 0$. Define z_n by (1.16.42), and obtain

$$\Delta^m z_n = -qy_{n-g} - py_{n+h} \quad \text{eventually.}$$

Clearly, $z_n > 0$ and $\Delta^i z_n$, $i = 0, 1, \cdots, m$ are eventually of one sign. Next, define w_n by (1.16.43) and get

$$\Delta^m w_n + qz_{n-g} + pz_{n+h} = 0$$

and

$$\Delta^m (w_n + aw_{n-\tau} + bw_{n+\sigma}) + qw_{n-g} + pw_{n+h} = 0. \qquad (1.16.47)$$

Here $w_n > 0$ for $n \ge n_1 \ge n_0$ and as in Theorem 1.16.6, the two cases $(\bar{I})$ and $(\overline{II})$ are considered:

$(\bar{I})$ Suppose $\Delta z_n > 0$ for $n \ge n_2 \ge n_1$. Then, $\Delta^i w_n < 0$, $i = m, m+1$ and $n \ge n_2$, and hence we see that $w_n < 0$ for $n \ge n_2$, a contradiction.

$(\overline{II})$ Suppose $\Delta z_n < 0$ for $n \ge n_2$. This is the case when m is odd. It is easy to check that $\Delta^{m+1} w_n > 0$, $\Delta^m w_n < 0$ and $\Delta^{m-1} w_n > 0$ for $n \ge n_3 \ge n_2$. Using this fact in (1.16.47), we have

$$(1 + a + b)\Delta^m w_{n+\sigma} + qw_{n-g} \le 0$$

or

$$\Delta^m w_n + \frac{q}{1+a+b} w_{n-g-\sigma} \le 0, \quad n \ge n_3. \qquad (1.16.48)$$

Now, if $\Delta w_n > 0$, $n \ge n_3$ then one can easily see that

$$0 < \Delta^{m-1} w_n \to -\infty \quad \text{as} \quad n \to \infty,$$

a contradiction. Thus, $\Delta w_n < 0$ and hence $(-1)^i \Delta^i w_n > 0$, $i = 0, 1, \cdots, m+1$ and $n \ge n_3$. Using this fact in (1.16.47), we obtain (1.16.48) and again we are led to a contradiction. ∎

Finally, we give the following two criteria for the oscillation of $(E_{42}; \delta)$.

Theorem 1.16.8. Let $a + b > 0$, $h > m$ and $g > \tau \ge 1$, and

$$p > \frac{m^m (h-m)^{h-m}}{h^h}. \qquad (1.16.49)$$

If

$$\frac{q}{a+b} > \frac{m^m (g-\tau)^{g-\tau}}{(m+g-\tau)^{m+g-\tau}}, \quad m \text{ is odd} \qquad (1.16.50)$$

and

$$q > \frac{m^m g^g}{(m+g)^{m+g}}, \quad m \text{ is even} \qquad (1.16.51)$$

then $(E_{42}; -1)$ is oscillatory.

Proof. Let $\{y_n\}$ be an eventually positive solution of $(E_{42}; -1)$, say $y_n > 0$ for $n \ge n_0 \ge 0$. Set

$$z_n = y_n - ay_{n-\tau} - by_{n+\sigma}. \qquad (1.16.52)$$

Then,

$$\Delta^m z_n = qy_{n-g} + py_{n+h} \ge 0 \quad \text{eventually}$$

and we see that $\Delta^i z_n$, $i = 0, 1, \cdots, m$ are eventually of one sign. There are two cases to consider:

(i) $z_n > 0$ eventually, and (ii) $z_n < 0$ eventually.

(i) Assume $z_n > 0$ for $n \ge n_1 \ge n_0$. Clearly, $y_n \ge z_n$ for $n \ge n_1$. Now, if $\Delta z_n > 0$ for $n \ge n_1$, then $\Delta^i z_n > 0$, $i = 0, 1, \cdots, m$, $n \ge n_2 \ge n_1$, and

$$\Delta^m z_n \ \ge \ p z_{n+h}, \qquad n \ge n_2.$$

By Lemma 1.16.1(i) and condition (1.16.47) we arrive at a contradiction. Next, if $\Delta z_n < 0$ for $n \ge n_1$, then m must be even, and $(-1)^i \Delta^i z_n > 0$, $i = 0, 1, \cdots, m$ and $n \ge n_2 \ge n_1$, also

$$\Delta^m z_n \ \ge \ q z_{n-g}, \qquad n \ge n_2.$$

Again by Lemma 1.16.1(ii) and condition (1.16.51) we obtain the desired contradiction.

(ii) Assume $z_n < 0$ for $n \ge n_1$. Set

$$0 \ < \ u_n \ = \ -z_n \ = \ a y_{n-\tau} + b y_{n+\sigma} - y_n. \tag{1.16.53}$$

Then,

$$\Delta^m u_n + q y_{n-g} + p y_{n+h} \ = \ 0.$$

Define

$$w_n \ = \ a u_{n-\tau} + b u_{n+\sigma} - u_n. \tag{1.16.54}$$

Then,

$$\Delta^m w_n + q u_{n-g} + p u_{n+h} \ = \ 0 \tag{1.16.55}$$

and

$$\Delta^m \left(a w_{n-\tau} + b w_{n+\sigma} - w_n \right) + q w_{n-g} + p w_{n+h} \ = \ 0. \tag{1.16.56}$$

Using the procedure of Case (i) above, one can easily see that $w_n > 0$ for $n \ge n_1$. Next, we consider the following two cases:

(I) $\Delta u_n > 0$ eventually, and (II) $\Delta u_n < 0$ eventually.

(I) Suppose $\Delta u_n > 0$ for $n \ge n_2 \ge n_1$. From (1.16.55), one can easily see that $\Delta^i w_n < 0$, $i = m, m+1$, $n \ge n_2$. Thus, $w_n < 0$ for $n \ge n_2$, a contradiction.

(II) Suppose $\Delta u_n < 0$ for $n \ge n_2$. This is the case when m is odd. From (1.16.55) we see that $\Delta^m w_n < 0$ and $\Delta^{m+1} w_n > 0$ for $n \ge n_2$. Next, we claim that $\Delta w_n < 0$ for $n \ge n_2$. Otherwise, $\Delta w_n > 0$ for $n \ge n_2$ and hence one can easily see that

$$0 \ < \ \Delta^{m-1} w_n \to -\infty \quad \text{as} \quad n \to \infty,$$

a contradiction. Thus, $\Delta w_n < 0$ for $n \geq n_2$, and hence we have $(-1)^i \Delta^i w_n > 0$, $i = 0, 1, \cdots, m$ and $n \geq n_2$.

From (1.16.56), one can easily obtain

$$(a+b)\Delta^m w_{n-\tau} + q w_{n-g} \leq 0$$

or

$$\Delta^m w_n + \frac{q}{a+b} w_{n-(g-\tau)} \leq 0 \quad \text{for} \quad n \geq n_2.$$

The rest of the proof is similar to that of Theorem 1.16.2(ii)–($\overline{II}$.) ∎

Theorem 1.16.9. Let $a+b > 0$, $h > \sigma + m$ and $g > \tau \geq 1$, and

$$\frac{p}{a+b} > \frac{m^m (h-\sigma-m)^{h-\sigma-m}}{(h-\sigma)^{h-\sigma}}. \tag{1.16.57}$$

If

$$q > \frac{m^m g^g}{(m+g)^{m+g}}, \quad m \text{ is odd} \tag{1.16.58}$$

and

$$\frac{q}{a+b} > \frac{m^m (g-\tau)^{g-\tau}}{(m+g-\tau)^{m+g-\tau}}, \quad m \text{ is even} \tag{1.16.59}$$

then $(E_{42}; 1)$ is oscillatory.

Proof. Let $\{y_n\}$ be an eventually positive solution of $(E_{42}; 1)$, say $y_n > 0$ for $n \geq n_0 \geq 0$. Define z_n by (1.16.52), and obtain

$$\Delta^m z_n = -q y_{n-g} - p y_{n+h} \leq 0 \quad \text{eventually,}$$

and hence, we conclude that $\Delta^i z_n$, $i = 0, 1, \cdots, m$ are eventually of one sign. As in the proof of Theorem 1.16.8, the two cases (i) and (ii) are considered:

(i) Assume $z_n > 0$ for $n \geq n_1 \geq n_0$. Then, $y_n \geq z_n$, $n \geq n_1$ and

$$\Delta^m z_n + q z_{n-g} \leq 0, \quad n \geq n_1.$$

Now, if $\Delta z_n > 0$, $n \geq n_1$ then one can easily see that

$$0 < \Delta^{m-1} z_n \to -\infty \quad \text{as} \quad n \to \infty,$$

a contradiction. Next, if $\Delta z_n < 0$, $n \geq n_1$ then m must be odd, and hence $(-1)^i \Delta^i z_n > 0$, $i = 0, 1, \cdots, m$ and $n \geq n_2 \geq n_1$, also

$$\Delta^m z_n + q z_{n-g} \leq 0 \quad \text{for} \quad n \geq n_2.$$

The rest of the proof is similar to that of Theorem 1.16.2(ii)–($\overline{II}$).

(ii) Assume $z_n < 0$ for $n \geq n_1$. Define u_n by (1.16.53), and obtain

$$\Delta^m u_n = q y_{n-g} + p y_{n+h}.$$

Next, define w_n by (1.16.54) and obtain

$$\Delta^m w_n = q u_{n-g} + p u_{n+h} \tag{1.16.60}$$

and

$$\Delta^m (a w_{n-\tau} + b w_{n+\sigma} - w_n) = q w_{n-g} + p w_{n+h}. \tag{1.16.61}$$

It is easy to check that $w_n > 0$ for $n \geq n_2 \geq n_1$, and as in the proof of Theorem 1.16.8, the two cases (I) and (II) are considered.

(I) Suppose $\Delta u_n > 0$ for $n \geq n_2$. From (1.16.60) we see that $\Delta^i w_n > 0$, $i = m, m+1$, $n \geq n_2$ and hence one can easily conclude that $\Delta^i w_n > 0$, $i = 0, 1, \cdots, m+1$ and $n \geq n_2$. From (1.16.61), we have

$$(a+b)\Delta^m w_{n+\sigma} > p w_{n+h}$$

or

$$\Delta^m w_n > \frac{p}{a+b} w_{n+(h-\sigma)}, \quad n \geq n_2.$$

By Lemma 1.16.1(i) and condition (1.16.57), we obtain the desired contradiction.

(II) Suppose $\Delta u_n < 0$ for $n \geq n_2$. This is the case when m is even, and we see that $\Delta^m w_n > 0$ and $\Delta^{m+1} w_n < 0$ for $n \geq n_2$. We claim that $\Delta w_n < 0$ for $n \geq n_2$. Otherwise, $\Delta w_n > 0$ for $n \geq n_2$ and hence one can easily conclude that $\Delta^i w_n > 0$, $i = 0, 1, \cdots, m$ and $n \geq n_2$. From (1.16.61) one can easily see that

$$(a+b)\Delta^m w_{n-\tau} > p w_{n+h}$$

or

$$(a+b)\Delta^m w_n > p w_{n+h+\tau} > p w_{n+h-\sigma}$$

and again we are led to a contradiction. Thus, $\Delta w_n < 0$ for $n \geq n_2$, and hence we conclude that $(-1)^i \Delta^i w_n > 0$, $i = 0, 1, \cdots, m$ and $n \geq n_2$. From (1.16.61), we have

$$(a+b)\Delta^m w_{n-\tau} \geq q w_{n-g}$$

or

$$\Delta^m w_n \geq \frac{q}{a+b} w_{n-(g-\tau)}, \quad n \geq n_3$$

and again we are led to a contradiction. ∎

Remark 1.16.2. From the proofs of the above results, we observe that when the coefficient $p \equiv 0$, or conditions on p are violated, the conclusions of the theorems presented may be replaced by "Every solution $\{y_n\}$ of each of the equation $(E_i;\delta)$, $i=39,40,41,42$ and $\delta=\pm1$ is oscillatory, or $\Delta^j y_n \to \infty$ monotonically as $n\to\infty$, $j=0,1,\cdots,m-1$".

As an example, we see that the equations

$$\Delta^m\left(y_n+e^\tau y_{n-\tau}-e^{-\sigma}y_{n+\sigma}\right) = \frac{(e-1)^m}{2}\left(e^g y_{n-g}+e^{-h}y_{n+h}\right) \qquad (E_{43})$$

and

$$\Delta^m\left(y_n+e^\tau y_{n-\tau}-e^{-\sigma}y_{n+\sigma}\right) = (e-1)^m e^g y_{n-g} \qquad (E_{44})$$

have a nonoscillatory solution $y_n=e^n\to\infty$ as $n\to\infty$.

Remark 1.16.3. Once again from the proofs of theorems presented above, we see that if $q=0$, or conditions on q are not satisfied, the conclusions of these theorems may be replaced by "Every solution $\{y_n\}$ of each of the equation $(E_i;\delta)$, $i=39,40,41,42$ and $\delta=\pm1$ is oscillatory, or $\Delta^j y_n\to 0$ monotonically as $n\to\infty$, $j=0,1,\cdots,m-1$".

As an example, we observe that the equations

$$\Delta^i\left(y_n+e^{-\tau}y_{n-\tau}-e^\sigma y_{n+\sigma}\right) = \frac{1}{2}\left(\frac{1}{e}-1\right)^m\left(e^{-g}y_{n-g}+e^h y_{n+h}\right) \qquad (E_{45})$$

and

$$\Delta^i\left(y_n+e^{-\tau}y_{n-\tau}-e^\sigma y_{n+\sigma}\right) = \left(\frac{1}{e}-1\right)^m e^h y_{n+h} \qquad (E_{46})$$

have a nonoscillatory solution $y_n=e^{-n}\to 0$ as $n\to\infty$.

Remark 1.16.4.

1. The above results are not applicable to equations $(E_i;\delta)$, $i=39,40,41,$ 42 and $\delta=\pm1$ when $p=0$ or $q=0$. However, these results are valid if $p=0$ or $q=0$ (but not $p=q=0$) provided that either $a=0$, $b=0$ or $a=b=0$.

2. By using the same technique presented above, one can easily obtain similar criteria for equations $(E_i;\delta)$, $i=39,40,41,42$ and $\delta=\pm1$ for different signs on τ and σ and also different signs on a and b.

1.17. Difference Equations Involving Quasi–differences

Let r_i, $0\le i\le m$ be functions such that $r_i:\mathbb{N}(n_0)\to\mathbb{R}_0$, where $n_0\in\mathbb{N}\cup\{0\}$. For an integer $k\in\mathbb{N}(0,m)$ and a function y defined on $\mathbb{N}(n_0)$, we define the *kth quasi–difference* of y, also known as the *kth r–difference* of y as

$$\Delta_r^k y = r_k(r_{k-1}(r_{k-2}(\cdots(r_1(r_0y)^\Delta)^\Delta\cdots)^\Delta)^\Delta)^\Delta, \qquad (1.17.1)$$

where $(\cdot)^\Delta \equiv \Delta(\cdot)$. It is clear from (1.17.1) that

$$\Delta_r^0 y = r_0\, y \tag{1.17.2}$$

and

$$\Delta_r^k y = r_k(\Delta_r^{k-1} y)^\Delta = r_k \Delta(\Delta_r^{k-1} y). \tag{1.17.3}$$

Further, if $r_i = 1,\ 0 \le i \le k,$ then $\Delta_r^k y = \Delta^k y$.

In this section we shall consider the difference equation

$$\Delta_r^m y(n) + g(n) F(y(a_1(n)), y(a_2(n)), \cdots, y(a_\zeta(n))) = h(n), \quad n \in \mathbb{N}(n_0) \tag{$E_{47};\zeta$}$$

where $F : \mathbb{R}^\zeta \to \mathbb{R}$ is continuous, $a_i : \mathbb{N}(n_0) \to \mathbf{Z} = \{0, \pm 1, \pm 2, \cdots\},\ 1 \le i \le \zeta$ and $g, h : \mathbb{N}(n_0) \to \mathbb{R}$. Throughout, it is assumed that

(A1) $r_m = 1$,

(A2) for each $1 \le i \le \zeta,\ \lim_{n\to\infty} a_i(n) = \infty$,

(A3) if $u_i > 0,\ 1 \le i \le \zeta,$ then $F(u_1, \cdots, u_\zeta) > 0,$ and if $u_i < 0,\ 1 \le i \le \zeta,$ then $F(u_1, \cdots, u_\zeta) < 0$.

Here we shall develop criteria which guarantee that a nonoscillatory solution of $(E_{47};\zeta)$ is bounded and/or tends to zero as $n \to \infty$. We begin with the following:

Let x be a function such that $x : \mathbb{N}(n_0) \to \mathbb{R}_0$, and let $T_i,\ 0 \le i \le m$ be defined by

$$T_m = x, \tag{1.17.4}$$

$$T_j(n+1) = r_j(n) \Delta T_{j+1}(n), \quad j = m-1, m-2, \cdots, 0. \tag{1.17.5}$$

Let $k \in \mathbb{N}(0, m-1)$. We say that x is *of type* $r[k]$ if

(B1) $T_j,\ k+1 \le j \le m$ are defined on $\mathbb{N}(n_0 + m - j)$,

(B2) T_{k+1} is a constant nonzero function on $\mathbb{N}(n_0 + m - k - 1)$,

(B3) if $k \le m-2,$ then for each $k+2 \le j \le m$,

$$\lim_{n\to\infty} T_j(n) \in \{0,\ \infty,\ -\infty\},$$

(B4) if $k \le m-3,$ then for each $k+2 \le j \le m-1,\ T_j(n) \ne 0$ for $n \ge n_1$ where n_1 is some integer.

Now we give some examples of functions of the type $r[k]$, which will be needed later. The verification is direct and hence is omitted.

Example 1.17.1. For $0 \le i \le j \le m-1$, define

$$\rho_{ij}(n) = \begin{cases} 1, & \text{if } i = j \\ \displaystyle\sum_{n_j=n_0}^{n-1} \frac{1}{r_j(n_j)} \sum_{n_{j-1}=n_0}^{n_j-1} \frac{1}{r_{j-1}(n_{j-1}+1)} \sum_{n_{j-2}=n_0}^{n_{j-1}-1} \frac{1}{r_{j-2}(n_{j-2}+2)} \cdots \\ \qquad \cdots \displaystyle\sum_{n_{i+1}=n_0}^{n_{i+2}-1} \frac{1}{r_{i+1}(n_{i+1}+j-i-1)}, & \text{if } i \le j-1. \end{cases} \tag{1.17.6}$$

In the case $k \le m-2$, suppose that

$$\sum^{\infty} \frac{1}{r_j(s)} = \infty, \quad k+1 \le j \le m-1. \tag{1.17.7}$$

Then, $\rho_{k,m-1}$ is of type $r[k]$ for any $k \in \mathbb{N}(0, m-1)$.

Example 1.17.2. For $0 \le k \le m-1$, define

$$\sigma_k(n) = \begin{cases} 1, \quad \text{if } k = m-1 \\ \displaystyle\sum_{n_{m-1}=n}^{\infty} \frac{1}{r_{m-1}(n_{m-1})} \sum_{n_{m-2}=n_{m-1}}^{\infty} \frac{1}{r_{m-2}(n_{m-2}+1)} \\ \displaystyle\sum_{n_{m-3}=n_{m-2}}^{\infty} \frac{1}{r_{m-3}(n_{m-3}+2)} \cdots \sum_{n_{k+1}=n_{k+2}}^{\infty} \frac{1}{r_{k+1}(n_{k+1}+m-k-2)}, \\ \qquad\qquad\qquad\qquad\qquad\qquad \text{if } k \le m-2. \end{cases} \tag{1.17.8}$$

In the case $k \le m-2$, suppose that

$$\sum^{\infty} \frac{1}{r_j(s)} < \infty, \quad k+1 \le j \le m-1. \tag{1.17.9}$$

Then, σ_k is of type $r[k]$ for any $k \in \mathbb{N}(0, m-1)$.

Example 1.17.3. Let $r_j = 1$ for $j \ne 0,\ m-1$ and let $r_{m-1} = r$. For $0 \le k \le m-2$, define

$$\theta_k(n) = \sum_{s=n}^{\infty} \frac{(s-n_0)^{(m-2-k)}}{r(s)}. \tag{1.17.10}$$

Suppose that

$$\sum^{\infty} \frac{s^{(m-2-k)}}{r(s)} < \infty. \tag{1.17.11}$$

Then, θ_k is of type $r[k]$ for any $k \in \mathbb{N}(0, m-2)$.

Example 1.17.4. Let $M \in \mathbb{N}(2, m-1)$ be fixed, $r_j = 1$ for $j \neq 0,\ m-M$ and $r_{m-M} = r$. For $0 \le k \le m-M-1$ and $1 \le j \le M-1$, define

$$\eta_{kj}(n) \;=\; \sum_{s=n_0}^{n-1} (n-s)^{(j-1)} \sum_{u=s}^{\infty} \frac{(u-n_0)^{(m-M-1-k)}}{r(u+1)}. \tag{1.17.12}$$

Suppose that

$$\sum^{\infty} \frac{s^{(m-M-1-k)}}{r(s+1)} \;<\; \infty \tag{1.17.13}$$

and

$$\lim_{n\to\infty} \eta_{kj}(n) \in \{0,\ \infty\}, \quad 1 \le j \le M-1. \tag{1.17.14}$$

Then, $\eta_{k,M-1}$ is of type $r[k]$ for any $k \in \mathbb{N}(0, m-M-1)$.

We shall also need the following two lemmas.

Lemma 1.17.1. Consider the difference equation

$$\Delta u(n) - \frac{\Delta T(n)}{T(n)}\, u(n) + \frac{\Delta T(n)}{T(n)}\, H(n) \;=\; 0, \tag{1.17.15}$$

where

(a) $T(n),\ \Delta T(n) \neq 0$ for $n \ge N$,
(b) $\lim_{n\to\infty} T(n) \in \{0,\ \infty,\ -\infty\}$.

Let y be the solution of (1.17.15) such that $y(N) = 0$. If $\lim_{n\to\infty} H(n)$ exists in $\mathbb{R}^*$, so does $\lim_{n\to\infty} y(n)$. Further, $\lim_{n\to\infty} |H(n)| = \infty$ implies $\lim_{n\to\infty} |y(n)| = \infty$.

Proof. It can be verified that the solution y is given by

$$y(n) \;=\; -T(n) \sum_{s=N}^{n-1} \frac{H(s)\Delta T(s)}{T(s)T(s+1)}. \tag{1.17.16}$$

Case 1 $\lim_{n\to\infty} T(n) = 0$. If $\lim_{n\to\infty} H(n) \neq 0$, then $H(n)$ is of fixed sign for large n. So it is obvious that

$$\sum_{s=N}^{\infty} \frac{H(s)\Delta T(s)}{T(s)T(s+1)} \;=\; \infty \quad \text{or} \quad -\infty. \tag{1.17.17}$$

Applying discrete l'Hospital's rule [2], we find

$$\lim_{n\to\infty} y(n) = \lim_{n\to\infty} \frac{-\sum_{s=N}^{n-1} \frac{H(s)\Delta T(s)}{T(s)T(s+1)}}{\frac{1}{T(n)}} = \lim_{n\to\infty} \frac{-\frac{H(n)\Delta T(n)}{T(n)T(n+1)}}{\frac{1}{T(n+1)} - \frac{1}{T(n)}} = \lim_{n\to\infty} H(n). \tag{1.17.18}$$

If $\lim_{n\to\infty} H(n) = 0$, then it is clear that $\sum_{s=N}^{\infty} \frac{H(s)\Delta T(s)}{T(s)T(s+1)}$ exists in $\mathbb{R}^*$. In the case that $\sum_{s=N}^{\infty} \frac{H(s)\Delta T(s)}{T(s)T(s+1)}$ is infinite, i.e. (1.17.17), we proceed as above and obtain (1.17.18). If $\sum_{s=N}^{\infty} \frac{H(s)\Delta T(s)}{T(s)T(s+1)}$ is finite, then it follows from (1.17.16) that

$$\lim_{n\to\infty} y(n) = -\lim_{n\to\infty} T(n) \cdot \sum_{s=N}^{\infty} \frac{H(s)\Delta T(s)}{T(s)T(s+1)} = 0 = \lim_{n\to\infty} H(n). \tag{1.17.19}$$

Case 2 $\lim_{n\to\infty} T(n) = \infty$ or $-\infty$. In view of condition (a), $T(n)$ and $\Delta T(n)$ are of the same fixed sign for large n. If $\lim_{n\to\infty} H(n) \neq 0$, then $H(n)$ is of fixed sign for large n. Hence, we see that $\sum_{s=N}^{\infty} \frac{H(s)\Delta T(s)}{T(s)T(s+1)}$ exists in $\mathbb{R}^*$. If $\lim_{n\to\infty} H(n) = 0$, then it is also obvious that $\sum_{s=N}^{\infty} \frac{H(s)\Delta T(s)}{T(s)T(s+1)}$ exists in $\mathbb{R}^*$.

Suppose that $\sum_{s=N}^{\infty} \frac{H(s)\Delta T(s)}{T(s)T(s+1)} \neq 0$. Then, it follows from (1.17.16) that

$$\lim_{n\to\infty} y(n) = -\lim_{n\to\infty} T(n) \cdot \sum_{s=N}^{\infty} \frac{H(s)\Delta T(s)}{T(s)T(s+1)} = \infty \text{ or } -\infty. \tag{1.17.20}$$

If $\sum_{s=N}^{\infty} \frac{H(s)\Delta T(s)}{T(s)T(s+1)} = 0$, then by using discrete l'Hospital's rule again we find (1.17.18).

In both cases, we have from (1.17.18) and (1.17.19),

$$\lim_{n\to\infty} y(n) = \lim_{n\to\infty} H(n)$$

or (1.17.20). Hence, the conclusion of the lemma follows. ∎

Lemma 1.17.2. Let x be a function of type $r[k]$ for some $k \in \mathbb{N}(0, m-1)$. If $\sum_{s=N}^{\infty} x(s+1)\Delta_r^m y(s)$ exists in $\mathbb{R}^*$, so does $\lim_{n\to\infty} \Delta_r^k y(n)$. Further, $\left|\sum_{s=N}^{\infty} x(s+1)\Delta_r^m y(s)\right| = \infty$ implies $\lim_{n\to\infty} |\Delta_r^k y(n)| = \infty$.

Proof. Case 1 $k = m-1$. By (1.17.4) and (B2), $x = T_m = T_{k+1} = c$ for some constant $c > 0$. It follows that

$$\lim_{n\to\infty} \Delta_r^{m-1}y(n) = \Delta_r^{m-1}y(N) + \frac{1}{c}\sum_{s=N}^{\infty} x(s+1)\Delta_r^m y(s)$$

from which the conclusion is immediate.

Case 2 $k \le m-2$. For $k+1 \le j \le m$, we define

$$q_j(n) = \sum_{s=N}^{n-1} T_j(s+1)\Delta(\Delta_r^{j-1}y(s)) = \sum_{s=N}^{n-1} T_j(s+1)\,\frac{\Delta_r^j y(s)}{r_j(s)}, \quad (1.17.21)$$

where T_j's are related to x as in the definition of a function of type $r[k]$.

On summing (1.17.21) by parts and using (1.17.5), it follows that

$$\begin{aligned} q_j(n) &= T_j(s)\Delta_r^{j-1}y(s)\Big|_{s=N}^{n} - \sum_{s=N}^{n-1} \Delta_r^{j-1}y(s)\Delta T_j(s) \\ &= \frac{T_j(n)}{\frac{T_{j-1}(n+1)}{r_{j-1}(n)}}\left[\frac{T_{j-1}(n+1)\Delta_r^{j-1}y(n)}{r_{j-1}(n)}\right] - T_j(N)\Delta_r^{j-1}y(N) \\ &\quad - \sum_{s=N}^{n-1} \Delta_r^{j-1}y(s)\,\frac{T_{j-1}(s+1)}{r_{j-1}(s)} \\ &= \frac{T_j(n)}{\Delta T_j(n)}\,\Delta q_{j-1}(n) - T_j(N)\Delta_r^{j-1}y(N) - q_{j-1}(n), \end{aligned}$$

or equivalently,

$$\Delta q_{j-1}(n) - \frac{\Delta T_j(n)}{T_j(n)}\,q_{j-1}(n) + \frac{\Delta T_j(n)}{T_j(n)}\,H_j(n) = 0,$$

where

$$H_j(n) = -q_j(n) - T_j(N)\Delta_r^{j-1}y(N). \quad (1.17.22)$$

Therefore, q_{j-1} is a solution of the difference equation $(1.17.15)|_{T=T_j,H=H_j}$ and we note that $q_{j-1}(N) = 0$. It is also obvious from (1.17.22) that if $\lim_{n\to\infty} q_j(n)$ exists in $\mathbb{R}^*$, so does $\lim_{n\to\infty} H_j(n)$. Applying Lemma 1.17.1, we see that $\lim_{n\to\infty} q_{j-1}(n)$ exists in $\mathbb{R}^*$. Further, if $\lim_{n\to\infty}|q_j(n)| = \infty$, then (1.17.22) provides $\lim_{n\to\infty}|H_j(n)| = \infty$. So by Lemma 1.17.1 again, we have $\lim_{n\to\infty}|q_{j-1}(n)| = \infty$.

Now, since

$$\lim_{n\to\infty} q_m(n) = \sum_{s=N}^{\infty} T_m(s+1)\,\frac{\Delta_r^m y(s)}{r_m(s)} = \sum_{s=N}^{\infty} x(s+1)\Delta_r^m y(s)$$

exists in $\mathbb{R}^*$ (given assumption), the above arguments can be applied for $j = m, m-1, \cdots, k+2$. Consequently, we have

$$\lim_{n\to\infty} q_{k+1}(n) \quad \text{exists in} \quad \mathbb{R}^*. \tag{1.17.23}$$

In addition,

$$\left|\sum_{s=N}^{\infty} x(s+1)\Delta_r^m y(s)\right| = \infty \quad \text{implies} \quad \lim_{n\to\infty} |q_{k+1}(n)| = \infty. \tag{1.17.24}$$

Next, since $T_{k+1} = c$ for some constant $c > 0$, from (1.17.21) we find

$$q_{k+1}(n) = \sum_{s=N}^{n-1} T_{k+1}(s+1)\Delta(\Delta_r^k y(s)) = c\left[\Delta_r^k y(n) - \Delta_r^k y(N)\right],$$

which leads to

$$\lim_{n\to\infty} \Delta_r^k y(n) = \frac{1}{c}\lim_{n\to\infty} q_{k+1}(n) + \Delta_r^k y(N). \tag{1.17.25}$$

Hence, in view of (1.17.23) and (1.17.24), it follows from (1.17.25) that $\lim_{n\to\infty} \Delta_r^k y(n)$ exists in $\mathbb{R}^*$ and $\left|\sum_{s=N}^{\infty} x(s+1)\Delta_r^m y(s)\right| = \infty$ implies $\lim_{n\to\infty} |\Delta_r^k y(n)| = \infty$. ∎

Theorem 1.17.3. Let

(C1) $\limsup_{n\to\infty} r_0(n) < \infty$.

Further, suppose that there exists a function x of type $r[k]$ for some $k \in \mathbb{N}(0, m-1)$ such that

(C2) if $k \geq 1$, then

$$\sum^{\infty} \frac{1}{r_i(n)} = \infty, \quad 1 \leq i \leq k,$$

(C3) for all $0 < \mu \leq 1$ and $\nu > 0$,

$$\sum^{\infty} x(n+1)[\mu g^+(n) - g^-(n) - \nu|h(n)|] = \infty$$

or

$$\sum^{\infty} x(n+1)[\mu g^-(n) - g^+(n) - \nu|h(n)|] = \infty,$$

where $g^+(n) = \max\{g(n), 0\}$ and $g^-(n) = \max\{-g(n), 0\}$. Then, any bounded solution y of $(E_{47};\zeta)$ satisfies $\liminf_{n\to\infty} |y(n)| = 0$.

Proof. Suppose the contrary and let y be a bounded solution of $(E_{47};\zeta)$ such that $\liminf_{n\to\infty} |y(n)| > 0$. Then, y is nonoscillatory. We can assume that y is eventually positive, since if otherwise, in view of (A3), the substitution $z = -y$ will transform $(E_{47};\zeta)$ into an equation of the same form satisfying the conditions of the theorem. So let $y(n) > 0$ for $n \ge n_1 \ge n_0$.

By (A2) and the boundedness of y, there exist positive constants M_1 and M_2 such that for some $N > n_1$,

$$M_1 \ \le\ y(a_i(n)) \ \le\ M_2, \quad 1 \le i \le \zeta, \quad n \ge N. \tag{1.17.26}$$

Then, in view of (A3) there exist positive constants c_1 and c_2 such that

$$c_1 \ \le\ F(y(a_1(n)), y(a_2(n)), \cdots, y(a_\zeta(n))) \ \le\ c_2, \quad n \ge N. \tag{1.17.27}$$

Now, we rewrite $(E_{47};\zeta)$ as

$$\begin{aligned} \Delta_r^m y(n) &= h(n) - g^+(n)F(y(a_1(n)), \cdots, y(a_\zeta(n))) \\ &\quad + g^-(n)F(y(a_1(n)), \cdots, y(a_\zeta(n))). \end{aligned} \tag{1.17.28}$$

Define $\mu = c_1 c_2^{-1} \in (0, 1]$ and $\nu = c_2^{-1} > 0$. Using (1.17.27), it follows from (1.17.28) that

$$\begin{aligned} x(n+1)\Delta_r^m y(n) &\ge x(n+1)[h(n) - c_2 g^+(n) + c_1 g^-(n)] \\ &= c_2 x(n+1)[\mu g^-(n) - g^+(n) + \nu h(n)], \ n \ge N \end{aligned} \tag{1.17.29}$$

and

$$\begin{aligned} x(n+1)\Delta_r^m y(n) &\le x(n+1)[h(n) - c_1 g^+(n) + c_2 g^-(n)] \\ &= -c_2 x(n+1)[\mu g^+(n) - g^-(n) - \nu h(n)], \ n \ge N. \end{aligned} \tag{1.17.30}$$

Noting (C3), inequalities (1.17.29) and (1.17.30) yield

$$\sum_{n=N}^{\infty} x(n+1)\Delta_r^m y(n) \ = \ \infty \quad \text{or} \quad -\infty. \tag{1.17.31}$$

By Lemma 1.17.2, relation (1.17.31) implies

$$\lim_{n\to\infty} \Delta_r^k y(n) \ = \ \infty \quad \text{or} \quad -\infty. \tag{1.17.32$_k$}$$

Finally, we shall show that

$$\lim_{n\to\infty} \Delta_r^0 y(n) \ = \ \infty \quad \text{or} \quad -\infty,$$

i.e. $(1.17.32)_0$ holds. This leads to a contradiction since by (C1) and the boundedness of y, $\Delta_r^0 y(n) = (r_0 y)(n)$ is bounded. The proof of the theorem is then complete. Obviously, $(1.17.32)_0$ is trivial if $k = 0$. Suppose that $k \geq 1$. In view of $(1.17.32)_k$, there exists $N_1 \geq N$ such that for $n \geq N_1$,

$$\Delta_r^k y(n) \geq 1 \quad \text{or} \quad \Delta_r^k y(n) \leq -1.$$

This is equivalent to

$$\Delta(\Delta_r^{k-1} y)(n) \geq \frac{1}{r_k(n)} \quad \text{or} \quad \Delta(\Delta_r^{k-1} y)(n) \leq -\frac{1}{r_k(n)}, \quad n \geq N_1. \tag{1.17.33}$$

Summing (1.17.33) from N_1 to $(n-1)$, we get

$$\Delta_r^{k-1} y(n) \geq \Delta_r^{k-1} y(N_1) + \sum_{s=N_1}^{n-1} \frac{1}{r_k(s)} \text{ or } \Delta_r^{k-1} y(n) \leq \Delta_r^{k-1} y(N_1) - \sum_{s=N_1}^{n-1} \frac{1}{r_k(s)}. \tag{1.17.34}$$

By (C2), letting $n \to \infty$ in (1.17.34) leads to

$$\lim_{n\to\infty} \Delta_r^{k-1} y(n) = \infty \quad \text{or} \quad -\infty,$$

which is the same as $(1.17.32)_{k-1}$. Hence, by repeating the above procedure we finally obtain $(1.17.32)_0$. ∎

Theorem 1.17.4. Let (C1) hold. Further, suppose that there exists a function x of type $r[0]$ such that

(C4) $\sum^{\infty} x(n+1)|h(n)| < \infty,$

(C5) $\sum^{\infty} x(n+1)g^+(n) = \infty$ and $\sum^{\infty} x(n+1)g^-(n) < \infty$

or

$\sum^{\infty} x(n+1)g^+(n) < \infty$ and $\sum^{\infty} x(n+1)g^-(n) = \infty.$

Then, any bounded nonoscillatory solution y of $(E_{47};\zeta)$ satisfies $\lim_{n\to\infty} r_0(n)$ $y(n) = 0$.

If, in addition

(C6) $\liminf_{n\to\infty} r_0(n) > 0,$

then any bounded nonoscillatory solution y of $(E_{47};\zeta)$ satisfies $\lim_{n\to\infty} y(n)$ $= 0$.

Proof. Let y be a bounded nonoscillatory solution of $(E_{47};\zeta)$. As in the proof of Theorem 1.17.3, we assume that y is eventually positive and

$y(n) > 0$ for $n \ge n_1 \ge n_0$. As before, we have (1.17.26) – (1.17.28). It follows from (1.17.28) that

$$\begin{aligned}\sum_{n=N}^{\tau} x(n+1)\Delta_r^m y(n) &= \sum_{n=N}^{\tau} x(n+1)h(n) - \sum_{n=N}^{\tau} x(n+1)g^+(n)F(y(a_1(n)),\cdots,y(a_\varsigma(n))) \\ &+ \sum_{n=N}^{\tau} x(n+1)g^-(n)F(y(a_1(n)),\cdots,y(a_\varsigma(n))), \quad \tau \ge N. \quad (1.17.35)\end{aligned}$$

Noting (1.17.27), (C4) and (C5), from (1.10.35) we see that $\sum_{n=N}^{\infty} x(n+1)\Delta_r^m y(n)$ exists in $\mathbb{R}^*$. So by Lemma 1.17.2, $\lim_{n\to\infty} \Delta_r^0 y(n)$ also exists in $\mathbb{R}^*$.

Next, since conditions (C4) and (C5) imply (C3), an application of Theorem 1.17.3 (with $k = 0$) provides

$$\liminf_{n\to\infty} y(n) = 0. \quad (1.17.36)$$

In view of (1.17.36) and (C1), we find

$$\liminf_{n\to\infty} \Delta_r^0 y(n) = \liminf_{n\to\infty} r_0(n)y(n) = \liminf_{n\to\infty} r_0(n) \cdot \liminf_{n\to\infty} y(n) = 0. \quad (1.17.37)$$

As $\lim_{n\to\infty} \Delta_r^0 y(n)$ exists, using (1.17.37) we immediately get

$$\lim_{n\to\infty} r_0(n)y(n) = \lim_{n\to\infty} \Delta_r^0 y(n) = \liminf_{n\to\infty} \Delta_r^0 y(n) = 0.$$

The last part of the theorem is obvious. ∎

The next result is for the difference equation $(E_{47}; 1)$. Here, we shall denote $a_1 = a$.

Theorem 1.17.5. Let (C1) hold and

(C7) $\displaystyle\sum^{\infty} \frac{1}{r_1(n)} = \infty,$

(C8) $\displaystyle\liminf_{n\to\infty} r_1(n) > 0.$

Suppose that there exists a function x of type $r[1]$ such that (C4) holds and

(C9) for some positive integer δ,

$$\liminf_{n\to\infty} \sum_{s=n}^{n+\delta} x(s+1)g^+(s) > 0 \quad \text{and} \quad \sum^{\infty} x(s+1)g^-(s) < \infty$$

or

$$\liminf_{n\to\infty}\sum_{s=n}^{n+\delta} x(s+1)g^-(s) > 0 \qquad \text{and} \qquad \sum^{\infty} x(s+1)g^+(s) < \infty.$$

Further, assume that a is nondecreasing and Δa is bounded on $\mathbb{N}(n_0)$. Then, any bounded nonoscillatory solution y of $(E_{47};1)$ satisfies

$$\lim_{n\to\infty} r_1(n)\Delta(r_0y)(n) \;=\; \lim_{n\to\infty}\Delta(r_0y)(n) \;=\; \lim_{n\to\infty} r_0(n)y(n) \;=\; 0.$$

Proof. Let y be a bounded nonoscillatory solution of $(E_{47};1)$. Once again y is assumed to be eventually positive and $y(n) > 0$ for $n \ge n_1 \ge n_0$. As before, we have (1.17.26) – (1.17.28) and (1.17.35) (with $\zeta = 1$). In view of (1.17.27), (C4) and (C9), from (1.17.35) we see that $\sum_{n=N}^{\infty} x(n+1)\Delta_r^m y(n)$ exists in $\mathbb{R}^*$. So by Lemma 1.17.2, $\lim_{n\to\infty}\Delta_r^1 y(n)$ also exists in $\mathbb{R}^*$. We shall first show that

$$\lim_{n\to\infty}\Delta_r^1 y(n) \;=\; \lim_{n\to\infty} r_1(n)\Delta(r_0y)(n) \;=\; 0. \tag{1.17.38}$$

Suppose the contrary. Then, there exists a positive constant c such that for some $N_1 \ge N$,

$$\Delta_r^1 y(n) \;\ge\; c \quad \text{or} \quad \Delta_r^1 y(n) \;\le\; -c, \qquad n \ge N_1.$$

This is the same as

$$\Delta(r_0y)(n) \;\ge\; \frac{c}{r_1(n)} \quad \text{or} \quad \Delta(r_0y)(n) \;\le\; -\frac{c}{r_1(n)}, \qquad n \ge N_1. \tag{1.17.39}$$

On summing (1.17.39) from N_1 to $(n-1)$ and taking (C7) into account, we get $\lim_{n\to\infty} r_0(n)y(n) = \infty$ or $-\infty$. This is a contradiction because r_0y is bounded by virtue of (C1) and the boundedness of y. Hence, (1.17.38) is proved.

Next, in view of (C8), relation (1.17.38) immediately provides

$$\lim_{n\to\infty}\Delta(r_0y)(n) \;=\; 0, \tag{1.17.40}$$

which is the second conclusion of the theorem.

It remains to show that

$$\lim_{n\to\infty}\Delta_r^0 y(n) \;=\; \lim_{n\to\infty} r_0(n)y(n) \;=\; 0. \tag{1.17.41}$$

For this, noting that (C7) is actually (C2) when $k = 1$, and also (C4) and (C9) imply (C3), we apply Theorem 1.17.3 (with $k = 1$) to get (1.17.36) which, in view of (C1), leads to (1.17.37). To complete the proof, we need to show further that

$$\limsup_{n\to\infty} \Delta_r^0 y(n) \;=\; 0. \tag{1.17.42}$$

Suppose (1.17.42) does not hold. Then, taking note of (A2) and (1.17.37) (which has been proved), we have

$$\liminf_{n\to\infty} \Delta_r^0 y(a(n)) \;=\; 0 \quad \text{and} \quad \limsup_{n\to\infty} \Delta_r^0 y(a(n)) \;\geq\; d,$$

where d is some positive constant. Now there exist three integer sequences $\{\alpha_j\}$, $\{\beta_j\}$, $\{\gamma_j\}$ with $\lim_{j\to\infty} \alpha_j = \infty$ such that for each $j \geq 1$,

$$N \;\leq\; \alpha_j \;<\; \gamma_j \;<\; \beta_j \;\leq\; \alpha_{j+1},$$

$$\Delta_r^0 y(a(\alpha_j)) \;=\; \frac{d}{2} \;=\; \Delta_r^0 y(a(\beta_j)), \tag{1.17.43}$$

$$\Delta_r^0 y(a(\gamma_j)) \;>\; d, \tag{1.17.44}$$

$$\Delta_r^0 y(a(n)) \;>\; \frac{d}{2}, \quad \alpha_j < n < \beta_j.$$

Next, by discrete mean value theorem [2], we have

$$\frac{\Delta_r^0 y(a(\gamma_j)) - \Delta_r^0 y(a(\alpha_j))}{a(\gamma_j) - a(\alpha_j)} \;\leq\; \Delta[\Delta_r^0 y(a(\xi_j))] \quad \text{or} \quad \nabla[\Delta_r^0 y(a(\xi_j))], \tag{1.17.45}$$

where $a(\xi_j) \in [a(\alpha_j) + 1, a(\gamma_j) - 1]$ and $\lim_{j\to\infty} \xi_j = \infty$. Using the fact that a is nondecreasing and also (1.17.43) and (1.17.44) in (1.17.45), we obtain

$$\frac{d - (d/2)}{a(\beta_j) - a(\alpha_j)} \;<\; \Delta[(r_0 y)(a(\xi_j))] \quad \text{or} \quad \nabla[(r_0 y)(a(\xi_j))]. \tag{1.17.46}$$

Since in view of (1.17.40), the right side of (1.17.46) tends to zero as $j \to \infty$, it is necessary that

$$\lim_{j\to\infty} [a(\beta_j) - a(\alpha_j)] \;=\; \infty. \tag{1.17.47}$$

Once again, by mean value theorem we have

$$a(\beta_j) - a(\alpha_j) \;\leq\; (\beta_j - \alpha_j)\Delta a(\theta_j) \quad \text{or} \quad (\beta_j - \alpha_j)\nabla a(\theta_j), \tag{1.17.48}$$

where $\theta_j \in [\alpha_j + 1, \beta_j - 1]$. Since Δa is bounded (so is ∇a), in view of (1.17.47), it is clear from (1.17.48) that

$$\lim_{j\to\infty} (\beta_j - \alpha_j) \;=\; \infty. \tag{1.17.49}$$

Next, we observe that for each $j \geq 1$,

$$\Delta_r^0 y(a(n)) \ \geq \ \frac{d}{2}, \quad n \in \mathbb{N}(\alpha_j, \beta_j).$$

Consequently, using (C1) we find

$$y(a(n)) \geq \frac{d}{2r_0(a(n))} \geq \frac{d}{2\sup_{n \geq n_0} r_0(n)} > 0, \quad n \in \mathbb{N}(\alpha_j, \beta_j).$$

Hence, $y(a(n))$ has a positive lower bound on the set $\cup_{j=1}^{\infty} [\alpha_j, \beta_j]$. So in view of (A3), there exists a positive constant L such that

$$F(y(a(n))) \geq L, \quad n \in \cup_{j=1}^{\infty}[\alpha_j, \beta_j].$$

Then, it is clear that

$$\begin{aligned} \sum_{s=N}^{\infty} x(s+1)g^{\pm}(s)F(y(a(s))) \ &\geq \ \sum_{j=1}^{\infty} \sum_{s=\alpha_j}^{\beta_j} x(s+1)g^{\pm}(s)F(y(a(s))) \\ &\geq \ L\sum_{j=1}^{\infty} \sum_{s=\alpha_j}^{\beta_j} x(s+1)g^{\pm}(s). \qquad (1.17.50) \end{aligned}$$

On the other hand, by (1.17.49) and (C9) we have

$$\sum_{j=1}^{\infty} \sum_{s=\alpha_j}^{\beta_j} x(s+1)g^{+}(s) \ = \ \infty \quad \text{or} \quad \sum_{j=1}^{\infty} \sum_{s=\alpha_j}^{\beta_j} x(s+1)g^{-}(s) = \infty.$$

Therefore, it follows from (1.17.50) that

$$\sum_{s=N}^{\infty} x(s+1)g^{+}(s)F(y(a(s))) = \infty \ \text{ or } \ \sum_{s=N}^{\infty} x(s+1)g^{-}(s)F(y(a(s))) = \infty. \qquad (1.17.51)$$

In view of (1.17.51) and (C4), we see from $(1.17.35)|_{\zeta=1}$ that $\sum_{s=N}^{\infty} x(s+1)\Delta_r^m y(s) = \infty$ or $-\infty$. By Lemma 1.17.2, this implies that $\lim_{n\to\infty} \Delta_r^1 y(n) = \infty$ or $-\infty$ which is a contradiction to (1.17.38). Hence, we have shown (1.17.42) and this completes the proof of the theorem. ∎

The next two results are for the difference equation $(E_{47}; \zeta)$ where g is nonnegative.

Theorem 1.17.6. Let (C1) and (C6) hold. Further, suppose that

(C10) $\displaystyle\sum^{\infty} \frac{1}{r_i(n)} = \infty, \ 1 \leq i \leq m-1,$

(C11) $g(n) \geq 0,\ n \geq n_0$ and $\sum^{\infty} g(n) = \infty$,

(C12) $\liminf_{\substack{u_i \to \infty \\ 1 \leq i \leq \zeta}} F(u_1, \cdots, u_\zeta) > 0$ and $\limsup_{\substack{u_i \to -\infty \\ 1 \leq i \leq \zeta}} F(u_1, \cdots, u_\zeta) < 0$,

(C13) $\sum^{\infty} x(n+1)|h(n)| < \infty$,

where x is a function of type $r[0]$ given by

$$\begin{aligned} x(n) &= \rho_{0,m-1}(n) \qquad \text{(see (1.17.6))} \\ &= \sum_{n_{m-1}=n_0}^{n-1} \frac{1}{r_{m-1}(n_{m-1})} \sum_{n_{m-2}=n_0}^{n_{m-1}-1} \frac{1}{r_{m-2}(n_{m-2}+1)} \\ &\quad \sum_{n_{m-3}=n_0}^{n_{m-2}-1} \frac{1}{r_{m-3}(n_{m-3}+2)} \cdots \sum_{n_1=n_0}^{n_2-1} \frac{1}{r_1(n_1+m-2)}. \end{aligned}$$

Then, every nonoscillatory solution y of $(E_{47};\zeta)$ is bounded and satisfies $\lim_{n\to\infty} y(n) = 0$.

Proof. Let y be a nonoscillatory solution of $(E_{47};\zeta)$. Again y is assumed to be eventually positive and $y(n) > 0$ for $n \geq n_1 \geq n_0$. Let N be such that for $n \geq N$, $a_i(n) \geq n_1$ for every $1 \leq i \leq \zeta$. It is clear from $(E_{47};\zeta)$ that

$$\begin{aligned} \sum_{n=N}^{T} x(n+1)\Delta_r^m y(n) &= \sum_{n=N}^{T} x(n+1)h(n) - \sum_{n=N}^{T} x(n+1)g(n) \times \\ &\quad F(y(a_1(n)), \cdots, y(a_\zeta(n))), \quad n \geq N. \quad (1.17.52) \end{aligned}$$

In view of (A3) and (C13), from (1.17.52) we see that $\sum_{n=N}^{\infty} x(n+1)\Delta_r^m y(n)$ exists in $\mathbb{R}^*$. It follows from Lemma 1.17.2 that $\lim_{n\to\infty} \Delta_r^0 y(n)$ also exists in $\mathbb{R}^*$. First, we shall show that

$$\lim_{n\to\infty} \Delta_r^0 y(n) = \lim_{n\to\infty} r_0(n) y(n) < \infty. \qquad (1.17.53)$$

Suppose the contrary, i.e.

$$\lim_{n\to\infty} \Delta_r^0 y(n) = \infty. \qquad (1.17.54)$$

Then, taking note of (C1) it is obvious that

$$\lim_{n\to\infty} y(n) = \lim_{n\to\infty} \frac{\Delta_r^0 y(n)}{r_0(n)} = \infty.$$

Consequently, by (A2) and (C12), the function $F(y(a_1(n)), \cdots, y(a_\zeta(n)))$ has a positive lower bound. This, together with (C11), yields

$$\sum^{\infty} g(n) F(y(a_1(n)), \cdots, y(a_\zeta(n))) = \infty. \qquad (1.17.55)$$

Further, it is clear that (C13) leads to

$$\sum^{\infty} |h(n)| \; < \; \infty. \tag{1.17.56}$$

Now, noting that $\Delta_r^m y = \Delta(\Delta_r^{m-1} y)$, we sum $(E_{47}; \zeta)$ from N to $(n-1)$ to get

$$\Delta_r^{m-1} y(n) \;=\; \Delta_r^{m-1} y(N) - \sum_{s=N}^{n-1} g(s) F(y(a_1(s)), \cdots, y(a_\zeta(s))) + \sum_{s=N}^{n-1} h(s). \tag{1.17.57}$$

Letting $n \to \infty$ in (1.17.52) and applying (1.17.55) and (1.17.56), we obtain

$$\lim_{n\to\infty} \Delta_r^{m-1} y(n) \;=\; -\infty. \tag{1.17.58}$$

Using (1.17.58) and (C10), a similar argument as in the proof of Theorem 1.17.3 gives

$$\lim_{n\to\infty} \Delta_r^0 y(n) \;=\; -\infty,$$

which is a contradiction to (1.17.54). Hence, we have proved (1.17.53).

Next, it follows from (1.17.53) and (C6) that

$$\lim_{n\to\infty} y(n) \;=\; \lim_{n\to\infty} \frac{\Delta_r^0 y(n)}{r_0(n)} \;<\; \infty,$$

i.e. y is bounded. Finally, noting that (C11) implies (C5) and (C13) is (C4), we apply Theorem 1.17.4 to get $\lim_{n\to\infty} y(n) = 0$. ∎

Theorem 1.17.7. Let (C1) and (C6) hold. Further, suppose that

(C14) $\displaystyle\sum^{\infty} \frac{1}{r_i(n)} < \infty, \; 1 \le i \le m-1,$

(C15) $\displaystyle\sum^{\infty} |h(n)| < \infty,$

(C16) $g(n) \ge 0, \; n \ge n_0$ and $\displaystyle\sum^{\infty} x(n+1) g(n) = \infty,$

where x is a function of type $r[0]$ given by

$$\begin{aligned} x(n) \;&=\; \sigma_0(n) \qquad \text{(see (1.17.8))} \\ &=\; \sum_{n_{m-1}=n}^{\infty} \frac{1}{r_{m-1}(n_{m-1})} \sum_{n_{m-2}=n_{m-1}}^{\infty} \frac{1}{r_{m-2}(n_{m-2}+1)} \\ &\qquad \sum_{n_{m-3}=n_{m-2}}^{\infty} \frac{1}{r_{m-3}(n_{m-3}+2)} \cdots \sum_{n_1=n_2}^{\infty} \frac{1}{r_1(n_1+m-2)}. \end{aligned}$$

Then, every nonoscillatory solution y of $(E_{47};\zeta)$ is bounded and satisfies $\lim_{n\to\infty} y(n) = 0$.

Proof. Let y be a nonoscillatory solution of $(E_{47};\zeta)$. We shall assume that y is eventually positive and $y(n) > 0$ for $n \ge n_1 \ge n_0$. From $(E_{47};\zeta)$ we have (1.17.57) which, in view of (A3) and (C15), leads to

$$\Delta_r^{m-1} y(n) \ \le\ L_{m-1}, \quad n \ge N \tag{1.17.59}$$

where L_{m-1} is some positive constant. Summing (1.17.59) from N to $(n-1)$ and taking into account (C14), we get

$$\Delta_r^{m-2} y(n) \ \le\ \Delta_r^{m-2} y(N) + M_{m-1} \sum_{s=N}^{n-1} \frac{1}{r_{m-1}(s)} \le L_{m-2}, \quad n \ge N$$

where once again L_{m-2} is some positive constant. By continuing the process, eventually we obtain $\Delta_r^0 y \ (= r_0 y)$ is bounded. By (C6), this implies that y is also bounded. Finally, noting that (C15) and (C16) respectively leads to (C4) and (C5), we use Theorem 1.17.4 to get $\lim_{n\to\infty} y(n) = 0$. ∎

To show the importance of Theorems 1.17.3 – 1.17.7 now we shall investigate the asymptotic behavior of nonoscillatory solutions of the difference equation

$$\Delta^M[r(n)\Delta^{m-M} y(n)] + g(n)F(y(a_1(n)), y(a_2(n)), \cdots, y(a_\zeta(n))) \ =\ h(n),$$
$$n \in \mathbb{N}(n_0) \qquad (E_{48};\zeta, M)$$

where $M \in \mathbb{N}(1, m-1)$ is fixed, $r : [n_0, \infty) \to \mathbb{R}_0$ and (A2), (A3) are satisfied.

We note that $(E_{48};\zeta, M)$ is a special case of $(E_{47};\zeta)$ when $r_j = 1$ for $j \ne m - M$ and $r_{m-M} = r$.

Corollary 1.17.8. Suppose that

$$\sum^{\infty} \frac{1}{r(n)} \ =\ \infty. \tag{1.17.60}$$

Further, there exists a real number p and an integer $k \in \mathbb{N}(0, M-1)$ such that for all $0 < \mu \le 1$ and $\nu > 0$,

$$\sum^{\infty} (n + 1 - p)^{(k)}[\mu g^+(n) - g^-(n) - \nu|h(n)|] \ =\ \infty$$

or (1.17.61)

$$\sum^{\infty} (n + 1 - p)^{(k)}[\mu g^-(n) - g^+(n) - \nu|h(n)|] \ =\ \infty.$$

Then, any bounded solution y of $(E_{48};\zeta,M)$ satisfies $\liminf_{n\to\infty}|y(n)| = 0$.

Proof. Without loss of generality, we assume that $n_0 \ge p+k$. It can be verified easily that the function $x(n) = (n-p)^{(k)}$ is of type $r[m-k-1]$. Noting that $1 \le m-M \le m-1$, we have $r_0 = 1$ and so (C1) is satisfied. Next, since $m-M \le m-k-1 \le m-1$, condition (C2) (with k replaced by $(m-k-1)$) reduces to (1.17.60). Finally, (1.17.61) is exactly (C3) with k replaced by $(m-k-1)$. Hence, the conclusion follows immediately from Theorem 1.17.3. ∎

Corollary 1.17.9. Suppose that there exists an integer $k \in \mathbb{N}(0,m-M-1)$ such that

$$\sum^{\infty} \frac{n^{(m-M-1-k)}}{r(n+M-1)} = \infty. \tag{1.17.62}$$

Further, let (C3) be satisfied with

$$x(n) = \sum_{s=n_0}^{n-1} \frac{(n-s-1)^{(M-1)}(s-n_0)^{(m-M-1-k)}}{r(s+M-1)}. \tag{1.17.63}$$

Then, any bounded solution y of $(E_{48};\zeta,M)$ satisfies $\liminf_{n\to\infty}|y(n)| = 0$.

Proof. By direct computation, we have

$$x(n) = (M-1)!(m-M-1-k)!\ \rho_{k,m-1}(n),$$

where $\rho_{k,m-1}$ is given in (1.17.6). Using (1.17.62), it can be verified that the function x is of type $r[k]$. Moreover, it is noted in the proof of Corollary 1.17.8 that (C1) is fulfilled. Finally, since $k \le m-M-1$, we observe that (C2) obviously holds. Hence, the conclusion is immediate from Theorem 1.17.3. ∎

Corollary 1.17.10. Suppose that there exists an integer $k \in \mathbb{N}(0,m-M-1)$ such that

(a) $\displaystyle\sum^{\infty} \frac{s^{(m-2-k)}}{r(s)} < \infty$ if $M=1$, and $\displaystyle\sum^{\infty} \frac{s^{(m-M-1-k)}}{r(s+1)} < \infty$ if $M>1$,

(b) if $M>1$, then $\lim_{n\to\infty}\eta_{kj}(n) \in \{0,\ \infty\}$ for every $1 \le j \le M-1$, and

(c) condition (C3) is satisfied with

$$x(n) = \begin{cases} \theta_k(n), & M=1 \\ \eta_{k,M-1}(n), & M>1 \end{cases}$$

where θ_k and η_{kj} are respectively given in (1.17.10) and (1.17.12). Then, any bounded solution y of $(E_{48};\zeta,M)$ satisfies $\liminf_{n\to\infty}|y(n)|=0$.

Proof. Clearly, the function x is of type $r[k]$. Moreover, as noted in the proof of Corollary 1.10.9, conditions (C1) and (C2) are satisfied. Therefore, the conclusion follows immediately from Theorem 1.17.3. ∎

Corollary 1.17.11. Suppose that there exists a real number p such that

$$\sum^{\infty}(n+1-p)^{(m-1)}|h(n)| \; < \; \infty, \tag{1.17.64}$$

$$\sum^{\infty}(n+1-p)^{(m-1)}g^{+}(n) \; = \; \infty \quad \textit{and} \quad \sum^{\infty}(n+1-p)^{(m-1)}g^{-}(n)<\infty$$

or (1.17.65)

$$\sum^{\infty}(n+1-p)^{(m-1)}g^{+}(n) \; < \; \infty \quad \textit{and} \quad \sum^{\infty}(n+1-p)^{(m-1)}g^{-}(n)=\infty.$$

Then, any bounded nonoscillatory solution y of $(E_{48};\zeta,M)$ satisfies $\lim_{n\to\infty}y(n)=0$.

Proof. Since $r_0=1$, conditions (C1) and (C6) are fulfilled. As in the proof of Corollary 1.17.8, it is noted that the function $x(n)=(n-p)^{(m-1)}$ is of type $r[0]$. Thus, (1.17.64) and (1.17.65) correspond to (C4) and (C5) respectively. We use Theorem 1.17.4 to get the conclusion. ∎

Corollary 1.17.12. Suppose that

$$\sum^{\infty}\frac{n^{(m-M-1)}}{r(n+M-1)} \; = \; \infty. \tag{1.17.66}$$

Further, let (C4) and (C5) be satisfied with

$$x(n)=\sum_{s=n_0}^{n-1}\frac{(n-s-1)^{(M-1)}(s-n_0)^{(m-M-1)}}{r(s+M-1)}. \tag{1.17.67}$$

Then, any bounded nonoscillatory solution y of $(E_{48};\zeta,M)$ satisfies $\lim_{n\to\infty}y(n)=0$.

Proof. Once again conditions (C1) and (C6) are satisfied. Next, we observe that $x(n)=(M-1)!(m-M-1)!\;\rho_{0,m-1}(n)$ where $\rho_{0,m-1}$ is given in (1.17.6), and also (1.17.66) ensures that x is of type $r[0]$. Theorem 1.17.4 is then used to get the conclusion. ∎

Corollary 1.17.13. Suppose that

(a) $\displaystyle\sum^{\infty}\frac{s^{(m-2)}}{r(s)}<\infty$ if $M=1$, and $\displaystyle\sum^{\infty}\frac{s^{(m-M-1)}}{r(s+1)}<\infty$ if $M>1$,

(b) if $M > 1$, then $\lim_{n\to\infty} \eta_{0j}(n) \in \{0, \infty\}$ for every $1 \le j \le M-1$, and

(c) conditions (C4) and (C5) are satisfied with

$$x(n) = \begin{cases} \theta_0(n), & M = 1 \\ \eta_{0,M-1}(n), & M > 1 \end{cases}$$

where θ_0 and η_{0j} are respectively given in (1.17.10) and (1.10.12). Then, any bounded nonoscillatory solution y of $(E_{48}; \zeta, M)$ satisfies $\lim_{n\to\infty} y(n) = 0$.

Proof. Since $r_0 = 1$, conditions (C1) and (C6) are satisfied. Next, noting that the function x is of type $r[0]$, the conclusion is clear from Theorem 1.17.4. ∎

Corollary 1.17.14. Consider the equation $(E_{48}; 1, m-1)$ and denote $a_1 = a$. Let

$$\sum^{\infty} \frac{1}{r(n)} = \infty, \tag{1.17.68}$$

$$\liminf_{n\to\infty} r(n) > 0 \tag{1.17.69}$$

and

$$\sum^{\infty} n^{(m-2)}|h(n)| < \infty. \tag{1.17.70}$$

Suppose that for some positive integer δ,

$$\liminf_{n\to\infty} \sum_{s=n}^{n+\delta} s^{(m-2)} g^+(s) > 0 \quad and \quad \sum^{\infty} s^{(m-2)} g^-(s) < \infty$$

or (1.17.71)

$$\liminf_{n\to\infty} \sum_{s=n}^{n+\delta} s^{(m-2)} g^-(s) > 0 \quad and \quad \sum^{\infty} s^{(m-2)} g^+(s) < \infty.$$

Further, assume that a is nondecreasing and Δa is bounded on $\mathbb{N}(n_0)$. Then, any bounded nonoscillatory solution y of $(E_{48}; 1, m-1)$ satisfies

$$\lim_{n\to\infty} r(n)\Delta y(n) = \lim_{n\to\infty} \Delta y(n) = \lim_{n\to\infty} y(n) = 0.$$

Proof. Here, $r_0 = 1$ and $r_1 = r$. Thus, (C1), (C7) (= (1.17.68)) and (C8) (= (1.17.69)) hold. Next, we note that the function $x(n) = (n - n_0)^{(m-2)}$ is of type $r[1]$. Subsequently, (1.17.70) and (1.17.71) respectively imply that (C4) and (C9) are satisfied for this x. The conclusion is now immediate from Theorem 1.17.5. ∎

Corollary 1.17.15. Consider the equation $(E_{48}; 1, M)$ where $M \le m-2$, and denote $a_1 = a$. Suppose that

$$\sum^{\infty} \frac{n^{(m-M-2)}}{r(n+M-1)} = \infty. \tag{1.17.72}$$

Let (C4) and (C9) be satisfied with

$$x(n) = \sum_{s=n_0}^{n-1} \frac{(n-s-1)^{(M-1)}(s-n_0)^{(m-M-2)}}{r(s+M-1)}. \tag{1.17.73}$$

Further, assume that a is nondecreasing and Δa is bounded on $\mathbb{N}(n_0)$. Then, any bounded nonoscillatory solution y of $(E_{48}; 1, M)$ satisfies

$$\lim_{n\to\infty} \Delta y(n) = \lim_{n\to\infty} y(n) = 0.$$

Proof. Since $2 \le m - M \le m - 1$, it is clear that $r_0 = r_1 = 1$. Therefore, (C1), (C7) and (C8) are fulfilled. Next, it is noted that $x(n) = (M-1)!(m-M-2)!\, \rho_{1,m-1}(n)$, where $\rho_{1,m-1}$ is given in (1.17.6). Using (1.17.72), it can be verified that the function x is of type $r[1]$. Hence, the conclusion is immediate from Theorem 1.17.5. ∎

Corollary 1.17.16. Consider the equation $(E_{48}; 1, M)$ where $M \le m-2$, and denote $a_1 = a$. Suppose that

(a) $\displaystyle\sum^{\infty} \frac{s^{(m-3)}}{r(s)} < \infty$ if $M = 1$, and $\displaystyle\sum^{\infty} \frac{s^{(m-M-2)}}{r(s+1)} < \infty$ if $M > 1$,

(b) if $M > 1$, then $\lim_{n\to\infty} \eta_{1j}(n) \in \{0, \infty\}$ for every $1 \le j \le M-1$, and

(c) conditions (C4) and (C9) are satisfied with

$$x(n) = \begin{cases} \theta_1(n), & M = 1 \\ \eta_{1,M-1}(n), & M > 1 \end{cases}$$

where θ_1 and η_{1j} are respectively given in (1.17.10) and (1.17.12). Further, assume that a is nondecreasing and Δa is bounded on $\mathbb{N}(n_0)$. Then, any bounded nonoscillatory solution y of $(E_{48}; 1, M)$ satisfies

$$\lim_{n\to\infty} \Delta y(n) = \lim_{n\to\infty} y(n) = 0.$$

Proof. As noted in the proof of Corollary 1.17.15, $r_0 = r_1 = 1$ and so (C1), (C7) and (C8) are satisfied. Moreover, the function x is of type $r[1]$. Thus, the conclusion follows immediately from Theorem 1.17.5. ∎

Corollary 1.17.17. Suppose that (1.17.60), (1.17.66), (C11) and (C12) hold. Further, let (C13) be satisfied with x defined in (1.17.67). Then, any nonoscillatory solution y of $(E_{48};\zeta, M)$ is bounded and satisfies $\lim_{n\to\infty} y(n) = 0$.

Proof. Once again conditions (C1) and (C6) are satisfied. Next, since $1 \le m - M \le m - 1$, condition (C10) reduces to (1.17.60). Noting that the function x given in (1.17.67) is of type $r[0]$, we apply Theorem 1.17.6 to get the conclusion. ∎

Corollary 1.17.18. Consider the equation $(E_{48};\zeta, 1)$ with $m = 2$, i.e.

$$\Delta[r(n)\Delta y(n)]+g(n)F(y(a_1(n)), y(a_2(n)), \cdots, y(a_\zeta(n))) = h(n), \quad n \in \mathbb{N}(n_0). \tag{1.17.74}$$

Let (C15) hold and let (C16) be satisfied with

$$x(n) \;=\; \sum_{s=n}^{\infty} \frac{1}{r(s)}. \tag{1.17.75}$$

Further, suppose that

$$\sum^{\infty} \frac{1}{r(n)} \;<\; \infty. \tag{1.17.76}$$

Then, any nonoscillatory solution y of (1.17.74) is bounded and satisfies $\lim_{n\to\infty} y(n) = 0$.

Proof. It is obvious that $r_0 = 1$ and $r_1 = r$. Thus, (C1) and (C6) are fulfilled, and (C14) reduces to (1.17.76). Noting that here $x = \sigma_0$, the conclusion is immediate from Theorem 1.17.7. ∎

Example 1.17.5. Consider the difference equation

$$\Delta^3[(n+2)\Delta y(n)] + \frac{1}{(n+1)(n+2)(n+3)} y(n+3) = \frac{7}{(n+1)(n+2)(n+3)(n+4)}, \quad n \in \mathbb{N}. \tag{$E_{48};1,3$}$$

Here, $m = 4$, $M = 3$, $\zeta = 1$, $n_0 = 0$, $r(n) = n + 2$, $g(n) = [(n+1)(n+2)(n+3)]^{-1}$, $h(n) = 7[(n+1)(n+2)(n+3)(n+4)]^{-1}$, $a(n) = n + 3$ and $F(u) = u$. Obviously, the hypothesis (A2) and (A3) are satisfied.

It can be easily verified that all the conditions of Corollaries 1.17.8 (take $k = 2$) and 1.17.14 are fulfilled. However, Corollaries 1.17.11 and 1.17.17 *cannot* be used as conditions (1.17.64) and (C11) are not met. Moreover, Corollaries 1.17.10 and 1.17.13 are also *not* applicable as their condition (a) is not satisfied. Consequently, we conclude from Corollaries 1.17.8 and 1.14.14 that

(a) if y_1 is a bounded solution of $(E_{48}; 1, 3)$, then $\liminf_{n\to\infty} |y_1(n)| = 0$,

(b) if y_2 is a bounded nonoscillatory solution of $(E_{48};\ 1,3)$, then $\lim_{n\to\infty}(n+2)\Delta y_2(n) = \lim_{n\to\infty}\ \Delta y_2(n) = \lim_{n\to\infty} y_2(n) = 0$.

In fact, a bounded as well as nonoscillatory solution is given by $y(n) = (n+1)^{-1}$ and we note that it does satisfy the above conclusions.

Example 1.17.6. Consider the difference equation

$$\Delta[n\Delta^2 y(n)] + 4y(n+3)^3 \;=\; -\frac{4(3n+7)}{(n+1)(n+2)(n+3)^3}, \quad n \in \mathbb{N}(1). \qquad (E_{48};1,1)$$

Here, $m = 3$, $M = 1$, $\zeta = 1$, $n_0 = 1$, $r(n) = n$, $g(n) = 4$, $h(n) = -4(3n+7)[(n+1)(n+2)(n+3)^3]^{-1}$, $a(n) = n+3$ and $F(u) = u^3$. Once again, the hypothesis (A2) and (A3) are satisfied.

We note that Corollaries 1.17.10, 1.17.13 and 1.17.16 are *not* applicable here as their condition (a) is violated. However, it can be easily checked that all the conditions of Corollary 1.17.15 are fulfilled. In fact, from (1.17.73) we have $x(n+1) = \sum_{s=1}^{n} 1/s$ and (C4) is satisfied as

$$\begin{aligned}
\sum^{\infty} x(n+1)|h(n)| &= \sum^{\infty} x(n+1)\frac{4(3n+7)}{(n+1)(n+2)(n+3)^3} \\
&< \sum^{\infty} x(n+1)\frac{28(n+1)}{(n+1)^5} \\
&< \sum^{\infty}\left(\sum_{s=1}^{n} 1\right)\frac{28}{(n+1)^4} \;<\; \infty.
\end{aligned}$$

It can also be verified that the conditions of Corollaries 1.17.8, 1.17.9 (take $k = 0$), 1.17.11, 1.17.12 and 1.17.17 are met. Hence, we conclude that

(a) if y_1 is a bounded solution of $(E_{48};1,1)$, then $\liminf_{n\to\infty}|y_1(n)| = 0$,

(b) if y_2 is a bounded nonoscillatory solution of $(E_{48};\ 1,1)$, then $\lim\limits_{n\to\infty} \Delta y_2(n) = \lim_{n\to\infty} y_2(n) = 0$,

(c) if y_3 is a nonoscillatory solution of $(E_{48};1,1)$, then it is bounded and satisfies $\lim_{n\to\infty} y_3(n) = 0$.

It is noted that a bounded as well as nonoscillatory solution is given by $y(n) = n^{-1}$ and indeed it does fulfill the above conclusions.

Example 1.17.7. Consider the difference equation

$$\Delta[n(n+1)\Delta y(n)] + n(n+1)(n+2)y(n)y(n+1)y(n+2)y(n+3)y(n+4) = \frac{1}{(n+3)(n+4)}, \quad n \in \mathbb{N}(1). \qquad (E_{48};5,1)$$

Here, $m = 2$, $M = 1$, $\zeta = 5$, $n_0 = 1$, $r(n) = n(n+1)$, $g(n) = n(n+1)(n+2)$, $h(n) = [(n+3)(n+4)]^{-1}$, $a_i(n) = n+i-1$, $1 \le i \le 5$

and $F(u_1, \cdots, u_5) = \prod_{i=1}^{5} u_i$. Once again, the hypothesis (A2) and (A3) are satisfied.

It can easily be verified that the conditions of Corollaries 1.17.10, 1.17.13 and 1.17.18 are fulfilled. On the other hand, conditions (1.17.60), (1.7.62), (1.17.64) and (1.17.66) are not met. Thus, we *cannot* employ Corollaries 1.17.8, 1.17.9, 1.17.11, 1.17.12 and 1.17.17 here. By Corollaries 1.17.10, 1.17.13 and 1.17.18, we conclude that

(a) if y_1 is a bounded solution of $(E_{48}; 5, 1)$, then $\liminf_{n\to\infty} |y_1(n)| = 0$,

(b) if y_2 is a bounded nonoscillatory solution of $(E_{48}; 5, 1)$, then $\lim\limits_{n\to\infty} y_2(n) = 0$,

(c) if y_3 is a nonoscillatory solution of $(E_{48}; 5, 1)$, then it is bounded and satisfies $\lim_{n\to\infty} y_3(n) = 0$.

In fact, a bounded as well as nonoscillatory solution is given by $y(n) = n^{-1}$ and we note that it does satisfy the above conclusions.

Example 1.17.8. Consider the difference equation

$$\Delta[(n+1)(n+2)\Delta^2 y(n)] + (n+2)(n+3)^2 \frac{1}{y(n+3)^5} = -\frac{n^3+15n^2+52n+54}{n(n+1)(n+3)^3},$$

$$n \in \mathbb{N}(1). \qquad (E_{48}; 1, 1)$$

Here, $m = 3$, $M = 1$, $\zeta = 1$, $n_0 = 1$, $r(n) = (n+1)(n+2)$, $g(n) = (n+2)(n+3)^2$, $h(n) = -(n^3 + 15n^2 + 52n + 54)[n(n+1)(n+3)^3]^{-1}$, $a(n) = n+3$ and $F(u) = u^{-5}$. Clearly, the hypothesis (A2) and (A3) are satisfied.

It can be checked that the conditions of Corollaries 1.17.10 (take $k = 1$) and 1.17.16 are met. However, conditions (1.17.60), (1.17.64), (1.17.72) and condition (a) of Corollary 1.17.13 are violated. As such, we *cannot* apply Corollaries 1.17.8, 1.17.11, 1.17.13, 1.17.15 and 1.17.17 here. By Corollaries 1.17.10 and 1.17.16, we conclude that

(a) if y_1 is a bounded solution of $(E_{48}; 1, 1)$, then $\liminf_{n\to\infty} |y_1(n)| = 0$,

(b) if y_2 is a bounded nonoscillatory solution of $(E_{48};\ 1, 1)$, then $\lim\limits_{n\to\infty} \Delta y_2(n) = \lim_{n\to\infty} y_2(n) = 0$.

Indeed, a bounded as well as nonoscillatory solution is given by $y(n) = n^{-1}$ and we note that it does fulfill the above conclusions.

Example 1.17.9. Consider the difference equation

$$\Delta[n(n+1)\Delta y(n)] + (n+2)(n+3)(n+4)y(n+1)y(n+2)^2$$

$$= (n+1)[2 + (n+2)^3(n+3)(n+4)], \quad n \in \mathbb{N}. \qquad (E_{48}; 2, 1)$$

Here, $m = 2$, $M = 1$, $\zeta = 2$, $n_0 = 0$, $r(n) = n(n+1)$, $g(n) = (n+2)(n+3)(n+4)$, $h(n) = (n+1)[2+(n+2)^3(n+3)(n+4)]$, $a_i(n) = n+i$, $i = 1, 2$ and $F(u_1, u_2) = u_1 u_2^2$. Obviously, the hypothesis (A2) and (A3) are satisfied.

It can be checked that (C11) and (C12) are fulfilled whereas (1.17.60), (1.17.66) and (C13) (with x defined in (1.17.67)) are not satisfied. Hence, the conditions of Corollary 1.17.17 are *violated*. Indeed, $(E_{48}; 2, 1)$ has a nonoscillatory solution given by $y(n) = n$ and contrary to the conclusion of Corollary 1.17.17, this solution is not bounded and $\lim_{n\to\infty} y(n) \neq 0$.

Moreover, (C16) (with x defined in (1.16.75)) and (1.17.76) are satisfied but (C15) is not met. Therefore, the conditions of Corollary 1.17.18 are *violated*. As noted above, $(E_{48}; 2, 1)$ has a nonoscillatory solution that contradicts the conclusion of Corollary 1.17.18.

1.18. Difference Equations with Distributed Deviating Arguments

Let $a, b\ (> a)$ be integers. Consider the difference inequalities

$$|\Delta^m[y(n) + g(n)y(n-\sigma)]|^{\alpha-1}\Delta^m[y(n) + g(n)y(n-\sigma)] + \sum_{\xi=a}^{b} T(n,\xi) f(y(h(n,\xi))) \leq 0, \quad n \in \mathbb{N}(n_0) \qquad (1.18.1)$$

and

$$|\Delta^m[y(n) + g(n)y(n-\sigma)]|^{\alpha-1}\Delta^m[y(n) + g(n)y(n-\sigma)] + \sum_{\xi=a}^{b} T(n,\xi) f(y(h(n,\xi))) \geq 0, \quad n \in \mathbb{N}(n_0) \qquad (1.18.2)$$

where $\alpha > 0$ and σ is a nonnegative integer. Unless otherwise stated, we shall assume that $T(n,\xi) \geq 0$ on $\mathbb{N}(n_0) \times \mathbb{N}(a,b)$, and $h : \mathbb{N}(n_0) \times \mathbb{N}(a,b) \to \mathbb{N}$ satisfies

$$\lim_{n\to\infty} \min_{\xi\in\mathbb{N}(a,b)} h(n,\xi) = \infty. \qquad (1.18.3)$$

By a *solution* of (1.18.1) ((1.18.2)), we mean a nontrivial sequence $\{y(n)\}$ defined for $n \in \mathbb{N}(n_0)$, $\Delta^m y(n)$ is not identically zero, and $y(n)$ fulfills (1.18.1) ((1.18.2)) for $n \geq n_0$. A solution $\{y(n)\}$ is said to be *nondecreasing* (*nonincreasing*) provided that $\Delta y(n)$ is nonnegative (nonpositive). If $\Delta^2 y(n)$ is positive (negative), then $\{y(n)\}$ is said to be *concave* (*convex*).

We shall develop nonexistence criteria for eventually positive (negative) solutions of (1.18.1) ((1.18.2)). These criteria lead to oscillation results for the difference equation

$$|\Delta^m[y(n)+g(n)y(n-\sigma)]|^{\alpha-1}\Delta^m[y(n)+g(n)y(n-\sigma)] + \sum_{\xi=a}^{b} T(n,\xi)f(y(h(n,\xi))) \ = \ 0, \quad n\in\mathbb{N}(n_0). \qquad (E_{49})$$

Next, we shall consider the partial difference equation

$$\Delta_n^m[y(i,n)+g(n)y(i,n-\sigma)] + \sum_{\xi=a}^{b}\Lambda(i,n,\xi)f(y(i,h(n,\xi))) = d(n)Ly(i,n) + \sum_{\ell=1}^{\mu} d_\ell(n)Ly(i,\tau_\ell(n)), \quad i\in\Omega,\ n\in\mathbb{N}(n_0) \quad (E_{50})$$

where $i=(i_1,\cdots,i_\wp)$,

$$\Omega \ \equiv \ \mathbb{N}(1,N_1)\times\cdots\times\mathbb{N}(1,N_\wp) \qquad (1.18.4)$$

and L is the discrete Laplacian on Ω defined as

$$Ly(i,n) \ = \ \sum_{\ell=1}^{\wp}\Delta_{i_\ell}^2\, y((i_1,\cdots,i_{\ell-1},i_\ell-1,i_{\ell+1},\cdots,i_\wp),n). \qquad (1.18.5)$$

Further, it is assumed that $\Lambda(i,n,\xi)$, $d(n)$, $d_\ell(n)\ge 0$, $1\le\ell\le\mu$ for $i\in\Omega$, $n\in\mathbb{N}(n_0)$, $\xi\in\mathbb{N}(a,b)$, and for each $1\le\ell\le\mu$, $\tau_\ell:\mathbb{N}(n_0)\to\mathbb{N}$ fulfills $\lim_{n\to\infty}\tau_\ell(n)=\infty$. Equation (E_{50}) is subjected to two types of boundary conditions, namely,

$$y(i,n) \ = \ 0, \quad i\in\partial\Omega,\ n\in\mathbb{N}(n_0) \qquad (1.18.6)$$

where

$$\partial\Omega \ \equiv \ \bigcup_{\ell=1}^{\wp}\Big\{(i_1,\cdots,i_{\ell-1},0,i_{\ell+1},\cdots,i_\wp),(i_1,\cdots,i_{\ell-1},N_\ell+1,i_{\ell+1},\cdots,i_\wp), \quad i_k\in\mathbb{N}(1,N_k),\quad 1\le k\le n\Big\}, \qquad (1.18.7)$$

and

$$\Delta_{i_\ell}y(i,n)\Big|_{i_\ell=0} - \theta_\ell(i,n)y(i,n)\Big|_{i_\ell=0} = 0,$$
$$\Delta_{i_\ell}y(i,n)\Big|_{i_\ell=N_\ell} + \theta_\ell(i,n)y(i,n)\Big|_{i_\ell=N_\ell+1} = 0,\ i\in\Omega,\ n\in\mathbb{N}(n_0),\ 1\le\ell\le\wp \qquad (1.18.8)$$

where for each $1 \le \ell \le \wp$, $\theta_\ell(i,n)|_{i_\ell=0,N_\ell+1} \ge 0$, $i \in \Omega$, $n \in \mathbb{N}(n_0)$.

A solution $\{y(i,n)\}$ of (E_{50}), (1.18.6) ((1.18.8)) is *eventually positive* (*negative*) if $y(i,n) > (<)\ 0$ for $i \in \Omega$ and all large n. A solution $\{y(i,n)\}$ is said to be *nondecreasing* (*nonincreasing*) provided $\Delta_n y(i,n)$ is nonnegative (nonpositive). If $\Delta_n^2 y(i,n)$ is positive (negative), then we say $\{y(i,n)\}$ is *concave* (*convex*). Using the results obtained earlier, we shall establish oscillation criteria for the boundary value problems (E_{50}), (1.18.6) and (E_{50}), (1.18.8).

We shall need the following:

Lemma 1.18.1. [28] Consider the eigenvalue problem

$$\begin{aligned} Ly(i) + \lambda y(i) &= 0, \quad i \in \Omega \\ y(i) &= 0, \quad i \in \partial\Omega \end{aligned} \tag{1.18.9}$$

where L, Ω and $\partial\Omega$ are defined in (1.18.5), (1.18.4) and (1.18.7), respectively. Let λ_0 be the least eigenvalue of (1.18.9) and $\psi(i)$ be the corresponding eigenfunction. Then, $\lambda_0 > 0$ and $\psi(i) > 0$, $i \in \Omega$.

Lemma 1.18.2. (Discrete Green's Theorem) Let $\{y(i)\}$ and $\{z(i)\}$ be sequences defined on $\bar{\Omega} \equiv \mathbb{N}(0, N_1+1) \times \cdots \times \mathbb{N}(0, N_\wp + 1)$. Then,

$$\sum_{i\in\Omega} z(i)Ly(i) - \sum_{i\in\Omega} y(i)Lz(i)$$
$$= \sum_{\ell=1}^{\wp} \left\{ \sum_{i_1=1}^{N_1} \cdots \sum_{i_{\ell-1}=1}^{N_{\ell-1}} \sum_{i_{\ell+1}=1}^{N_{\ell+1}} \cdots \sum_{i_\wp=1}^{N_\wp} \left[z(i)\Delta_{i_\ell} y(i) - y(i)\Delta_{i_\ell} z(i)\right]_{i_\ell=0}^{N_\ell} \right\}, \tag{1.18.10}$$

where L and Ω are defined in (1.18.5) and (1.18.4) respectively.

Proof. For the one–dimensional case ($\wp = 1$), from [159] we have

$$\sum_{i\in\mathbb{N}(1,N)} z(i)Ly(i) = \sum_{i\in\mathbb{N}(1,N)} y(i)Lz(i) + [z(i)\Delta y(i) - y(i)\Delta z(i)]_{i=0}^{N}, \tag{1.18.11}$$

where $Ly(i) = \Delta^2 y(i-1)$. It is noted that (1.8.11) coincides with (1.18.10) when $\wp = 1$.

Consider the case when $\wp = 2$. Using (1.18.11), we find

$$\sum_{i_1\in\mathbb{N}(1,N_1)} \sum_{i_2\in\mathbb{N}(1,N_2)} z(i_1,i_2)Ly(i_1,i_2)$$
$$= \sum_{i_1\in\mathbb{N}(1,N_1)} \sum_{i_2\in\mathbb{N}(1,N_2)} z(i_1,i_2)\left[\Delta_{i_1}^2\, y(i_1-1,i_2) + \Delta_{i_2}^2\, y(i_1,i_2-1)\right]$$

$$
\begin{aligned}
&= \sum_{i_2 \in \mathbb{N}(1,N_2)} \Bigg\{ \sum_{i_1 \in \mathbb{N}(1,N_1)} y(i_1,i_2)\Delta_{i_1}^2\, z(i_1-1,i_2) \\
&\quad + [z(i_1,i_2)\Delta_{i_1} y(i_1,i_2) - y(i_1,i_2)\Delta_{i_1} z(i_1,i_2)]_{i_1=0}^{N_1} \Bigg\} \\
&\quad + \sum_{i_1 \in \mathbb{N}(1,N_1)} \Bigg\{ \sum_{i_2 \in \mathbb{N}(1,N_2)} y(i_1,i_2)\Delta_{i_2}^2\, z(i_1,i_2-1) \\
&\quad + [z(i_1,i_2)\Delta_{i_2} y(i_1,i_2) - y(i_1,i_2)\Delta_{i_2} z(i_1,i_2)]_{i_2=0}^{N_2} \Bigg\} \\
&= \sum_{i_1 \in \mathbb{N}(1,N_1)} \sum_{i_2 \in \mathbb{N}(1,N_2)} y(i_1,i_2) Lz(i_1,i_2) \\
&\quad + \sum_{i_2 \in \mathbb{N}(1,N_2)} [z(i_1,i_2)\Delta_{i_1} y(i_1,i_2) - y(i_1,i_2)\Delta_{i_1} z(i_1,i_2)]_{i_1=0}^{N_1} \\
&\quad + \sum_{i_1 \in \mathbb{N}(1,N_1)} [z(i_1,i_2)\Delta_{i_2} y(i_1,i_2) - y(i_1,i_2)\Delta_{i_2} z(i_1,i_2)]_{i_2=0}^{N_2},
\end{aligned}
$$

which is exactly (1.18.10). Now, by induction we can show that (1.18.10) is true. ∎

Lemma 1.18.3. [217] (Discrete Jansen's Inequality) Let f be a positive and convex function on $\mathbb{R}_0$. Then, for any $y(i),\ \psi(i) > 0,\ i \in \Omega$,

$$\sum_{i\in\Omega} f(y(i))\psi(i) \ \ge\ \rho\, f\left(\frac{1}{\rho}\sum_{i\in\Omega} y(i)\psi(i)\right),$$

where $\rho = \sum_{i\in\Omega}\psi(i)$.

Theorem 1.18.4. Suppose that

$$0 \le g(n) \le 1, \quad n \ge n_0 \tag{1.18.12}$$

$$f \text{ is an odd function, i.e. } f(-u) = -f(u), \tag{1.18.13}$$

$$\text{for } u \ne 0,\ \frac{f(u)}{u} \ge q \text{ for some } q > 0 \tag{1.18.14}$$

and

$$\sum_{n=c}^{\infty} \left\{ \sum_{\xi=a}^{b} T(n,\xi)[1 - g(h(n,\xi))] \right\}^{1/\alpha} = \infty. \tag{1.18.15}$$

(a) If m is even, then (1.18.1) ((1.18.2)) has no eventually positive (negative) solution. Further, all solutions of (E_{49}) are oscillatory.

(b) If m is odd, then (1.18.1) ((1.18.2)) has no eventually positive (negative) and nondecreasing (nonincreasing) solution. Further, (E_{49}) has neither eventually positive and nondecreasing nor eventually negative and nonincreasing solutions. Hence, all nonoscillatory solutions of (E_{49}) are bounded.

Proof. Let $\{y(n)\}$ be an eventually positive solution of (1.18.1). We shall consider only this case because by (1.18.13) the sequence $\{-y(n)\}$ is an eventually negative solution of (1.18.2). In view of (1.18.3), there exists $n_1 > n_0$ such that

$$y(n) > 0, \quad y(n-\sigma) > 0, \quad y(h(n,\xi)) > 0, \quad n \in \mathbb{N}(n_1),\ \xi \in \mathbb{N}(a,b). \tag{1.18.16}$$

We define

$$x(n) \ = \ y(n) + g(n)y(n-\sigma). \tag{1.18.17}$$

Due to condition (1.18.12), $x(n)$ is eventually positive and $x(n) \ge y(n)$. Further, it follows from (1.18.1), (1.18.14) and (1.18.16) that

$$\begin{aligned} |\Delta^m x(n)|^{\alpha-1}\Delta^m x(n) \ &\le \ -\sum_{\xi=a}^{b} T(n,\xi)f(y(h(n,\xi))) \\ &\le \ -q\sum_{\xi=a}^{b} T(n,\xi)y(h(n,\xi)) \ \le \ 0, \quad n \in \mathbb{N}(n_1). \end{aligned} \tag{1.18.18}$$

Hence, we have

$$\Delta^m x(n) \ \le \ 0, \quad n \in \mathbb{N}(n_1). \tag{1.18.19}$$

Case (a) m is even

In view of (1.18.19), from Lemma 1.13.1 (here p is odd and $1 \le p \le m-1$) it follows that

$$\Delta^{m-1}x(n) \ > \ 0, \quad n \in \mathbb{N}(n_1) \tag{1.18.20}$$

and

$$\Delta x(n) \ > \ 0, \quad n \in \mathbb{N}(n_1). \tag{1.18.21}$$

It is clear from (1.18.21) that for some $n_2 > n_1$,

$$x(h(n,\xi)) \ \ge \ x(h(n,\xi)-\sigma) \ \ge \ y(h(n,\xi)-\sigma), \quad n \in \mathbb{N}(n_2),\ \xi \in \mathbb{N}(a,b). \tag{1.18.22}$$

Further, noting (1.18.3) we may choose $n_3 > n_2$ so that $h(n,\xi) \ge v$, $n \ge n_3$, $\xi \in \mathbb{N}(a,b)$, where v is sufficiently large so that $x(v) > 0$. Subsequently, in view of (1.18.21) we have

$$x(h(n,\xi)) \ \ge \ x(v) \ > \ 0, \quad n \in \mathbb{N}(n_3),\ \xi \in \mathbb{N}(a,b). \tag{1.18.23}$$

Now, using (1.18.19), (1.18.12), (1.18.22) and (1.18.23) successively, from (1.18.18) we find for $n \in \mathbb{N}(n_3)$,

$$\begin{aligned} |\Delta^m x(n)|^\alpha &\geq q\sum_{\xi=a}^{b} T(n,\xi)y(h(n,\xi)) \\ &= q\sum_{\xi=a}^{b} T(n,\xi)[x(h(n,\xi)) - g(h(n,\xi))y(h(n,\xi)-\sigma)] \\ &\geq q\sum_{\xi=a}^{b} T(n,\xi)[1-g(h(n,\xi))]x(h(n,\xi)) \\ &\geq q\sum_{\xi=a}^{b} T(n,\xi)[1-g(h(n,\xi))]x(v), \end{aligned}$$

which is the same as

$$-\Delta^m x(n) = |\Delta^m x(n)| \geq \left\{ q\sum_{\xi=a}^{b} T(n,\xi)[1-g(h(n,\xi))]x(v) \right\}^{1/\alpha}, \quad n \in \mathbb{N}(n_3). \quad (1.18.24)$$

Summing (1.18.24) from n_3 to $(N-1)$, we obtain

$$\Delta^{m-1}x(N) - \Delta^{m-1}x(n_3) + [q\,x(v)]^{1/\alpha} \sum_{n=n_3}^{N-1} \left\{ \sum_{\xi=a}^{b} T(n,\xi)[1-g(h(n,\xi))] \right\}^{1/\alpha} \leq 0. \quad (1.18.25)$$

By (1.18.15) and (1.18.20), the right side of (1.18.25) tends to ∞ as $N \to \infty$. Thus, we get a contradiction.

Case (b) m is odd

Here, let $\{y(n)\}$ be an eventually positive and nondecreasing solution of (1.18.1). Again we shall only consider this case because (1.18.13) implies that $\{-y(n)\}$ is an eventually negative and nonincreasing solution of (1.18.2). In view of (1.18.19), in Lemma 1.13.1 we have p is even and $0 \leq p \leq m-1$. Therefore, (1.18.20) still holds but we cannot conclude that (1.18.21) is true.

Since $y(n)$ is eventually nondecreasing, for sufficiently large n_2 we have

$$x(h(n,\xi)) \geq y(h(n,\xi)) \geq y(h(n,\xi)-\sigma), \quad n \geq n_2, \ \xi \in \mathbb{N}(a,b)$$

which is equivalent to (1.18.22).

Further, the fact that $y(n)$ is eventually positive and nondecreasing leads to $\liminf_{n\to\infty} y(n) > 0$. Since $x(n) \ge y(n)$, it follows that

$$\liminf_{n\to\infty} x(n) \;\ge\; \liminf_{n\to\infty} y(n) \;>\; 0.$$

Therefore, in view of (1.18.3) there exists $n_3 > n_2$ so that for a given $w > 0$,

$$x(h(n,\xi)) \;\ge\; w \;>\; 0, \quad n \in \mathbb{N}(n_3),\ \xi \in \mathbb{N}(a,b). \tag{1.18.26}$$

Using a similar technique as in Case (a) and also applying relation (1.18.26) (instead of (1.18.23)), we obtain the following inequality which corresponds to (1.18.24)

$$-\Delta^m x(n) = |\Delta^m x(n)| \ge \left\{ q\sum_{\xi=a}^{b} T(n,\xi)[1-g(h(n,\xi))]w \right\}^{1/\alpha}, \quad n \in \mathbb{N}(n_3). \tag{1.18.27}$$

The rest of the argument follows as in Case (a). ∎

Theorem 1.18.5. Suppose that (1.18.12) – (1.18.14) hold, and there exists an eventually nondecreasing function $z : \mathbb{N} \to \mathbb{R}_0$ such that for any $\delta > 0$,

$$\sum_{n=c}^{\infty} \left\{ \left(\sum_{\xi=a}^{b} T(n,\xi)[1-g(h(n,\xi))] \right)^{1/\alpha} - \delta\Delta z(n) \right\} \;=\; \infty. \tag{1.18.28}$$

Then, the conclusions of Theorem 1.18.4 hold.

Proof. Case (a) m is even

Let $\{y(n)\}$ be an eventually positive solution of (1.18.1). As in the proof of Theorem 1.18.4, we have (1.18.16) – (1.18.24). We define

$$\phi(n) \;=\; z(n)\Delta^{m-1}x(n).$$

Then, it is clear from (1.18.20) that $\phi(n) > 0$, $n \ge n_1$. Further, using (1.18.24) as well as the fact that $z(n)$ is nondecreasing and $\Delta^{m-1}x(n)$ is nonincreasing (see (1.18.19)), we find for $n \ge n_3$,

$$\Delta\phi(n) = z(n+1)\Delta^m x(n) + \Delta^{m-1}x(n)\Delta z(n)$$

$$\le -z(n+1)\left\{ q\sum_{\xi=a}^{b} T(n,\xi)[1-g(h(n,\xi))]x(v) \right\}^{1/\alpha} + \Delta^{m-1}x(n)\Delta z(n)$$

$$\leq -z(n_3+1)\left\{q\sum_{\xi=a}^{b}T(n,\xi)[1-g(h(n,\xi))]x(v)\right\}^{1/\alpha} +\Delta^{m-1}x(n_3)\Delta z(n)$$

$$= -[q\,x(v)]^{1/\alpha}z(n_3+1)\left\{\left(\sum_{\xi=a}^{b}T(n,\xi)[1-g(h(n,\xi))]\right)^{1/\alpha} - \delta'\Delta z(n)\right\} \tag{1.18.29}$$

where $\delta' = [q\ x(v)]^{-1/\alpha}\Delta^{m-1}x(n_3)/z(n_3+1) > 0$. A summation of (1.18.29) from n_3 to $(N-1)$ yields

$$\phi(N) \leq \phi(n_3)-[qx(v)]^{1/\alpha}z(n_3+1)\sum_{n=n_3}^{N-1}\left\{\left(\sum_{\xi=a}^{b}T(n,\xi)[1-g(h(n,\xi))]\right)^{1/\alpha} -\delta'\Delta z(n)\right\}. \tag{1.18.30}$$

Letting $N \to \infty$ in (1.18.30), in view of (1.18.28) we get $\limsup_{N\to\infty}\phi(N) = -\infty$. This contradicts the fact that $\phi(n)$ is eventually positive.

Case (b) m is odd

Here, let $\{y(n)\}$ be an eventually positive and nondecreasing solution of (1.18.1). As in the proof of Theorem 1.18.4(b), we find that (1.18.27) holds. Now, using (1.18.27) instead of (1.18.24), a similar argument as in Case (a) leads to the following inequality which corresponds to (1.18.29)

$$\Delta\phi(n) \leq -(qw)^{1/\alpha}z(n_3+1)\left\{\left(\sum_{\xi=a}^{b}T(n,\xi)[1-g(h(n,\xi))]\right)^{1/\alpha} - \delta''\Delta z(n)\right\},$$

$$n \in \mathbb{N}(n_3)$$

where $\delta'' = (qw)^{-1/\alpha}\Delta^{m-1}x(n_3)/z(n_3+1) > 0$. The rest of the proof follows as in Case (a). ∎

Theorem 1.18.6. Suppose that (1.18.12), (1.18.13) hold,

$$\text{for } u \neq 0,\ uf(u) > 0, \tag{1.18.31}$$

$$f \text{ is nondecreasing}, \tag{1.18.32}$$

$$h(n,\xi) \text{ is nondecreasing in } n \tag{1.18.33}$$

and for any $\delta > 0$,

$$\sum_{n=c}^{\infty} \left\{ \sum_{\xi=a}^{b} T(n,\xi) f([1 - g(h(n,\xi))]\delta) \right\}^{1/\alpha} = \infty. \qquad (1.18.34)$$

(a) If m is even, then the conclusion of Theorem 1.18.4(a) holds.

(b) If m is odd and $g(n)$ is eventually nondecreasing, then the conclusion of Theorem 1.18.4(b) holds.

Proof. Once again let $\{y(n)\}$ be an eventually positive solution of (1.18.1). As before, we have (1.18.16) and (1.18.17). In view of (1.18.31), inequality (1.18.1) provides

$$|\Delta^m x(n)|^{\alpha-1} \Delta^m x(n) \le -\sum_{\xi=a}^{b} T(n,\xi) f(y(h(n,\xi))) \le 0, \quad n \in \mathbb{N}(n_1)$$

which gives rise to (1.18.19).

Case (a) m is even

By Lemma 1.13.1, we obtain (1.18.20) and (1.18.21). Therefore, it follows that

$$x(n-\sigma) \le x(n) = y(n)+g(n)y(n-\sigma) \le y(n)+g(n)x(n-\sigma), \quad n \in \mathbb{N}(n_2)$$

for some $n_2 > n_1$. The above inequality is equivalent to

$$[1 - g(n)]x(n-\sigma) \le y(n), \quad n \in \mathbb{N}(n_2). \qquad (1.18.35)$$

Next, in view of (1.18.3) we may choose $n_3 > n_2$ such that

$$h(n,\xi) \ge n_2, \quad x(h(n,\xi) - \sigma) > 0, \quad n \in \mathbb{N}(n_3), \ \xi \in \mathbb{N}(a,b). \quad (1.18.36)$$

Using the fact that f is nondecreasing as well as (1.18.35), (1.18.36), (1.18.21) and (1.18.33), from (1.18.1) we find for $n \in \mathbb{N}(n_3)$,

$$\begin{aligned} 0 &\ge -|\Delta^m x(n)|^{\alpha} + \sum_{\xi=a}^{b} T(n,\xi) f(y(h(n,\xi))) \\ &\ge -|\Delta^m x(n)|^{\alpha} + \sum_{\xi=a}^{b} T(n,\xi) f([1 - g(h(n,\xi))]x(h(n,\xi) - \sigma)) \\ &\ge -|\Delta^m x(n)|^{\alpha} + \sum_{\xi=a}^{b} T(n,\xi) f([1 - g(h(n,\xi))]\delta'), \end{aligned}$$

where $\delta' = x(h(n_3,\xi)-\sigma) > 0$. Hence, it follows that

$$-\Delta^m x(n) = |\Delta^m x(n)| \geq \left\{\sum_{\xi=a}^{b} T(n,\xi) f([1-g(h(n,\xi))]\delta')\right\}^{1/\alpha}, \quad n \in \mathbb{N}(n_3). \tag{1.18.37}$$

Summing (1.18.37) from n_3 to $(N-1)$ gives

$$\Delta^{m-1}x(N) - \Delta^{m-1}x(n_3) + \sum_{n=n_3}^{N-1}\left\{\sum_{\xi=a}^{b} T(n,\xi) f([1-g(h(n,\xi))]\delta')\right\}^{1/\alpha} \leq 0. \tag{1.18.38}$$

By (1.18.34) and (1.18.20), the right side of (1.18.38) tends to ∞ as $N \to \infty$. Therefore, we get a contradiction.

Case (b) m is odd

Once again let $\{y(n)\}$ be an eventually positive and nondecreasing solution of (1.18.1). In this case, from Lemma 1.13.1 we obtain (1.18.20) but not (1.18.21). However, from definition (1.18.17) we note that

$$\Delta x(n) = \Delta y(n) + g(n+1)\Delta y(n-\sigma) + y(n-\sigma)\Delta g(n). \tag{1.18.39}$$

Since $y(n)$ and $g(n)$ (≥ 0) are eventually nondecreasing, it is clear from (1.18.39) that (1.18.21) holds. The proof then follows as in Case (a). ∎

Theorem 1.18.7. Let $\alpha \geq 1$. Suppose that (1.18.13), (1.18.31), (1.18.32) hold,

$$\text{for } u,v > 0, \ f(u+v) \leq f(u) + f(v), \tag{1.18.40}$$

$$\text{for } \delta, u > 0, \ f(\delta u) \leq \delta \ f(u), \tag{1.18.41}$$

$$g(n) = g = \text{constant}\begin{cases} \geq 1, & \alpha > 1 \\ > 0, & \alpha = 1 \end{cases} \tag{1.18.42}$$

$$T(n,\xi) \text{ is periodic in } n \text{ with period } \sigma, \tag{1.18.43}$$

$$\text{for } \delta > 0, \ h(n-\delta,\xi) = h(n,\xi) - \delta \tag{1.18.44}$$

and

$$\sum_{n=c}^{\infty}\left[\sum_{\xi=a}^{b} T(n,\xi)\right]^{1/\alpha} = \infty. \tag{1.18.45}$$

Then, the conclusions of Theorem 1.18.4 hold.

Proof. Again suppose that $\{y(n)\}$ is an eventually positive solution of (1.18.1). As in the proof of Theorem 1.18.6, we have (1.18.16), (1.18.17) and (1.18.19).

Case (a) m is even

In this case, from Lemma 1.13.1 we obtain (1.18.20) and (1.18.21). Let

$$\phi(n) \;=\; x(n) + g\, x(n-\sigma).$$

Clearly, $\phi(n)$ is eventually positive. Further, in view of (1.18.20) and (1.18.21) respectively, there exists $n_2 > n_1$ such that

$$\Delta^{m-1}\phi(n) \;=\; \Delta^{m-1}x(n) + g\Delta^{m-1}x(n-\sigma) \;>\; 0, \quad n \in \mathbb{N}(n_2) \quad (1.18.46)$$

and

$$\Delta\phi(n) \;=\; \Delta x(n) + g\Delta x(n-\sigma) > 0, \quad n \in \mathbb{N}(n_2). \qquad (1.18.47)$$

Next, on noting (1.18.19) the difference inequality (1.18.1) is the same as

$$-|\Delta^m x(n)|^\alpha \;\le\; -\sum_{\xi=a}^{b} T(n,\xi) f(y(h(n,\xi)))$$

or

$$\Delta^m x(n) \;\le\; -\left[\sum_{\xi=a}^{b} T(n,\xi) f(y(h(n,\xi)))\right]^{1/\alpha}, \quad n \in \mathbb{N}(n_1). \qquad (1.18.48)$$

Using (1.18.48), we find for $n \ge n_3$ $(> n_2)$,

$$\Delta^m\phi(n) = \Delta^m x(n) + g\Delta^m x(n-\sigma)$$

$$\le -\left[\sum_{\xi=a}^{b} T(n,\xi) f(y(h(n,\xi)))\right]^{1/\alpha} - g\left[\sum_{\xi=a}^{b} T(n-\sigma,\xi) f(y(h(n-\sigma,\xi)))\right]^{1/\alpha}. \qquad (1.18.49)$$

Choose $n_4 > n_3$ such that $y(h(n-\sigma,\xi)),\ y(h(n-2\sigma,\xi)) > 0$, $n \in \mathbb{N}(n_4)$. Then, applying (1.18.49) and (1.18.43) we get

$$\Delta^m\phi(n) + g\Delta^m\phi(n-\sigma)$$

$$\le -\left[\sum_{\xi=a}^{b} T(n,\xi) f(y(h(n,\xi)))\right]^{1/\alpha} - 2g\left[\sum_{\xi=a}^{b} T(n-\sigma,\xi) f(y(h(n-\sigma,\xi)))\right]^{1/\alpha}$$

$$-\; g^2\left[\sum_{\xi=a}^{b} T(n-2\sigma,\xi) f(y(h(n-2\sigma,\xi)))\right]^{1/\alpha}$$

$$= -\left[\sum_{\xi=a}^{b} T(n,\xi) f(y(h(n,\xi)))\right]^{1/\alpha} - 2g\left[\sum_{\xi=a}^{b} T(n,\xi) f(y(h(n-\sigma,\xi)))\right]^{1/\alpha}$$

$$- \; g^2 \left[\sum_{\xi=a}^{b} T(n,\xi) f(y(h(n-2\sigma,\xi)))\right]^{1/\alpha}, \quad n \in \mathbb{N}(n_4). \qquad (1.18.50)$$

Now, from definition (1.18.17) we have

$$\phi(n) \;=\; x(n) + g\, x(n-\sigma) \;=\; y(n) + 2g\, y(n-\sigma) + g^2 y(n-2\sigma). \quad (1.18.51)$$

Hence, in view of (1.18.40), (1.18.41) and (1.18.44) it follows that

$$\begin{aligned}
&\sum_{\xi=a}^{b} T(n,\xi) f(\phi(h(n,\xi))) \\
&= \sum_{\xi=a}^{b} T(n,\xi) f(y(h(n,\xi)) + 2g\, y(h(n,\xi)-\sigma) + g^2 y(h(n,\xi)-2\sigma)) \\
&\le \sum_{\xi=a}^{b} T(n,\xi) \left[f(y(h(n,\xi)) + 2g f(y(h(n-\sigma,\xi)) + g^2 f(y(h(n-2\sigma,\xi)))\right], \\
&\hspace{20em} n \in \mathbb{N}(n_4). \quad (1.18.52)
\end{aligned}$$

Denoting $A = \sum_{\xi=a}^{b} T(n,\xi) f(y(h(n,\xi))$, $B = \sum_{\xi=a}^{b} T(n,\xi) f(y(h(n-\sigma,\xi))$ and $C = \sum_{\xi=a}^{b} T(n,\xi) f(y(h(n-2\sigma,\xi))$, from (1.18.50) and (1.18.52) we find for $n \in \mathbb{N}(n_4)$,

$$\begin{aligned}
&\left[\sum_{\xi=a}^{b} T(n,\xi) f(\phi(h(n,\xi)))\right]^{1/\alpha} + \Delta^m \phi(n) + g \Delta^m \phi(n-\sigma) \\
&\qquad \le (A + 2gB + g^2 C)^{1/\alpha} - A^{1/\alpha} - 2gB^{1/\alpha} - g^2 C^{1/\alpha} \le 0, \quad (1.18.53)
\end{aligned}$$

where we have also made used of condition (1.18.42) in the last inequality.

Let v and $n_5 > n_4$ be such that $h(n,\xi) > v$, $n \ge n_5$, $\xi \in \mathbb{N}(a,b)$ and $\phi(v) > 0$. Then, it is clear from (1.18.47), (1.18.32) and (1.18.31) that

$$\phi(h(n,\xi)) \;>\; \phi(v), \quad f(\phi(h(n,\xi))) \;\ge\; f(\phi(v)) \;>0, \quad n \in \mathbb{N}(n_5),\ \xi \in \mathbb{N}(a,b).$$

Therefore, it is immediate from (1.18.53) that

$$\left[\sum_{\xi=a}^{b} T(n,\xi) f(\phi(v))\right]^{1/\alpha} + \Delta^m \phi(n) + g \Delta^m \phi(n-\sigma) \;\le\; 0, \quad n \in \mathbb{N}(n_5).$$

Summing the above relation from n_5 to $(N-1)$, we obtain

$$[f(\phi(v))]^{1/\alpha} \sum_{n=n_5}^{N-1} \left[\sum_{\xi=a}^{b} T(n,\xi)\right]^{1/\alpha} + \Delta^{m-1}\phi(N) - \Delta^{m-1}\phi(n_5)$$

$$+\ g\Delta^{m-1}\phi(n-\sigma) - g\Delta^{m-1}\phi(n_5-\sigma)\ \le\ 0.$$

In view of (1.18.46) and also (1.18.45), the left side of the above inequality tends to ∞ as $N \to \infty$. Hence, we get a contradiction.

Case (b) m is odd

Let $\{y(n)\}$ be an eventually positive and nondecreasing solution of (1.18.1). Here, from Lemma 1.13.1 we cannot obtain (1.18.21) and hence (1.18.47). However, using the fact that $\Delta y(n)$ is eventually positive in expression (1.18.51) immediately yields the relation (1.18.47). Therefore, the proof proceeds as in Case (a). ∎

Theorem 1.18.8. Suppose that (1.18.13), (1.18.31), (1.18.32), (1.18.45) hold, and

$$-1 \le g(n) \le 0, \quad n \in \mathbb{N}(n_0). \tag{1.18.54}$$

(a) If m is even, then (1.18.1) ((1.18.2)) has no eventually positive (negative) and unbounded solution. Further, all nonoscillatory solutions of (E_{49}) are bounded, and all unbounded solutions of (E_{49}) are oscillatory.

(b) If m is odd and $g(n)$ is eventually nondecreasing, then (1.18.1) ((1.18.2)) has no eventually positive (negative) and concave (convex) solution. Further, (E_{49}) has neither eventually positive and concave nor eventually negative and convex solutions.

Proof. Let $\{y(n)\}$ be an eventually positive and unbounded solution of (1.18.1). As in the proof of Theorem 1.18.6, we have (1.18.16), (1.18.17) and (1.18.19). Further, it is noted that $x(n)$ is eventually positive, for if $x(n) \le 0$, $n \ge n_1$, then it follows from (1.18.17) and (1.18.54) that

$$y(n)\ \le\ -g(n)y(n-\sigma)\ \le\ y(n-\sigma), \quad n \in \mathbb{N}(n_1)$$

and hence $y(n)$ will not be unbounded.

Case (a) m is even

In view of (1.18.19) and the fact that $x(n)$ is eventually positive, from Lemma 1.13.1 we get (1.18.20) and (1.18.21).

Next, noting (1.18.3) we may choose $n_2 > n_1$ so that $h(n,\xi) \ge v$, $n \ge n_2$, $\xi \in \mathbb{N}(a,b)$, where v is sufficiently large enough such that $x(v) > 0$. Subsequently, in view of (1.18.21) and (1.18.32) we have

$$f(x(h(n,\xi)))\ \ge\ f(x(v)) > 0, \quad n \in \mathbb{N}(n_2),\ \xi \in \mathbb{N}(a,b). \tag{1.18.55}$$

Using the fact that $y(n) \ge x(n)$, f is nondecreasing and also (1.18.55), it follows from (1.18.1) that for $n \geq n_2$,

$$\begin{aligned} -|\Delta^m x(n)|^\alpha &\leq -\sum_{\xi=a}^{b} T(n,\xi) f(y(h(n,\xi))) \\ &\leq -\sum_{\xi=a}^{b} T(n,\xi) f(x(h(n,\xi))) \leq -\sum_{\xi=a}^{b} T(n,\xi) f(x(v)) \end{aligned}$$

or

$$\Delta^m x(n) \leq -\left[f(x(v)) \sum_{\xi=a}^{b} T(n,\xi)\right]^{1/\alpha}, \quad n \in \mathbb{N}(n_2). \tag{1.18.56}$$

Summing (1.18.56) from n_2 to $(N-1)$ yields

$$\Delta^{m-1}x(N) - \Delta^{m-1}x(n_2) + [f(x(v))]^{1/\alpha} \sum_{n=n_2}^{N-1} \left[\sum_{\xi=a}^{b} T(n,\xi)\right]^{1/\alpha} \leq 0.$$

Clearly, by (1.18.20) and (1.18.45) the right side of the above inequality tends to ∞ as $N \to \infty$. Hence, we get a contradiction.

Case (b) m is odd

Let $\{y(n)\}$ be an eventually positive and concave solution of (1.18.1), i.e. $\Delta^2 y(n) > 0$ for all large n. Here, from Lemma 1.13.1 we cannot conclude that (1.18.21) is true. However, it follows from expression (1.18.39) that (1.18.21) holds if

$$\Delta y(n) + g(n+1)\Delta y(n-\sigma) > 0 \tag{1.18.57}$$

eventually. Since $y(n)$ is eventually concave and $-1 \leq g(n+1) \leq 0$, we see that (1.18.57) is fulfilled. Hence, we get (1.18.21). The rest of the proof proceeds as in Case (a). ∎

Theorem 1.18.9. Let $\alpha \leq 1$. Suppose that (1.18.12) – (1.18.14), (1.18.33) hold,

$$h(n,\xi) \text{ is monotone in } \xi, \tag{1.18.58}$$

and there exists an eventually nondecreasing function $z : \mathbb{N} \to \mathbb{R}_0$ such that for any $\delta, \gamma > 0$,

$$\sum_{n=c}^{\infty} \left\{ \left(\gamma \sum_{\xi=a}^{b} T(n,\xi)[1 - g(h(n,\xi))]\right)^{1/\alpha} z(n+1) - \delta \Delta z(n) \right\} = \infty. \tag{1.18.59}$$

Then, the conclusions of Theorem 1.18.6 hold.

Proof. Case (a) m is even

Let $\{y(n)\}$ be an eventually positive solution of (1.18.1). As in the proof of Theorem 1.18.4, we have (1.18.16) – (1.18.22). Further, it is clear from (1.18.58) that

$$h(n,\xi) \ \geq \ \min\{h(n,a),\ h(n,b)\} \ \equiv \ H(n), \quad \xi \in \mathbb{N}(a,b)$$

which in view of (1.18.21) leads to

$$x(h(n,\xi)) \ \geq \ x(H(n)) \ > \ 0, \quad n \in \mathbb{N}(n_3),\ \xi \in \mathbb{N}(a,b)$$

for some $n_3 > n_2$. Using the above inequality together with (1.18.19), (1.18.12) and (1.18.22) in relation (1.18.18), we get for $n \geq n_3$,

$$\begin{aligned} |\Delta^m x(n)|^\alpha \ &\geq \ q\sum_{\xi=a}^{b} T(n,\xi)y(h(n,\xi)) \\ &= \ q\sum_{\xi=a}^{b} T(n,\xi)[x(h(n,\xi)) - g(h(n,\xi))y(h(n,\xi)-\sigma)] \\ &\geq \ q\sum_{\xi=a}^{b} T(n,\xi)[1-g(h(n,\xi))]x(h(n,\xi)) \\ &\geq \ q\sum_{\xi=a}^{b} T(n,\xi)[1-g(h(n,\xi))]x(H(n)) \end{aligned}$$

or

$$\Delta^m x(n) \ \leq \ -\left\{q\ x(H(n))\sum_{\xi=a}^{b} T(n,\xi)[1-g(h(n,\xi))]\right\}^{1/\alpha}, \quad n \in \mathbb{N}(n_3). \tag{1.18.60}$$

Now, we define

$$\phi(n) \ = \ \frac{z(n)\Delta^{m-1}x(n)}{x(H(n-1))}.$$

Then, it is clear from (1.18.20) that $\phi(n)$ is eventually positive. In view of (1.18.20), (1.18.21) and also (1.18.60), we find for $n \geq n_4$ $(> n_3)$,

$$\begin{aligned} \Delta\phi(n) &= \frac{\begin{array}{r}\{x(H(n-1))\left[\Delta^{m-1}x(n)\Delta z(n) + z(n+1)\Delta^m x(n)\right] \\ -z(n)\Delta^{m-1}x(n)\Delta x(H(n-1))\}\end{array}}{x(H(n-1))\ x(H(n))} \\ &\leq \frac{\Delta^{m-1}x(n)\Delta z(n) + z(n+1)\Delta^m x(n)}{x(H(n))} \end{aligned}$$

$$\leq \frac{\Delta^{m-1}x(n)\Delta z(n)}{x(H(n))} - \frac{\left\{qx(H(n))\sum_{\xi=a}^{b}T(n,\xi)[1-g(h(n,\xi))]\right\}^{1/\alpha} \times z(n+1)}{x(H(n))} \tag{1.18.61}$$

Further, using (1.18.21), (1.18.33) as well as the fact that $\Delta^{m-1}x(n)$ is nonincreasing (see (1.18.19)), for $n \geq n_4$ we have

$$\begin{aligned} \Delta^{m-1}x(n) \ \leq \ \Delta^{m-1}x(n_4), \quad x(H(n)) \ \geq \ x(H(n_4)), \\ [x(H(n))]^{1/\alpha-1} \ \geq \ [x(H(n_4))]^{1/\alpha-1}. \end{aligned} \tag{1.18.62}$$

An application of (1.18.62) in (1.18.61) provides

$$\begin{aligned} \Delta\phi(n) \ \leq \ [x(H(n_4))]^{-1} \Bigg\{ \delta'\Delta z(n) \\ - \left(\gamma' \sum_{\xi=a}^{b} T(n,\xi)[1-g(h(n,\xi))]\right)^{1/\alpha} z(n+1) \Bigg\}, \\ n \in \mathbb{N}(n_4) \end{aligned} \tag{1.18.63}$$

where $\delta' = \Delta^{m-1}x(n_4)$, $\gamma' = q\, x(H(n_4)) > 0$.

Summing (1.18.63) from n_4 to $(N-1)$ yields

$$\begin{aligned} \phi(N) \ \leq \ \phi(n_4) - [x(H(n_4))]^{-1} \sum_{n=n_4}^{N-1} \Bigg\{ \delta'\Delta z(n) \\ - \left(\gamma' \sum_{\xi=a}^{b} T(n,\xi)[1-g(h(n,\xi))]\right)^{1/\alpha} z(n+1) \Bigg\}. \end{aligned}$$

By (1.18.59), the right side of the above inequality tends to $-\infty$ as $N \to \infty$. This contradicts the fact that $\phi(n)$ is eventually positive.

Case (b) m is odd

The proof is similar to that of Theorem 1.18.6(b). ∎

Theorem 1.18.10. Let $\alpha = 1$. Suppose that (1.18.14), (1.18.33) hold,

$$h(n,\xi) \leq n, \quad \xi \in \mathbb{N}(a,b) \tag{1.18.64}$$

$$g(n) \equiv 0, \quad n \in \mathbb{N}(n_0) \tag{1.18.65}$$

$$(-1)^{m+1}T(n,\xi) \geq 0, \quad n \in \mathbb{N}(n_0),\ \xi \in \mathbb{N}(a,b) \tag{1.18.66}$$

and

$$(-1)^{m+1} q \liminf_{\ell\to\infty} \sum_{n=h^*(\ell)}^{\ell} \sum_{\xi=a}^{b} T(n,\xi) \frac{(h(\ell,\xi)-h(n,\xi)+m-2)^{(m-1)}}{(m-1)!} \ge 1, \tag{1.18.67}$$

where $h^*(\ell) = \max_{\xi\in\mathbb{N}(a,b)} h(\ell,\xi)$. Then, (1.18.1) ((1.18.2)) has no eventually positive (negative) and bounded solution. Further, all nonoscillatory solutions of (E_{49}) are unbounded, and all bounded solutions of (E_{49}) are oscillatory.

Proof. Let $\{y(n)\}$ be an eventually positive and bounded solution of (1.18.1). Then, there exists $n_1 > n_0$ such that $y(h(n,\xi)) > 0,\ n \ge n_1,\ \xi \in \mathbb{N}(a,b)$. Let m be odd. We shall only provide the proof for this case as the arguments are similar when m is even.

Using (1.18.66) and (1.18.14), from (1.18.1) we find

$$\Delta^m y(n) \le -\sum_{\xi=a}^{b} T(n,\xi) f(y(h(n,\xi))) \le -q\sum_{\xi=a}^{b} T(n,\xi) y(h(n,\xi)) \le 0,$$

$$n \in \mathbb{N}(n_1). \tag{1.18.68}$$

Consequently, it follows from Lemma 1.13.1 that

$$(-1)^i \Delta^i y(n) \ge 0, \quad n \in \mathbb{N}(n_1),\ 1 \le i \le m-1. \tag{1.18.69}$$

Let $r \ge s \ge n_1$. Then, by discrete Taylor's formula [2] and (1.18.68) we have

$$y(s) = \sum_{i=0}^{m-1} \frac{(r-s+i-1)^{(i)}}{i!} (-1)^i \Delta^i y(r) - \sum_{\ell=s}^{r-1} \frac{(\ell-s+m-1)^{(m-1)}}{(m-1)!} (-1)^{m-1} \Delta^m y(\ell)$$

$$\ge \sum_{i=0}^{m-1} \frac{(r-s+i-1)^{(i)}}{i!} (-1)^i \Delta^i y(r). \tag{1.18.70}$$

Now, let $n, \ell\ (\ge n \ge n_1)$ be sufficiently large such that

$$r \equiv h(\ell,\xi) \ge h(n,\xi) \equiv s \ge n_1, \quad \xi \in \mathbb{N}(a,b).$$

Then, applying (1.18.70) in (1.18.68) we find

$$\Delta^m y(n) \le -q\sum_{\xi=a}^{b} T(n,\xi) y(h(n,\xi))$$

$$\le -q\sum_{\xi=a}^{b} T(n,\xi) \left[\sum_{i=0}^{m-1} \frac{(h(\ell,\xi)-h(n,\xi)+i-1)^{(i)}}{i!} (-1)^i \Delta^i y(h(\ell,\xi))\right],$$

which on summing from $h^*(\ell)$ to ℓ ($\ge h^*(\ell)$ by (1.18.64)) gives

$$\begin{aligned}
&\Delta^{m-1}y(\ell+1) - \Delta^{m-1}y(h^*(\ell))\\
&\le -q\sum_{n=h^*(\ell)}^{\ell}\sum_{\xi=a}^{b}T(n,\xi)\left[\sum_{i=0}^{m-1}\frac{(h(\ell,\xi)-h(n,\xi)+i-1)^{(i)}}{i!}(-1)^i\Delta^i y(h(\ell,\xi))\right]\\
&\le -q\sum_{n=h^*(\ell)}^{\ell}\sum_{\xi=a}^{b}T(n,\xi)\left[\sum_{i=0}^{m-2}\frac{(h(\ell,\xi)-h(n,\xi)+i-1)^{(i)}}{i!}(-1)^i\Delta^i y(h(\ell,\xi))\right]\\
&\quad -q\sum_{n=h^*(\ell)}^{\ell}\sum_{\xi=a}^{b}T(n,\xi)\frac{(h(\ell,\xi)-h(n,\xi)+m-2)^{(m-1)}}{(m-1)!}(-1)^{m-1}\Delta^{m-1}y(h^*(\ell))
\end{aligned} \tag{1.18.71}$$

where in the last inequality we have also used the fact that $\Delta^{m-1}y(n)$ is nonincreasing (by (1.18.68)) and so

$$(-1)^{m-1}\Delta^{m-1}y(h^*(\ell)) \le (-1)^{m-1}\Delta^{m-1}y(h(\ell,\xi)).$$

Relation (1.18.71) is actually equivalent to

$$\begin{aligned}
&\Delta^{m-1}y(\ell+1)\\
&\le -q\sum_{n=h^*(\ell)}^{\ell}\sum_{\xi=a}^{b}T(n,\xi)\left[\sum_{i=0}^{m-2}\frac{(h(\ell,\xi)-h(n,\xi)+i-1)^{(i)}}{i!}(-1)^i\Delta^i y(h(\ell,\xi))\right]\\
&\quad -(-1)^{m-1}\Delta^{m-1}y(h^*(\ell))\times\\
&\quad \left[q\sum_{n=h^*(\ell)}^{\ell}\sum_{\xi=a}^{b}T(n,\xi)\frac{(h(\ell,\xi)-h(n,\xi)+m-2)^{(m-1)}}{(m-1)!}-1\right].
\end{aligned} \tag{1.18.72}$$

By (1.18.69) and (1.18.67), all the terms in the right side of (1.18.72) are nonpositive. However, the left side of (1.18.72) is nonnegative by (1.18.69). Hence, we get a contradiction. ∎

Theorem 1.18.11. Suppose that $y(i,n)$ is an eventually positive solution of (E_{50}), (1.18.6) and

$$f \text{ is a positive and convex function on } \mathbb{R}_0. \tag{1.18.73}$$

Then, the inequality

$$\Delta_n^m[u(n)+g(n)u(n-\sigma)]+\sum_{\xi=a}^{b}\left[\min_{i\in\Omega}\Lambda(i,n,\xi)\right]f(u(h(n,\xi)))\le 0,\quad n\in\mathbb{N}(n_0) \tag{1.18.74}$$

has an eventually positive solution given by

$$u(n) = \frac{1}{\rho}\sum_{i\in\Omega}y(i,n)\psi(i) \tag{1.18.75}$$

where $\psi(i)$ is defined in Lemma 1.18.1 and $\rho = \sum_{i\in\Omega}\psi(i)$.

Proof. Since $y(i,n)$ is eventually positive, there exists $n_1 > n_0$ such that

$$y(i,n) > 0, \; y(i,n-\sigma) > 0, \; y(i,h(n,\xi)) > 0, \; y(i,\tau_\ell(n)) > 0,$$
$$i \in \Omega, \; n \geq n_1, \; \xi \in \mathbb{N}(a,b), \; 1 \leq \ell \leq \mu. \quad (1.18.76)$$

Multiplying (E_{50}) by the eigenfunction $\psi(i)$ (defined in Lemma 1.18.1) and then summing over i, we get

$$\Delta_n^m\left[\sum_{i\in\Omega} y(i,n)\psi(i) + g(n)\sum_{i\in\Omega} y(i,n-\sigma)\psi(i)\right]$$
$$+\sum_{i\in\Omega}\sum_{\xi=a}^{b}\Lambda(i,n,\xi)f(y(i,h(n,\xi)))\psi(i)$$
$$= d(n)\sum_{i\in\Omega}\psi(i)Ly(i,n) + \sum_{\ell=1}^{\mu}d_\ell(n)\sum_{i\in\Omega}\psi(i)Ly(i,\tau_\ell(n)), \quad n \in \mathbb{N}(n_1).$$
$$(1.18.77)$$

Now, it is clear from Lemma 1.18.2 that

$$\sum_{i\in\Omega}\psi(i)Ly(i,n) - \sum_{i\in\Omega}y(i,n)L\psi(i)$$
$$= \sum_{\ell=1}^{\wp}\left\{\sum_{i_1=1}^{N_1}\cdots\sum_{i_{\ell-1}=1}^{N_{\ell-1}}\sum_{i_{\ell+1}=1}^{N_{\ell+1}}\cdots\sum_{i_\wp=1}^{N_\wp}\left[\psi(i)\Delta_{i_\ell}y(i,n) - y(i,n)\Delta_{i_\ell}\psi(i)\right]_{i_\ell=0}^{N_\ell}\right\}$$
$$= \sum_{\ell=1}^{\wp}\left\{\sum_{i_1=1}^{N_1}\cdots\sum_{i_{\ell-1}=1}^{N_{\ell-1}}\sum_{i_{\ell+1}=1}^{N_{\ell+1}}\cdots\sum_{i_\wp=1}^{N_\wp}\left(\psi(i_1,\cdots,i_{\ell-1},N_\ell,i_{\ell+1},\cdots,i_\wp)\times\right.\right.$$
$$\left[-y\left((i_1,\cdots,i_{\ell-1},N_\ell,i_{\ell+1},\cdots,i_\wp),n\right)\right]$$
$$\left.\left.-y\left((i_1,\cdots,i_{\ell-1},N_\ell,i_{\ell+1},\cdots,i_\wp),n\right)\left[-\psi(i_1,\cdots,i_{\ell-1},N_\ell,i_{\ell+1},\cdots,i_\wp)\right]\right)\right\}$$
$$= 0, \quad (1.18.78)$$

where we have used the boundary condition (1.18.6) in the second last equality. Subsequently, in view of Lemma 1.18.1 relation (1.18.78) further reduces to

$$\sum_{i\in\Omega}\psi(i)Ly(i,n) = \sum_{i\in\Omega}y(i,n)L\psi(i) = -\lambda_0\sum_{i\in\Omega}y(i,n)\psi(i). \quad (1.18.79)$$

Similarly, for each $1 \le \ell \le \mu$ we find that

$$\sum_{i\in\Omega} \psi(i) L y(i,\tau_\ell(n)) = -\lambda_0 \sum_{i\in\Omega} y(i,\tau_\ell(n))\psi(i). \qquad (1.18.80)$$

Further, an application of Lemma 1.18.3 provides

$$\sum_{i\in\Omega} f(y(i,h(n,\xi)))\psi(i) \ge \rho f\left(\frac{1}{\rho}\sum_{i\in\Omega} y(i,h(n,\xi))\psi(i)\right), \quad n \in \mathbb{N}(n_1). \qquad (1.18.81)$$

Using (1.18.81), we find

$$\begin{aligned} &\sum_{i\in\Omega}\sum_{\xi=a}^{b} \Lambda(i,n,\xi) f(y(i,h(n,\xi)))\psi(i) \\ &\ge \sum_{\xi=a}^{b}\left[\min_{i\in\Omega}\Lambda(i,n,\xi)\right]\sum_{i\in\Omega} f(y(i,h(n,\xi)))\psi(i) \\ &\ge \rho\sum_{\xi=a}^{b}\left[\min_{i\in\Omega}\Lambda(i,n,\xi)\right] f\left(\frac{1}{\rho}\sum_{i\in\Omega} y(i,h(n,\xi))\psi(i)\right), \quad n \in \mathbb{N}(n_1). \end{aligned} \qquad (1.18.82)$$

In view of (1.18.82), (1.18.79) and (1.18.80), it now follows from (1.18.77) that

$$\begin{aligned} &\Delta_n^m\left[\sum_{i\in\Omega} y(i,n)\psi(i) + g(n)\sum_{i\in\Omega} y(i,n-\sigma)\psi(i)\right] \\ &+\rho\sum_{\xi=a}^{b}\left[\min_{i\in\Omega}\Lambda(i,n,\xi)\right] f\left(\frac{1}{\rho}\sum_{i\in\Omega} y(i,h(n,\xi))\psi(i)\right) \\ &\le d(n)\left[-\lambda_0\sum_{i\in\Omega} y(i,n)\psi(i)\right] + \sum_{\ell=1}^{\mu} d_\ell(n)\left[-\lambda_0\sum_{i\in\Omega} y(i,\tau_\ell(n))\psi(i)\right] \le 0, \\ &\qquad n \in \mathbb{N}(n_1). \end{aligned}$$

It is clear from the above relation that the sequence defined in (1.18.75) is an eventually positive solution of (1.18.74). ∎

Theorem 1.18.12. Suppose that (1.18.73) holds and $y(i,n)$ is an eventually positive solution of (E_{50}), (1.18.8). Then, the inequality (1.18.74) has an eventually positive solution given by

$$u(n) = \frac{1}{|\Omega|}\sum_{i\in\Omega} y(i,n), \qquad (1.18.83)$$

where $|\Omega| = \sum_{i\in\Omega} 1 = N_1 N_2 \cdots N_{\wp}$.

Proof. Since $y(i,n)$ is eventually positive, there exists $n_1 > n_0$ such that (1.18.76) holds.

A summation of (E_{50}) over i provides

$$\Delta_n^m \left[\sum_{i\in\Omega} y(i,n) + g(n)\sum_{i\in\Omega} y(i,n-\sigma)\right] + \sum_{i\in\Omega}\sum_{\xi=a}^{b} \Lambda(i,n,\xi) f(y(i,h(n,\xi)))$$

$$= d(n)\sum_{i\in\Omega} Ly(i,n) + \sum_{\ell=1}^{\mu} d_\ell(n)\sum_{i\in\Omega} Ly(i,\tau_\ell(n)), \quad n \in \mathbb{N}(n_1). \tag{1.18.84}$$

Next, applying Lemma 1.18.2 and also the boundary condition (1.18.8), we find

$$\sum_{i\in\Omega} Ly(i,n) = \sum_{\ell=1}^{\wp}\left\{\sum_{i_1=1}^{N_1}\cdots\sum_{i_{\ell-1}=1}^{N_{\ell-1}}\sum_{i_{\ell+1}=1}^{N_{\ell+1}}\cdots\sum_{i_\wp=1}^{N_\wp}\left[\Delta_{i_\ell} y(i,n)\right]_{i_\ell=0}^{N_\ell}\right\}$$

$$= \sum_{\ell=1}^{\wp}\left\{\sum_{i_1=1}^{N_1}\cdots\sum_{i_{\ell-1}=1}^{N_{\ell-1}}\sum_{i_{\ell+1}=1}^{N_{\ell+1}}\cdots\sum_{i_\wp=1}^{N_\wp}\left[-\theta_\ell(i,n)y(i,n)\big|_{i_\ell=N_\ell+1}\right.\right.$$

$$\left.\left. -\ \theta_\ell(i,n)y(i,n)\big|_{i_\ell=0}\right]\right\} \equiv A(n). \tag{1.18.85}$$

Likewise, for each $1 \le \ell \le \mu$ we have

$$\sum_{i\in\Omega} Ly(i,\tau_\ell(n)) = A(\tau_\ell(n)). \tag{1.18.86}$$

Further, it follows from Lemma 1.18.3 that for $n \ge n_1$,

$$\sum_{i\in\Omega} f(y(i,h(n,\xi)))\cdot 1 \ \ge\ |\Omega|\, f\left(\frac{1}{|\Omega|}\sum_{i\in\Omega} y(i,h(n,\xi))\right). \tag{1.18.87}$$

In view of (1.18.87), we find that

$$\sum_{i\in\Omega}\sum_{\xi=a}^{b} \Lambda(i,n,\xi) f(y(i,h(n,\xi)))$$

$$\ge \sum_{\xi=a}^{b}\left[\min_{i\in\Omega} \Lambda(i,n,\xi)\right]\sum_{i\in\Omega} f(y(i,h(n,\xi)))$$

$$\geq \; |\Omega| \sum_{\xi=a}^{b} \left[\min_{i\in\Omega} \Lambda(i,n,\xi)\right] f\left(\frac{1}{|\Omega|}\sum_{i\in\Omega} y(i,h(n,\xi))\right), \quad n \in \mathbb{N}(n_1). \quad (1.18.88)$$

Noting (1.18.88), (1.18.85) and (1.18.86), from (1.18.84) we obtain

$$\Delta_n^m \left[\sum_{i\in\Omega} y(i,n) + g(n)\sum_{i\in\Omega} y(i,n-\sigma)\right]$$
$$+ |\Omega| \sum_{\xi=a}^{b} \left[\min_{i\in\Omega} \Lambda(i,n,\xi)\right] f\left(\frac{1}{|\Omega|}\sum_{i\in\Omega} y(i,h(n,\xi))\right)$$
$$\leq \; d(n)A(n) + \sum_{\ell=1}^{\mu} d_\ell(n)A(\tau_\ell(n)) \; \leq \; 0, \quad n \in \mathbb{N}(n_1).$$

Hence, it is clear from the above relation that the sequence defined in (1.18.83) is an eventually positive solution of (1.18.74). ∎

Theorem 1.18.13. Suppose that (1.18.73), (1.18.12) - (1.18.14) hold, and

$$\sum_{n=c}^{\infty}\sum_{\xi=a}^{b} \left[\min_{i\in\Omega} \Lambda(i,n,\xi)\right] [1 - g(h(n,\xi))] \; = \; \infty. \qquad (1.18.89)$$

(a) If m is even, then all solutions of (E_{50}), (1.18.6) and (E_{50}), (1.18.8) are oscillatory.

(b) If m is odd, then (E_{50}), (1.18.6) and (E_{50}), (1.18.8) have no eventually positive (negative) and nondecreasing (nonincreasing) solution. Hence, all nonoscillatory solutions of (E_{50}), (1.18.6) and (E_{50}), (1.18.8) are bounded.

Proof. Case (a) m is even

Suppose that $\{y(i,n)\}$ is a nonoscillatory solution of (E_{50}), (1.18.6), say $y(i,n)$ is eventually positive. We shall consider only this case because if $y(i,n)$ is eventually negative, then by (1.18.13) the sequence $\{-y(i,n)\}$ is an eventually positive solution of (E_{50}), (1.18.6). It follows from Theorem 1.18.11 that the sequence defined in (1.18.75) is an eventually positive solution of (1.18.74). However, by Theorem 1.18.4(a) the inequality (1.18.74) has no eventually positive solution. Hence, we get a contradiction.

Case (b) m is odd

In this case, let $y(i,n)$ be an eventually positive and nondecreasing solution of (E_{50}), (1.18.6). Then, once again by Theorem 1.18.11 the sequence defined in (1.18.75) is an eventually positive and nondecreasing solution of (1.18.74). However, by Theorem 1.18.4(b) the inequality (1.18.74) has no eventually positive and nondecreasing solution.

This completes the proof for the boundary value problem (E_{50}), (1.18.6). The proof for (E_{50}), (1.18.8) is similar and makes use of Theorem 1.18.12 instead of Theorem 1.18.11. ∎

Using Theorems 1.18.5 – 1.18.12 and a similar argument as in the proof of Theorem 1.18.13, we obtain the following results.

Theorem 1.18.14. Suppose that (1.18.73), (1.18.12) – (1.18.14) hold, and there exists an eventually nondecreasing function $z : \mathbb{N} \to \mathbb{R}_0$ such that for any $\delta > 0$,

$$\sum_{n=c}^{\infty} \left\{ \sum_{\xi=a}^{b} \left[\min_{i \in \Omega} \Lambda(i, n, \xi) \right] [1 - g(h(n, \xi))] - \delta \Delta z(n) \right\} = \infty. \qquad (1.18.90)$$

Then, the conclusions of Theorem 1.18.13 hold.

Theorem 1.18.15. Suppose that (1.18.73), (1.18.12), (1.18.13), (1.18.31) – (1.18.33) hold and for any $\delta > 0$,

$$\sum_{n=c}^{\infty} \sum_{\xi=a}^{b} \left[\min_{i \in \Omega} \Lambda(i, n, \xi) \right] f([1 - g(h(n, \xi))]\delta) = \infty. \qquad (1.18.91)$$

(a) If m is even, then the conclusion of Theorem 1.18.13(a) holds.

(b) If m is odd and $g(n)$ is eventually nondecreasing, then the conclusion of Theorem 1.18.13(b) holds.

Theorem 1.18.16. Suppose that (1.18.73), (1.18.13), (1.18.31), (1.18.32), (1.18.40), (1.18.41), (1.18.44) hold,

$$g(n) = g = \text{constant} > 0, \qquad (1.18.92)$$

$$\Lambda(i, n, \xi) \text{ is periodic in } n \text{ with period } \sigma \qquad (1.18.93)$$

and

$$\sum_{n=c}^{\infty} \sum_{\xi=a}^{b} \left[\min_{i \in \Omega} \Lambda(i, n, \xi) \right] = \infty. \qquad (1.18.94)$$

Then, the conclusions of Theorem 1.18.13 hold.

Theorem 1.18.17. Suppose that (1.18.73), (1.18.13), (1.18.31), (1.18.32), (1.18.54) and (1.18.94) hold.

(a) If m is even, then all nonoscillatory solutions of (E_{50}), (1.18.6) and (E_{50}), (1.18.8) are bounded, and all unbounded solutions of (E_{50}), (1.18.6) and (E_{50}), (1.18.8) are oscillatory.

(b) If m is odd and $g(n)$ is eventually nondecreasing, then (E_{50}), (1.18.6) and (E_{50}), (1.18.8) have no eventually positive (negative) and concave (convex) solution.

Theorem 1.18.18. Suppose that (1.18.73), (1.18.12) – (1.18.14), (1.18.33), (1.18.58) hold, and there exists an eventually nondecreasing function $z : \mathbb{N} \to \mathbb{R}_0$ such that for any $\delta, \gamma > 0$,

$$\sum_{n=c}^{\infty} \left\{ \gamma \sum_{\xi=a}^{b} \left[\min_{i \in \Omega} \Lambda(i,n,\xi) \right] [1 - g(h(n,\xi))] z(n+1) - \delta \Delta z(n) \right\} = \infty. \tag{1.18.95}$$

Then, the conclusions of Theorem 1.18.15 hold.

Theorem 1.18.19. Let m be odd. Suppose that (1.18.73), (1.18.14), (1.18.33), (1.18.64), (1.18.65) hold, and

$$q \liminf_{\ell \to \infty} \sum_{n=h^*(\ell)}^{\ell} \sum_{\xi=a}^{b} \left[\min_{i \in \Omega} \Lambda(i,n,\xi) \right] \frac{(h(\ell,\xi) - h(n,\xi) + m - 2)^{(m-1)}}{(m-1)!} \geq 1, \tag{1.18.96}$$

where $h^*(\ell) = \max_{\xi \in \mathbb{N}(a,b)} h(\ell,\xi)$. Then, all nonoscillatory solutions of (E_{50}), (1.18.6) and (E_{50}), (1.18.8) are unbounded, and all bounded solutions of (E_{50}), (1.18.6) and (E_{50}), (1.18.8) are oscillatory.

The following examples illustrate the importance of the results of this section.

Example 1.18.1. Consider the difference equation

$$|\Delta^2 y(n)| \Delta^2 y(n) + 16(n+1) y(n+1) = 0, \quad n \in \mathbb{N}. \tag{E_{51}}$$

Here, $m = \alpha = 2$, $g(n) = 0$, $f(u) = u$, $T(n,\xi) = 16(n+1)$, $h(n,\xi) = n+1$ and $a = b = 1$.

It can be easily checked that the conditions of Theorems 1.18.4, 1.18.5 (take $z(n) = 1$) and 1.18.6 are fulfilled. Hence, it follows that all solutions of (E_{51}) are oscillatory. In fact, one such solution is given by $\{y(n)\} = \{(-1)^n n\}$.

Further, we note that the conditions of Theorem 1.18.8 are also satisfied. Thus, all unbounded solutions of (E_{51}) are oscillatory. Indeed, the unbounded solution $\{y(n)\} = \{(-1)^n n\}$ *is* oscillatory.

Example 1.18.2. Consider the difference equation

$$|\Delta^4 y(n)|^2 \Delta^4 y(n) + \frac{3^{12}}{2^9} [y(n+1)]^3 = 0, \quad n \in \mathbb{N}. \tag{E_{52}}$$

With $m = 4,\ \alpha = 3,\ g(n) = 0,\ f(u) = u^3,\ T(n,\xi) = 3^{12}/2^9,\ h(n,\xi) = n+1$ and $a = b = 1$, we see that the conditions of Theorems 1.18.4 – 1.18.6 and 1.18.8 are satisfied. Therefore, all solutions of (E_{52}) are oscillatory. One such solution is given by $\{y(n)\} = \{(-1)^n/2^n\}$.

Example 1.18.3. Consider the difference equation

$$|\Delta[y(n) + y(n-\sigma)]|^{\alpha-1}\Delta[y(n) + y(n-\sigma)] + \frac{4^\alpha}{b-a+1}\sum_{\xi=a}^{b} y(n \pm 2\xi) \ = \ 0,$$

$$n \in \mathbb{N}(2b) \qquad (E_{53})$$

where $\alpha \geq 1$, σ is a nonnegative even integer and a, b are nonnegative integers.

Here, we have $m = 1,\ g(n) = 1,\ f(u) = u,\ T(n,\xi) = 4^\alpha/(b-a+1)$ and $h(n,\xi) = n \pm 2\xi$. Clearly, the conditions of Theorem 1.18.7 are fulfilled. We conclude that equation (E_{53}) has no eventually positive (negative) and nondecreasing (nonincreasing) solution. In fact, (E_{53}) has an oscillatory solution given by $\{y(n)\} = \{(-1)^n\}$.

Example 1.18.4. Consider the difference equation

$$\Delta^m y(n) + \frac{2^m}{b-a+1}\sum_{\xi=a}^{b} y(n-2\xi) = 0, \quad n \in \mathbb{N}(2b) \qquad (E_{54})$$

where m is odd and a, b are positive integers.

In this example, $\alpha = 1,\ g(n) = 0,\ f(u) = u,\ T(n,\xi) = 2^m/(b-a+1)$ and $h(n,\xi) = n - 2\xi$. It can be checked that the conditions of Theorem 1.18.4 – 1.18.6, 1.18.8 and 1.18.9 are satisfied. It follows that (E_{54}) has no eventually positive (negative) and nondecreasing (nonincreasing) solution. In fact, (E_{54}) has an oscillatory solution given by $\{y(n)\} = \{(-1)^n\}$.

Further, with $h^*(\ell) = \max_{\xi\in\mathbb{N}(a,b)} h(\ell,\xi) = \ell - 2a$, the right side of (1.18.67) reduces to

$$\begin{aligned}
&\liminf_{\ell\to\infty} \sum_{n=\ell-2a}^{\ell}\sum_{\xi=a}^{b} \frac{2^m}{b-a+1}\,\frac{(\ell-n+m-2)^{(m-1)}}{(m-1)!} \\
&= \frac{2^m}{(m-1)!}\liminf_{\ell\to\infty}\sum_{n=\ell-2a}^{\ell}(\ell-n+m-2)^{(m-1)} \\
&> \frac{2^m}{(m-1)!}(2a+m-2)^{(m-1)} > 1.
\end{aligned}$$

Hence, the conditions of Theorem 1.18.10 are fulfilled and all bounded solutions of (E_{54}) are oscillatory. It is noted that the bounded solution $\{y(n)\} = \{(-1)^n\}$ is indeed oscillatory.

Example 1.18.5. Consider the difference equation

$$\Delta^m y(n) - \frac{2^m}{b-a+1}\sum_{\xi=a}^{b} y(n-2\xi) \;=\; 0, \quad n \in \mathbb{N}(2b) \qquad (E_{55})$$

where m is even and a, b are positive integers.

It can be verified that all the conditions of Theorem 1.18.10 are satisfied. Hence, all bounded solutions of (E_{55}) are oscillatory. In fact, (E_{55}) has a bounded solution given by $\{y(n)\} = \{(-1)^n\}$ which is also oscillatory.

Example 1.18.6. Consider the partial difference equation

$$\Delta_n^m[y(i,n) + g\; y(i,n-3)] + \sum_{\xi=a}^{b}(n^2+\xi)[y(i,n+\xi)]^{1/3}$$

$$= \;(n+1)Ly(i,n) + (n+2)Ly(i,n-1), \quad i \in \mathbb{N}(1,N),\; n \in \mathbb{N}(3) \quad (E_{56})$$

together with the following two types of boundary conditions

$$y(0,n) \;=\; y(N+1,n) \;=\; 0, \quad n \in \mathbb{N}(3) \qquad (1.18.97)$$

$$\begin{aligned} \Delta_i y(i,n)\Big|_{i=0} - \theta(i,n)y(i,n)\Big|_{i=0} &= 0,\\ \Delta_i y(i,n)\Big|_{i=N} + \theta(i,n)y(i,n)\Big|_{i=N+1} &= 0, \quad n \in \mathbb{N}(3). \end{aligned} \qquad (1.18.98)$$

Here, it is assumed that $0 \le g < 1$, a, b are nonnegative integers and $\theta(0,n)$, $\theta(N+1,n) \ge 0$, $n \ge 3$.

With $f(u) = u^{1/3}$, $g(n) = g$, $h(n,\xi) = n+\xi$ and $\Lambda(i,n,\xi) = n^2+\xi$, we find that for any $\delta > 0$,

$$\sum_{n=c}^{\infty}\sum_{\xi=a}^{b}\left[\min_{i\in\Omega}\Lambda(i,n,\xi)\right] f([1-g(h(n,\xi))]\delta) = \sum_{n=c}^{\infty}\sum_{\xi=a}^{b}(n^2+\xi)[(1-g)\delta]^{1/3} = \infty.$$

Thus, the conditions of Theorem 1.18.15 are fulfilled and the conclusions of Theorem 1.18.15 hold for the boundary value problems (E_{56}), (1.18.97) and (E_{56}), (1.18.98).

For the special case $g = 0$, it can be easily checked that the conditions of Theorem 1.18.17 are satisfied. Hence, the conclusions of Theorem 1.18.17 hold for the boundary value problems (E_{56}), (1.18.97) and (E_{56}), (1.18.98).

1.19. Oscillation of Systems of Higher Order Difference Equations

Here we shall consider the following linear system of difference equations

$$(-1)^{m+1}\Delta^m y_i(n)+\sum_{j=1}^{N} q_{ij}y_j(n-\tau_{jj}) \;=\; 0, \quad m\ge 1, \quad i=1,\cdots,N \quad (E_{57})$$

where q_{ij} are real numbers, and τ_{jj} are positive integers.

We say that a solution $y(n)=[y_1(n),\cdots,y_N(n)]^T$ of (E_{57}) *oscillates* if for some $i\in\{1,\cdots,N\}$, and for every integer $n_0>0$ there exists $n>n_0$ such that $y_i(n)y_i(n+1)<0$. A solution $y(n)$ of (E_{57}) is said to be *nonoscillatory* if there exists an integer $n_0>0$ such that for each $i=1,\cdots,N$, $y_i(n)\neq 0$ for $n\ge n_0$.

We shall establish sufficient conditions for the oscillation of (E_{57}), and investigate the oscillatory behavior of the neutral system of difference equations of the type

$$(-1)^{m+1}\Delta^m\left(y_i(n)+cy_i(n-\delta\sigma)\right)+\sum_{j=1}^{N} q_{ij}y_j(n-\tau) \;=\; 0,$$

$$m\ge 1, \quad i=1,\cdots,N \qquad (E_{58};\delta)$$

where c and q_{ij} are real number, $\delta=\pm 1$, and σ and τ are positive integers.

The following result provides sufficient conditions for the oscillation of all bounded solutions of (E_{57}).

Theorem 1.19.1. Let in the system (E_{57}), q_{ij} be real numbers and τ_{jj} be positive integers. If every bounded solution of the equation

$$(-1)^{m+1}\Delta^m z(n)+qz(n-\tau) \;=\; 0 \qquad (1.19.1)$$

oscillates, where

$$q \;=\; \min_{1\le i\le N}\left\{q_{ii}-\sum_{j=1,j\neq i}^{N}|q_{ji}|\right\} \;>\; 0 \quad \text{and} \quad \tau=\min_{1\le i\le N}\{\tau_{ii}\}, \qquad (1.19.2)$$

then every bounded solution of (E_{57}) is oscillatory.

Proof. Suppose that (E_{57}) has a nonoscillatory, bounded and eventually positive solution $y(n)=[y_1(n),\cdots,y_N(n)]^T$. There exists an integer $n_0\ge 0$ such that $y_i(n)>0$ for $n\ge n_0$, $i=1,\cdots,N$. If we let $w(n)=\sum_{j=1}^{N}y_j(n)$, then

$$\begin{aligned}(-1)^{m+1}\Delta^m w(n) &= -\sum_{j=1}^{N} q_{ii}y_i(n-\tau_{ii}) - \sum_{i=1}^{N}\sum_{j=1,j\neq i}^{N} q_{ij}y_j(n-\tau_{jj}) \\ &\le -\sum_{i=1}^{N} q_{ii}y_i(n-\tau_{ii}) + \sum_{i=1}^{N}\sum_{j=1,j\neq i}^{N} |q_{ji}|y_i(n-\tau_{ii}).\end{aligned}$$

It follows from the above inequality that

$$(-1)^{m+1}\Delta^m w(n) + \sum_{i=1}^{N}\left[q_{ii} - \sum_{j=1,j\neq i}^{N} |q_{ji}|\right] y_i(n-\tau_{ii}) \le 0$$

or

$$(-1)^{m+1}\Delta^m w(n) + q\left(\sum_{i=1}^{N} y_i(n-\tau_{ii})\right) \le 0 \quad \text{for} \quad n \ge n_1 \ge n_0. \tag{1.19.3}$$

From the boundedness, nonoscillation and eventual positivity of $y_1(n), \cdots, y_N(n)$, we see that $w(n)$ is bounded and eventually positive. From the fact that $(-1)^{m+1}\Delta^m w(n) \le 0$ eventually and by Lemma 1.13.1, the sequence $\{w(n)\}$ is eventually decreasing and satisfies

$$(-1)^k\Delta^k w(n) > 0 \quad \text{eventually}, \quad k = 0, 1, \cdots, m. \tag{1.19.4}$$

Thus, we conclude that $y_i(n)$ converges as $n \to \infty$, $i = 1, \cdots, N$. We let

$$\lim_{n\to\infty} y_i(n) = \alpha_i > 0, \quad i = 1, \cdots, N.$$

We claim that $\alpha_i = 0$, $i = 1, \cdots, N$. If not, then there exists an integer $n_1 > t_0 + \tau^*$ where $\tau^* = \max_{1\le i\le N}\{\tau_{ii}\}$ such that

$$y_i(n-\tau_{ii}) \ge \frac{\alpha_i}{2} \quad \text{for} \quad n \ge n_1 + \tau^*.$$

We have from (1.19.3) that

$$(-1)^{m+1}\Delta^m w(n) \le -\left(\frac{1}{2}q\right)\sum_{i=1}^{N}\alpha_i \quad \text{for} \quad n \ge n_1 + \tau^*,$$

which leads to

$$\begin{aligned}(-1)^{m+1}\Delta^{m-1} w(n) &\le (-1)^{m+1}\Delta^{m-1} w(n_1+\tau^*) \\ &\quad -\left(\frac{1}{2}q\right)\left(\sum_{i=1}^{N}\alpha_i\right)(n-n_1-\tau^*)\end{aligned}$$

implying that $(-1)^{m+1}\Delta^{m-1}w(n)$ can become negative for all sufficiently large n; but this is impossible. Thus, we have $\sum_{i=1}^{N}\alpha_i = 0$ and hence $\alpha_i = 0$, $i = 1, \cdots, N$.

Thus,

$$\lim_{n\to\infty} y_i(n) = 0, \quad i = 1, \cdots, N$$

and hence

$$\lim_{n\to\infty} w(n) = 0.$$

Summing both sides of (1.19.3) from $n \geq n_2 \geq n_1 + \tau^*$ to $K \geq n$ repeatedly and letting $K \to \infty$, we get

$$\begin{aligned}
w(n) &\geq \sum_{j_1=n}^{\infty}\sum_{j_2=j_1}^{\infty}\cdots\sum_{j_m=j_{m-1}}^{\infty} q\left(\sum_{i=1}^{N} y_i(j_m - \tau_{ii})\right) \\
&= \sum_{i=1}^{N} q\left(\sum_{j_1=n}^{\infty}\sum_{j_2=j_1}^{\infty}\cdots\sum_{j_m=j_{m-1}}^{\infty} y_i(j_m - \tau_{ii})\right) \\
&= \sum_{i=1}^{N} q\left(\sum_{j=n-\tau_{ii}}^{\infty}\sum_{j_2=j_1-\tau_{ii}}^{\infty}\cdots\sum_{j_m=j_{m-1}-\tau_{ii}} y_i(j_m)\right) \\
&\geq \sum_{i=1}^{N} q\left(\sum_{j=n-\tau}^{\infty}\sum_{j_2=j_1-\tau}^{\infty}\cdots\sum_{j_m=j_{m-1}-\tau}^{\infty} y_i(j_m)\right) \\
&= q\left(\sum_{j_1=n}^{\infty}\sum_{j_2=j_1}^{\infty}\cdots\sum_{j_m=j_{m-1}}^{\infty}\left(\sum_{i=1}^{N} y_i(j_m - \tau)\right)\right)
\end{aligned}$$

or

$$w(n) \geq q\left(\sum_{j_1=n}^{\infty}\sum_{j_2=j_1}^{\infty}\cdots\sum_{j_m=j_{m-1}}^{\infty} w(j_m - \tau)\right), \quad n \geq n_3 \geq n_2 \quad (1.19.5)$$

where n is sufficiently large and $\{w(n)\}$ is a decreasing positive sequence which tends to zero as $n \to \infty$. Then by Lemma 5 in [194] we conclude that the equation

$$q\left(\sum_{j_1=n}^{\infty}\sum_{j_2=j_1}^{\infty}\cdots\sum_{j_m=j_{m-1}}^{\infty} z(j_m - \tau)\right) = z(n)$$

has a positive solution $\{z(n)\}$ for $n \geq n_3$. It is now easy to see that $\{z(n)\}$ is also a positive bounded solution of the equation

$$(-1)^{m+1}\Delta^m z(n) + qz(n-\tau) = 0.$$

This contradiction to our hypothesis completes the proof. ∎

The following corollary is immediate from Lemma 1.16.1(ii) and Theorem 1.19.1.

Corollary 1.19.2. Let q and τ be as in (1.19.2). If

$$q > \frac{m^m \tau^\tau}{(m+\tau)^{m+\tau}}, \tag{1.19.6}$$

then all bounded solutions of (E_{57}) are oscillatory.

Example 1.19.1. Consider the system of difference equations

$$\begin{aligned}(-1)^{m+1}\Delta^m y_1(n) + 2y_1(n-m/2) - y_2(n-m) = 0\\ (-1)^{m+1}\Delta^m y_2(n) - y_1(n-m/2) + 2y_1(n-m) = 0.\end{aligned} \tag{E_{59}}$$

All conditions of Corollary 1.19.2 are satisfied and hence all bounded solutions of (E_{59}) are oscillatory.

Now, for the equation $(E_{58};\delta)$ we state and prove the following results.

Theorem 1.19.3. Let q_{ij} be real numbers, τ and σ be positive integers, and $0 \le c < 1$ be a real number. If every bounded solution of the difference equation

$$(-1)^{m+1}\Delta^m V(n) + q(1-c)V(n-\tau) \ = \ 0 \tag{1.19.7}$$

is oscillatory, where q is defined in (1.19.2), then every bounded solution of $(E_{58};-1)$ is oscillatory.

Theorem 1.19.4. Let q_{ij} be real numbers, τ and σ be positive integers such that $\tau > \sigma$, and $c > 1$ be a real number. If every bounded solution of the difference equation

$$(-1)^{m+1}\Delta^m U(n) + q\left(\frac{c-1}{c^2}\right) U(n-(\tau-\sigma)) \ = \ 0 \tag{1.19.8}$$

is oscillatory, where q is defined in (1.19.2), then every bounded solution of $(E_{58};1)$ is oscillatory.

Theorem 1.19.5. Let q_{ij} be real numbers, τ and σ be positive integers, and $0 < c^* = -c \le 1$ be a real number. If every bounded solution of the difference equation

$$(-1)^{m+1}\Delta^m W(n) + qW(n-\tau) \ = \ 0 \tag{1.19.9}$$

is oscillatory, where q is defined in (1.19.2), then every bounded solution of $(E_{58};1)$ is oscillatory.

Proof of Theorems 1.19.3 – 1.19.5. Let $y(n) = [y_1(n), \cdots, y_N(n)]^T$ be a nonoscillatory, bounded eventually positive solution of $(E_{58};\delta)$, $\delta = \pm 1$.

Then, there exists an integer $n_0 \ge 0$ such that $y_i(n) > 0$ for $n \ge n_0$, $i = 1, \cdots, N$. We let

$$z(n) = \sum_{i=1}^{N} y_i(n) + c\sum_{i=1}^{N} y_i(n-\delta\tau) \tag{1.19.10}$$

and again $w(n) = \sum_{i=1}^{N} y_i(n)$. Then,

$$(-1)^{m+1}\Delta^m z(n) + \sum_{i=1}^{N}\sum_{j=1}^{N} q_{ij}y_j(n-\tau) = 0.$$

Further, as in Theorem 1.19.1 we have

$$(-1)^{m+1}\Delta^m z(n) + qw(n-\tau) \le 0 \quad \text{for } n \ge n_1 \ge n_0, \tag{1.19.11}$$

where q is defined in (1.19.2). Clearly $\{z(n)\}$ and $\{w(n)\}$ are positive and bounded sequences, and

$$(-1)^{m+1}\Delta^m z(n) \le 0 \quad \text{for } n \ge n_1. \tag{1.19.12}$$

Thus, as in Theorem 1.19.1, we find that $\lim_{n\to\infty} z(n) = 0$ and $\{z(n)\}$ satisfies

$$(-1)^k\Delta^k z(n) > 0 \quad \text{eventually for } k = 0, 1, \cdots, m. \tag{1.19.13}$$

Using this fact, we see from (1.19.10) that if $\delta = -1$ and $0 \le c < 1$, then eventually $z(n) = w(n) + cw(n+\tau)$, and hence

$$w(n) = z(n) - cw(n+\tau) = z(n) - c(z(n+\tau) - cz(n+2\tau)) \ge (1-c)z(n) \tag{1.19.14}$$

and if $\delta = 1$ and $c > 1$, then

$$\begin{aligned} w(n) &= \frac{1}{c}(z(n+\tau) - w(n+\tau)) \\ &= \frac{1}{c}z(n+\tau) - \frac{1}{c^2}(z(n+2\tau) - w(n+2\tau)) \\ &\ge \left(\frac{c-1}{c^2}\right) z(n+\tau). \end{aligned} \tag{1.19.15}$$

Next we consider the case when $\delta = 1$ and $0 \le c^* = -c < 1$. Clearly, $\{w(n)\}$ is bounded and eventually positive. Since (1.19.12) holds, we find that $\{z(n)\}$ is either eventually positive or eventually negative. If $z(n) < 0$ eventually, there is a sequence $\{n_k\}$ such that $\lim_{n\to\infty} n_k = \infty$ and $\lim_{k\to\infty} w(n_k) = \limsup_{n\to\infty} w(n)$. Without any loss of generality, we assume that $\{w(n_k - \tau)\}$ is convergent. Then,

$$0 > \lim_{k\to\infty} z(n_k) \ge \limsup_{n\to\infty} w(n)(1-c^*) \ge 0$$

and hence we conclude that $z(n) < 0$ eventually is impossible. Thus, as in Theorem 1.19.1 we have $\lim_{n\to\infty} z(n) = 0$, and (1.19.13) holds eventually, and

$$z(n) \le w(n) \quad \text{eventually.} \tag{1.19.16}$$

From the above we have the following:

(I) Suppose $\delta = -1$ and $0 \le c < 1$. From (1.19.11) and (1.19.14), we obtain

$$(-1)^{m+1}\Delta^m z(n) + q(1-c)z(n-\tau) \le 0 \quad \text{eventually.} \tag{1.19.17}$$

(II) Supose $\delta = 1$ and $c > 1$. From (1.19.11) and (1.19.15), we have

$$(-1)^{m+1}\Delta^m z(n) + q\left(\frac{c-1}{c^2}\right) z(n-(\tau-\sigma)) \le 0 \quad \text{eventually.} \tag{1.19.18}$$

(III) Suppose $\delta = 1$ and $0 < c^* = -c \le 1$. From (1.19.11) and (1.19.16), we find

$$(-1)^{m+1}\Delta^m z(n) + qz(n-\tau) \le 0 \quad \text{eventually.} \tag{1.19.19}$$

The rest of the proof is similar to that of Theorem 1.19.1. ∎

The following corollaries are immediate.

Corollary 1.19.6. Let q_{ij} be real numbers, τ and σ be positive integers, $0 \le c < 1$ be a real number, and

$$(1-c)q > \frac{m^m \tau^\tau}{(m+\tau)^{m+\tau}}, \tag{1.19.20}$$

where q is defined in (1.19.2), then every bounded solution of $(E_{58}; -1)$ is oscillatory.

Corollary 1.19.7. Let q_{ij} be real numbers, τ and σ be positive integers such that $\tau > \sigma$, $c > 1$ be a real number, and

$$\left(\frac{c-1}{c^2}\right) q > \frac{m^m(\tau-\sigma)^{\tau-\sigma}}{(m+\tau-\sigma)^{m+\tau-\sigma}}, \tag{1.19.21}$$

where q is defined in (1.19.2), then every bounded solution of $(E_{58}; 1)$ is oscillatory.

Corollary 1.19.8. Let q_{ij} be real numbers, τ and σ be positive integers, $0 < c^* = -c \le 1$ be a real number, and condition (1.19.6) holds. Then every bounded solution of $(E_{58}; 1)$ is oscillatory.

Example 1.19.2. Consider the system of difference equations

$$(-1)^{m+1}\Delta^m (y_1(n) + cy_1(n-\delta\sigma)) + 2y_1(n-m) - y_2(n-m) = 0$$
$$(-1)^{m+1}\Delta^m (y_2(n) + cy_2(n-\delta\sigma)) - y_1(n-m) + 2y_2(n-m) = 0, \qquad (E_{60};\delta)$$

where $\delta = \pm 1$, σ is a positive integer.

One can easily conclude the following:

(i) If $0 \le c < 1$ and $(1-c) > (1/2)^{2m}$, then all conditions of Coroallry 1.19.6 are satisfied and hence all bounded solutions of $(E_{60};-1)$ are oscillatory.

(ii) If $c > 1$, $m > \sigma$ and

$$\left(\frac{c-1}{c^2}\right) > \frac{m^m(m-\sigma)^{m-\sigma}}{(2m-\sigma)^{2m-\sigma}},$$

then all conditions of Corollary 1.19.7 are satisfied and hence all bounded solutions of $(E_{60};1)$ are oscillatory.

(iii) If $-1 \le c < 0$, then all conditions of Coroallry 1.19.8 are satisfied and hence all bounded solutions of $(E_{60};1)$ are oscillatory.

1.20. Partial Difference Equations with Continuous Variables

Consider the partial difference equation with continuous variables

$$p_1 z(s+a, t+b) + p_2 z(s+a, t) + p_3 z(s, t+b) - p_4 z(s,t) + P(s,t) z(s-\tau, t-\sigma) = 0, \qquad (E_{61})$$

where $P \in C(\mathbb{R}_0 \times \mathbb{R}_0, \mathbb{R}_+)$, a, b, τ, σ are real numbers, and p_i, $i = 1,2,3,4$ are nonnegative constants.

By a solution of (E_{61}), we mean a continuous function $z(s,t)$ which satisfies (E_{61}) for $s \ge s_0 \ge 0$, $t \ge t_0 \ge 0$. As usual a solution $z(s,t)$ of (E_{61}) is said to be *eventually positive* if $z(s,t) > 0$ for all large s and t, and *eventually negative* if $z(s,t) < 0$ for all large s and t. It is said to be *oscillatory* if it neither eventually positive nor eventually negative.

In what follows we shall assume that

(i) $p_1 \ge 0$, p_2, $p_3 \ge p_4 > 0$, $P(s,t) > 0$, and $a\tau > 0$, $b\sigma > 0$,

(ii) $\tau = ka + \theta$, $\sigma = \ell b + \eta$, where k, ℓ are nonnegative integers, $\theta \in [0,a)$ for $a > 0$, and $\theta \in (a, 0]$ for $a < 0$, $\eta \in [0,b)$ for $b > 0$, and $\eta \in (b, 0]$ for $b < 0$,

(iii)

$$Q(s,t) = \begin{cases} \min\{P(u,v) : s \le u \le s+a, t \le v \le t+b\} & \text{if } a>0, b>0 \\ \min\{P(u,v) : s+a \le u \le s, t \le v \le t+b\} & \text{if } a<0, b>0 \\ \min\{P(u,v) : s \le u \le s+a, t+b \le v \le t\} & \text{if } a>0, b<0 \\ \min\{P(u,v) : s+a \le u \le s, t+b \le v \le t\} & \text{if } a<0, b<0 \end{cases} \tag{1.20.1}$$

and

$$\limsup_{s,t\to\infty} Q(s,t) \ > \ 0.$$

Now, we define a set E by

$$E \ = \ \{\lambda > 0 \ : \ p_4 - \lambda Q(s,t) > 0 \ \text{ eventually}\}\,. \tag{1.20.2}$$

Lemma 1.20.1. Assume that (E_{61}) has an eventually positive solution. Then, the difference inequality

$$p_1 w(s+a,t+b) + p_2 w(s+a,t) + p_3 w(s,t+b) - p_4 w(s,t)$$
$$+ \, Q(s,t) w(s-ka, t-\ell b) \ \le \ 0 \tag{1.20.3}$$

has an eventually positive solution.

Proof. Let $z(s,t)$ be an eventually positive solution of (E_{61}). Then, we have

$$p_4(z(s+a,t) + z(s,t+b) - z(s,t))$$
$$< \ p_1 z(s+a,t+b) + p_2 z(s+a,t) + p_3 z(s,t+b) - p_4 z(s,t) \ < \ 0$$

eventually. We consider the following four cases:

Case 1 $a > 0,\ b > 0$. Let

$$w(s,t) \ = \ \int_s^{s+a}\int_t^{t+b} z(u,v)dudv. \tag{1.20.4}$$

Then,

$$\frac{\partial w(s,t)}{\partial s} \ = \ \int_t^{t+b} (z(s+a,v) - z(s,v))dv \ < \ 0$$

and

$$\frac{\partial w(s,t)}{\partial t} \ = \ \int_s^{s+a} (z(u,t+b) - z(u,t))du < 0.$$

Integrating (E_{61}), we get

$$p_1 \int_s^{s+a} \int_t^{t+b} z(u+a, v+b)dudv + p_2 \int_s^{s+a} \int_t^{t+b} z(u+a, v)dudv$$

$$+p_3 \int_s^{s+a} \int_t^{t+b} z(u, v+b)dudv - p_4 \int_s^{s+a} \int_t^{t+b} z(u, v)dudv$$

$$+ \int_s^{s+a} \int_t^{t+b} P(u,v)z(u-\tau, v-\sigma)dudv \ = \ 0.$$

By (1.20.1), (1.20.4) and the above equality, we obtain

$$p_1 w(s+a, t+b) + p_2 w(s+a, t) + p_3 w(s, t+b) - p_4 w(s,t)$$
$$+ Q(s,t)w(s-\tau, t-\sigma) \ \leq \ 0.$$

Since $\dfrac{\partial w}{\partial s}$ and $\dfrac{\partial w}{\partial t} < 0$, we have $w(s-\tau, t-\sigma) = w(s-(ka+\theta), t-(\ell b+\eta)) \geq w(s-ka, t-\ell b)$. Therefore, (1.20.3) holds.

Case 2 $a < 0,\ b > 0$. Let

$$w(s,t) \ = \ \int_{s+a}^{s} \int_t^{t+b} z(u,v)dudv. \tag{1.20.5}$$

Integrating (E_{61}) and using (1.20.1) and (1.20.5), again we get

$$p_1 w(s+a, t+b) + p_2 w(s+a, t) + p_3 w(s, t+b) - p_4 w(s,t)$$
$$+ Q(s,t)w(s-\tau, t-\sigma) \ \leq \ 0.$$

Since $\dfrac{\partial w}{\partial s} > 0$ and $\dfrac{\partial w}{\partial t} < 0$, we have $w(s-\tau, t-\sigma) = w(s-(ka+\theta), t-(\ell b+\eta)) \geq w(s-ka, t-(\ell b+\eta)) \geq w(s-ka, t-\ell b)$. Therefore, (1.20.3) holds.

Case 3 $a > 0,\ b < 0$. Let

$$w(s,t) \ = \ \int_s^{s+a} \int_{t+b}^{t} z(u,v)dudv. \tag{1.20.6}$$

Then, $\dfrac{\partial w}{\partial s} < 0$ and $\dfrac{\partial w}{\partial t} > 0$. Now as in Case 1, (1.20.3) holds.

Case 4 $a < 0,\ b < 0$. Let

$$w(s,t) \ = \ \int_{s+a}^{s} \int_{t+b}^{t} z(u,v)dudv. \tag{1.20.7}$$

Then, $\dfrac{\partial w}{\partial s} > 0$ and $\dfrac{\partial w}{\partial t} > 0$. Now as in Case 1, (1.20.3) holds. ∎

Theorem 1.20.2. Assume that there exist $S \geq s_0$, $T \geq t_0$ such that if $k > \ell > 0$

$$\sup_{\lambda \in E, s \geq S, t \geq T} \lambda \prod_{i=1}^{\ell} (p_4 - \lambda Q(s - ia, t - ib)) \prod_{j=1}^{k-\ell} (p_4 - \lambda Q(s - \ell a - ja, t - \ell b))$$

$$< \left(p_1 + \frac{2p_2 p_3}{p_4} \right)^{\ell} p_2^{k-\ell} \tag{1.20.8}$$

and if $\ell > k > 0$

$$\sup_{\lambda \in E, s \geq S, t \geq T} \lambda \prod_{i=1}^{k} (p_4 - \lambda Q(s - ia, t - ib)) \prod_{j=1}^{\ell-k} (p_4 - \lambda Q(s - ka, t - kb - jb))$$

$$< \left(p_1 + \frac{2p_2 p_3}{p_4} \right)^{k} p_3^{\ell-k}. \tag{1.20.9}$$

Then, every solution of (E_{61}) is oscillatory.

Proof. Suppose to the contrary $z(s,t)$ is an eventually positive solution. Let $w(s,t)$ be as in Lemma 1.20.1. We define a subset $\mathcal{S}$ of the positive numbers as follows:

$$\begin{aligned} \mathcal{S}(\lambda) \; &= \{\lambda > 0 \; : \; p_1 w(s+a, t+b) + p_2 w(s+a, t) + p_3 w(s, t+b) \\ &\qquad -(p_4 - \lambda Q(s,t)) w(s,t) \leq 0, \quad \text{eventually}\}. \end{aligned}$$

From (1.20.3) and the definition of $w(s,t)$ in Lemma 1.20.3, we have

$$p_1 w(s+a, t+b) + p_2 w(s+a, t) + p_3 w(s, t+b) - (p_4 - Q(s,t)) w(s,t) \; \leq \; 0,$$

which implies that $1 \in \mathcal{S}(\lambda)$. Hence, $\mathcal{S}(\lambda)$ is nonempty. For $\lambda \in \mathcal{S}$, we have eventually that $p_4 - \lambda Q(s,t) > 0$, which implies that $\mathcal{S} \subset E$. From the condition (i), the set E is bounded, and hence $\mathcal{S}(\lambda)$ is also bounded. From (1.20.3), we have

$$p_1 w(s+a, t+b) + p_2 w(s+a, t) + p_3 w(s, t+b) \; < \; p_4 w(s,t),$$

and so

$$w(s+a, t+b) \; \leq \; \frac{p_4}{p_2} w(s, t+b) \quad \text{and} \quad w(s+a, t+b) \; \leq \; \frac{p_4}{p_3} w(s+a, t). \tag{1.20.10}$$

Let $\mu \in \mathcal{S}(\lambda)$. Then, we have

$$\begin{aligned} \left(p_1 + \frac{2p_2 p_3}{p_4} \right) w(s+a, t+b) \; &\leq \; p_1 w(s+a, t+b) + p_2 w(s+a, t) \\ &\qquad + p_3 w(s, t+b) \\ &\leq \; (p_4 - \mu Q(s,t)) w(s,t). \end{aligned}$$

If $k > \ell$, then it follows that

$$w(s,t) \le \left(p_1 + \frac{2p_2p_3}{p_4}\right)^{-\ell} \prod_{i=1}^{\ell}(p_4 - \mu Q(s-ia, t-ib))w(s-\ell a, t-\ell b) \tag{1.20.11}$$

and

$$\begin{aligned} w(s-\ell a, t-\ell b) &\le \frac{1}{p_2}(p_4 - \mu Q(s-\ell a - a, t-\ell b))w(s-\ell a - a, t-\ell b) \\ &\le \dots \le \left(\frac{1}{p_2}\right)^{k-\ell} \prod_{j=1}^{k-\ell}(p_4 - \mu Q(s-\ell a - ja, t-\ell b)) \\ &\qquad \times w(s-ka, t-\ell b). \end{aligned}$$

Hence

$$\begin{aligned} w(s,t) \le& \left(p_1 + \frac{2p_2p_3}{p_4}\right)^{-\ell} p_2^{\ell-k} \prod_{i=1}^{\ell}(p_4 - \mu Q(s-ia, t-ib)) \\ &\times \prod_{j=1}^{k-\ell}(p_4 - \mu Q(s-\ell a - ja, t-\ell b))w(s-ka, t-\ell b). \end{aligned} \tag{1.20.12}$$

Similarly, if $\ell > k$, then

$$\begin{aligned} w(s,t) \le& \left(p_1 + \frac{2p_2p_3}{p_4}\right)^{-k} p_3^{k-\ell} \prod_{i=1}^{k}(p_4 - \mu Q(s-ia, t-ib)) \\ &\times \prod_{j=1}^{\ell-k}(p_4 - \mu Q(s-ka, t-kb-jb))w(s-ka, t-\ell b). \end{aligned} \tag{1.20.13}$$

Substituting (1.20.12) and (1.20.13) into (1.20.3), we get

$$\begin{aligned} & p_1 w(s+a, t+b) + p_2 w(s+a, t) + p_3 w(s, t+b) - p_4 w(s,t) \\ & + Q(s,t)\left(p_1 + \frac{2p_2p_3}{p_4}\right)^{\ell} p_2^{k-\ell} \left(\prod_{i=1}^{\ell}(p_4 - \mu Q(s-ia, t-ib)) \right. \\ & \left. \times \prod_{j=1}^{k-\ell}(p_4 - \mu Q(s-\ell a - ja, t-\ell b))\right)^{-1} w(s,t) \le 0, \quad \text{for } k > \ell \end{aligned} \tag{1.20.14}$$

and

$$\begin{aligned} & p_1 w(s+a, t+b) + p_2 w(s+a, t) + p_3 w(s, t+b) - p_4 w(s,t) \\ & + Q(s,t)\left(p_1 + \frac{2p_2p_3}{p_4}\right)^{k} . \, p_3^{\ell-k} \left(\prod_{i=1}^{k}(p_4 - \mu Q(s-ia, t-ib)) \right. \end{aligned}$$

$$\times \prod_{j=1}^{\ell-k}(p_4 - \mu Q(s-ka, t-kb-jb))\Bigg)^{-1} w(s,t) \le 0, \quad \text{for } \ell > k. \tag{1.20.15}$$

Therefore, we have

$$p_1 w(s+a,t+b) + p_2 w(s+a,t) + p_3 w(s,t+b)$$
$$- \left\{ p_4 - Q(s,t)\left(p_1 + \frac{2p_2p_3}{p_4}\right)^{\ell} p_2^{k-\ell} \sup_{s\ge S, t\ge T}\left[\left(\prod_{i=1}^{\ell}(p_4 - \mu Q(s-ia, t-ib))\right.\right.\right.$$
$$\left.\left.\left.\times \prod_{j=1}^{k-\ell}(p_4 - \mu Q(s-\ell a - ja, t-\ell b))\right)^{-1}\right]\right\} w(s,t) \le 0, \quad \text{for } k > \ell \tag{1.20.16}$$

and

$$p_1 w(s+a,t+b) + p_2 w(s+a,t) + p_3 w(s,t+b)$$
$$- \left\{ p_4 - Q(s,t)\left(p_1 + \frac{2p_2p_3}{p_4}\right)^{k} p_3^{\ell-k} \sup_{s\ge S, t\ge T}\left[\left(\prod_{i=1}^{k}(p_4 - \mu Q(s-ia, t-ib))\right.\right.\right.$$
$$\left.\left.\left.\times \prod_{j=1}^{\ell-k}(p_4 - \mu Q(s-ka, t-kb-jb))\right)^{-1}\right]\right\} w(s,t) \le 0, \quad \text{for } \ell > k. \tag{1.20.17}$$

From (1.20.16) and (1.20.17), we find

$$\left(p_1 + \frac{2p_2p_3}{p_4}\right)^{\ell} p_2^{k-\ell} \sup_{s\ge S, t\ge T}\left[\left(\prod_{i=1}^{\ell}(p_4 - \mu Q(s-ia, t-ib))\right.\right.$$
$$\left.\left.\times \prod_{j=1}^{k-\ell}(p_4 - \mu Q(s-\ell a-ja, t-\ell b))\right)^{-1}\right] \in \mathcal{S}, \quad \text{for } k > \ell \tag{1.20.18}$$

and

$$\left(p_1 + \frac{2p_2p_3}{p_4}\right)^{k} p_3^{\ell-k} \sup_{s\ge S, t\ge T}\left[\left(\prod_{i=1}^{k}(p_4 - \mu Q(s-ia, t-ib))\right.\right.$$
$$\left.\left.\times \prod_{j=1}^{\ell-k}(p_4 - \mu Q(s-ka, t-kb-jb))\right)^{-1}\right] \in \mathcal{S}, \quad \text{for } \ell > k. \tag{1.20.19}$$

On the other hand (1.20.8) implies that there exists $\alpha_1 \in (0,1)$ such that for $k > \ell$

$$\sup_{\lambda \in E, s \geq S, t \geq T} \lambda \prod_{i=1}^{\ell} (p_4 - \lambda Q(s-ia, t-ib)) \prod_{j=1}^{k-\ell} (p_4 - \lambda Q(s - \ell a - ja, t - \ell b))$$

$$\leq \alpha_1 \left(p_1 + \frac{2p_2p_3}{p_4} \right)^{\ell} p_2^{k-\ell} \tag{1.20.20}$$

and (1.20.9) implies that there exists $\alpha_1 \in (0,1)$ (we can choose the same) such that for $\ell > k$

$$\sup_{\lambda \in E, s \geq S, t \geq T} \lambda \prod_{i=1}^{k} (p_4 - \lambda Q(s-ia, t-ib)) \prod_{j=1}^{\ell-k} (p_4 - \lambda Q(s - ka, t - kb - jb))$$

$$\leq \alpha_1 \left(p_1 + \frac{2p_2p_3}{p_4} \right)^{k} p_3^{\ell-k}. \tag{1.20.21}$$

In particular, (1.20.20) and (1.20.21) lead to (when $\lambda = \mu$), respectively

$$\left(p_1 + \frac{2p_2p_3}{p_4} \right)^{\ell} p_2^{k-\ell} \sup_{s \geq S, t \geq T} \left[\left(\prod_{i=1}^{\ell} (p_4 - \mu Q(s-ia, t-ib)) \times \prod_{j=1}^{k-\ell} (p_4 - \mu Q(s - \ell a - ja, t - \ell b)) \right)^{-1} \right] \geq \frac{\mu}{\alpha_1}, \quad \text{for } k > \ell \tag{1.20.22}$$

and

$$\left(p_1 + \frac{2p_2p_3}{p_4} \right)^{k} p_3^{\ell-k} \sup_{s \geq S, t \geq T} \left[\left(\prod_{i=1}^{k} (p_4 - \mu Q(s-ia, t-ib)) \times \prod_{j=1}^{\ell-k} (p_4 - \mu Q(s - ka, t - kb - jb)) \right)^{-1} \right] \geq \frac{\mu}{\alpha_1}, \quad \text{for } \ell > k. \tag{1.20.23}$$

Since $\mu \in \mathcal{S}$ and $\mu' \leq \mu$ implies that $\mu' \in \mathcal{S}$, it follows from (1.20.18) and (1.20.22) for $k > \ell$, (1.20.19) and (1.20.23) for $\ell > k$ that $\mu/\alpha_1 \in \mathcal{S}$. Repeating the above arguments with μ replaced by μ/α_1, we get $\mu/(\alpha_1\alpha_2) \in \mathcal{S}$, where $\alpha_2 \in (0,1)$. Continuing in this way, we obtain $\mu \big/ \left(\prod_{i=1}^{\infty} \alpha_i\right) \in \mathcal{S}$, where $\alpha_i \in (0,1)$. But, this contradicts the boundedness of $\mathcal{S}$. ∎

Corollary 1.20.3. Assume that for $k > \ell > 0$

$$\liminf_{s,t \to \infty} Q(s,t) = q > p_4^{k+1} \left(p_1 + \frac{2p_2p_3}{p_4} \right)^{-\ell} p_2^{\ell-k} \frac{k^k}{(k+1)^{k+1}} \tag{1.20.24}$$

and for $\ell > k > 0$

$$\liminf_{s,t\to\infty} Q(s,t) = q > p_4^{\ell+1}\left(p_1 + \frac{2p_2p_3}{p_4}\right)^{-k} p_3^{k-\ell}\frac{\ell^\ell}{(\ell+1)^{\ell+1}}. \quad (1.20.25)$$

Then, every solution of (E_{61}) is oscillatory.

Proof. We note that

$$\max_{p_4/q>\lambda>0} \lambda(p_4 - \lambda q)^k = \frac{p_4^{k+1}k^k}{q(k+1)^{k+1}}.$$

Hence (1.20.24) and (1.20.25) imply that (1.20.8) and (1.20.9) hold. Now by Theorem 1.20.1 every solution of (E_{61}) oscillates. ∎

Theorem 1.20.4. Assume that there exist $S \geq s_0$, $T \geq t_0$ such that if $k > \ell > 0$

$$\sup_{\lambda\in E, s\geq S, t\geq T} \lambda\left[\prod_{j=1}^{k-\ell}\prod_{i=1}^{\ell}(p_4 - \lambda Q(s-ia-ja, t-ib))\right]^{1/(k-\ell)}$$
$$< \left(p_1 + \frac{2p_2p_3}{p_4}\right)^{\ell}\left(\frac{p_2}{p_4}\right)^{(k-\ell+1)/2} \quad (1.20.26)$$

and if $\ell > k > 0$

$$\sup_{\lambda\in E, s\geq S, t\geq T} \lambda\left[\prod_{j=1}^{\ell-k}\prod_{i=1}^{k}(p_4 - \lambda Q(s-ia, t-ib-jb))\right]^{1/(\ell-k)}$$
$$< \left(p_1 + \frac{2p_2p_3}{p_4}\right)^{k}\left(\frac{p_4}{p_3}\right)^{(\ell-k+1)/2} \quad (1.20.27)$$

Then, every solution of (E_{61}) is oscillatory.

Proof. If $k > \ell$, then from (1.20.11) for $j = 1, 2, \cdots, k-\ell$ we have

$$\begin{aligned} &w(s-ja,t) \\ &\leq \left(p_1 + \frac{2p_2p_3}{p_4}\right)^{-\ell}\prod_{i=1}^{\ell}(p_4 - \mu Q(s-ia-ja, t-ib))w(s-\ell a-ja, t-\ell b) \\ &\leq \left[\left(p_1 + \frac{2p_2p_3}{p_4}\right)^{-\ell}\prod_{i=1}^{\ell}(p_4 - \mu Q(s-ia-ja, t-ib))\right] w(s-ka, t-\ell b). \end{aligned} \quad (1.20.28)$$

Now, from (1.20.10) and (1.20.28) it follows that

$$\begin{aligned} & w^{k-\ell}(s,t) \\ & \le \prod_{j=1}^{k-\ell} \left(\frac{p_4}{p_2}\right)^j w(s-ja,t) \\ & \le \prod_{j=1}^{k-\ell} \left\{ \left(\frac{p_4}{p_2}\right)^j \left[\left(p_1 + \frac{2p_2p_3}{p_4}\right)^{-\ell} \prod_{i=1}^{\ell} (p_4 - \mu Q(s-ia-ja, t-ib)) \right] \right. \\ & \qquad\qquad \left. \times\ w(s-ka, t-\ell b) \right\} \\ & = \left(p_1 + \frac{2p_2p_3}{p_4}\right)^{-\ell(k-\ell)} \left(\frac{p_4}{p_2}\right)^{(k-\ell+1)(k-\ell)/2} \\ & \quad \times \left[\prod_{j=1}^{k-\ell} \prod_{i=1}^{\ell} (p_4 - \mu Q(s-ia-ja, t-ib)) \right] w^{k-\ell}(s-ka, t-\ell b), \end{aligned} \tag{1.20.29}$$

i.e.

$$\begin{aligned} w(s,t) \ \le\ & \left[\left(p_1 + \frac{2p_2p_3}{p_4}\right)^{-\ell(k-\ell)} \left(\frac{p_4}{p_2}\right)^{(k-\ell+1)(k-\ell)/2} \right. \\ & \left. \times \prod_{j=1}^{k-\ell} \prod_{i=1}^{\ell} (p_4 - \mu Q(s-ia-ja, t-ib)) \right]^{1/(k-\ell)} w(s-ka, t-\ell b). \end{aligned}$$

Similarly, if $\ell > k$ then

$$\begin{aligned} w(s,t) \ \le\ & \left[\left(p_1 + \frac{2p_2p_3}{p_4}\right)^{-k(\ell-k)} \left(\frac{p_4}{p_3}\right)^{(\ell-k+1)(\ell-k)/2} \right. \\ & \left. \times \prod_{j=1}^{\ell-k} \prod_{i=1}^{k} (p_4 - \mu Q(s-ia, t-ib-jb)) \right]^{1/(\ell-k)} w(s-ka, t-\ell b). \end{aligned}$$

The rest of the proof is similar to that of Theorem 1.20.2. ∎

Corollary 1.20.5. Assume that for $k > \ell > 0$

$$\begin{aligned} & \liminf_{s,t\to\infty} \frac{1}{(k-\ell)\ell} \sum_{j=1}^{k-\ell} \sum_{i=1}^{\ell} Q(s-ia-ja, t-ib) \\ & \qquad > \frac{p_4^{\ell+1}\ell^\ell}{(\ell+1)^{\ell+1}} \left(p_1 + \frac{2p_2p_3}{p_4}\right)^{-\ell} \left(\frac{p_4}{p_2}\right)^{(k-\ell+1)/2} \end{aligned} \tag{1.20.30}$$

and for $\ell > k > 0$

$$\liminf_{s,t\to\infty} \frac{1}{(\ell-k)k}\sum_{j=1}^{\ell-k}\sum_{i=1}^{k} Q(s-ia, t-ib-jb)$$

$$> \frac{p_4^{k+1}k^k}{(k+1)^{k+1}}\left(p_1+\frac{2p_2p_3}{p_4}\right)^{-k}\left(\frac{p_4}{p_3}\right)^{(\ell-k+1)/2} \tag{1.20.31}$$

Then, every solution of (E_{61}) is oscillatory.

Proof. We note that

$$\max_{p_4/c>\lambda>0} \lambda(p_4-\lambda c)^\ell = \frac{p_4^{\ell+1}\ell^\ell}{c(\ell+1)^{\ell+1}}.$$

We shall use this for

$$c = \frac{1}{(k-\ell)\ell}\sum_{j=1}^{k-\ell}\sum_{i=1}^{\ell} Q(s-ia-ja, t-ib).$$

Clearly,

$$\lambda\left[\prod_{j=1}^{k-\ell}\prod_{i=1}^{\ell}(p_4-\lambda Q(s-ia-ja,t-ib))\right]^{1/(k-\ell)}$$

$$\le \lambda\left[\frac{\lambda}{(k-\ell)\ell}\sum_{j=1}^{k-\ell}\sum_{i=1}^{\ell}(p_4-\lambda Q(s-ia-ja,t-ib))\right]^\ell$$

$$\le \lambda\left[p_4-\frac{\lambda}{(k-\ell)\ell}\sum_{j=1}^{k-\ell}\sum_{i=1}^{\ell}Q(s-ia-ja,t-ib)\right]^\ell$$

$$\le p_4^{\ell+1}\frac{\ell^\ell}{(\ell+1)^{\ell+1}}\left[\frac{1}{(k-\ell)\ell}\sum_{j=1}^{k-\ell}\sum_{i=1}^{\ell}Q(s-ia-ja,t-ib)\right]^{-1}$$

$$< \left(p_1+\frac{2p_2p_3}{p_4}\right)^\ell\left(\frac{p_4}{p_2}\right)^{(k-\ell+1)/2}.$$

Similarly, we have

$$\lambda\left[\prod_{j=1}^{\ell-k}\prod_{i=1}^{k}(p_4-\lambda Q(s-ia,t-ib-jb))\right]^{1/(\ell-k)}$$

$$< \left(p_1+\frac{2p_2p_3}{p_4}\right)^k\left(\frac{p_4}{p_3}\right)^{(\ell-k+1)/2}$$

Hence (1.20.30) and (1.20.31) imply that (1.20.26) and (1.20.27) hold. Now, by Theorem 1.20.4 every solution of (E_{61}) oscillates. ■

Theorem 1.20.6. Assume that there exist $S \geq s_0$, $T \geq t_0$ such that if $k = \ell > 0$

$$\sup_{\lambda \in E, s \geq S, t \geq T} \lambda \prod_{i=1}^{k} (p_4 - \lambda Q(s - ia, t - ib)) < \left(p_1 + \frac{2p_2p_3}{p_4}\right)^k. \quad (1.20.32)$$

Then, every solution of (E_{61}) is oscillatory.

Proof. Let $\mu \in \mathcal{S}$. Then, from (1.20.10) we have

$$w(s,t) \leq \left(p_1 + \frac{2p_2p_3}{p_4}\right)^{-k} \prod_{i=1}^{k} (p_4 - \mu Q(s - ia, t - ib)) w(s - ka, t - kb).$$

The rest of the proof is similar to that of Theorem 1.20.2. ■

Corollary 1.20.7. Assume that $k = \ell > 0$, and

$$\liminf_{s,t \to \infty} Q(s,t) = q > \frac{p_4^{k+1} k^k}{(k+1)^{k+1}} \left(p_1 + \frac{2p_2p_3}{p_4}\right)^{-k}.$$

Then, every solution of (E_{61}) is oscillatory.

Proof. It suffices to note that

$$\max_{p_4/q > \lambda > 0} \lambda (p_4 - \lambda q)^k = \frac{p_4^{k+1} k^k}{q(k+1)^{k+1}}. \quad ■$$

Theorem 1.20.8. Assume that there exist $S \geq s_0$, $T \geq t_0$ such that if $k,\ \ell > 0$

$$\sup_{\lambda \in E, s \geq S, t \geq T} \lambda \prod_{i=1}^{k} (p_4 - \lambda Q(s - ia, t)) \prod_{j=1}^{\ell} (p_4 - \lambda Q(s - ka, t - jb)) < p_2^k p_3^\ell. \quad (1.20.33)$$

Then, every solution of (E_{61}) is oscillatory.

Proof. Let $\mu \in \mathcal{S}$. Then, we have

$$\begin{aligned} &w(s,t) \\ &\leq \frac{1}{p_2}(p_4 - \mu Q(s - a, t)) w(s - a, t) \\ &\leq \left(\frac{1}{p_2}\right)^k \prod_{i=1}^{k} (p_4 - \mu Q(s - ia, t)) w(s - ka, t) \end{aligned}$$

$$\leq \left(\frac{1}{p_2}\right)^k \left(\frac{1}{p_3}\right) \prod_{i=1}^{k}(p_4 - \mu Q(s-ia,t))(p_4 - \mu Q(s-ka,t-b)) \times w(s-ka,t-b)$$

$$\leq \left(\frac{1}{p_2}\right)^k \left(\frac{1}{p_3}\right)^\ell \prod_{i=1}^{k}(p_4 - \mu Q(s-ia,t)) \prod_{j=1}^{\ell}(p_4 - \mu Q(s-ka,t-jb)) \times w(s-ka,t-\ell b).$$

The rest of the proof is similar to that of Theorem 1.20.2. ∎

Corollary 1.20.9. Assume that $k,\ \ell > 0$, and

$$\liminf_{s,t\to\infty} Q(s,t) \ = \ q \ > \ \frac{p_4^{k+\ell+1}(k+\ell)^{k+\ell}}{p_2^k p_3^\ell (k+\ell+1)^{k+\ell+1}}. \tag{1.20.34}$$

Then, every solution of (E_{61}) is oscillatory.

Proof. It suffices to note that

$$\max_{p_4/q>\lambda>0} \lambda(p_4 - \lambda q)^{k+\ell} \ = \ \frac{p_4^{k+\ell+1}(k+\ell)^{k+\ell}}{q(k+\ell+1)^{k+\ell+1}}.$$ ∎

Theorem 1.20.10. Assume that there exist $S \geq s_0,\ T \geq t_0$ such that if $k,\ \ell > 0$

$$\sup_{\lambda\in E, s\geq S, t\geq T} \lambda \left[\prod_{j=1}^{\ell}\prod_{i=1}^{k}(p_4 - \lambda Q(s-ia,t-jb))\right]^{1/\ell} \ < \ p_2^k \left(\frac{p_3}{p_4}\right)^{(\ell+1)/2} \tag{1.20.35}$$

or

$$\sup_{\lambda\in E, s\geq S, t\geq T} \lambda \left[\prod_{i=1}^{k}\prod_{j=1}^{\ell}(p_4 - \lambda Q(s-ia,t-jb))\right]^{1/k} \ < \ p_3^\ell \left(\frac{p_2}{p_4}\right)^{(k+1)/2} \tag{1.20.36}$$

Then, every solution of (E_{61}) is oscillatory.

Proof. Let $\mu \in \mathcal{S}$. Then, eventually

$$p_2 w(s+a,t) \ \leq \ (p_4 - \mu Q(s,t))w(s,t) \tag{1.20.37}$$

and

$$p_3 w(s,t+b) \ \leq \ (p_4 - \mu Q(s,t))w(s,t). \tag{1.20.38}$$

Using (1.20.37), we get

$$\begin{aligned} w(s,t) &\le \frac{1}{p_2}(p_4 - \mu Q(s-a,t))w(s-a,t) \\ &\le \cdots \le \left(\frac{1}{p_2}\right)^k \prod_{i=1}^{k}(p_4 - \mu Q(s-ia,t))w(s-ka,t). \end{aligned}$$

Hence, in view of (1.20.10) and the fact that $p_3 \ge p_4$, for $j = 1, 2, \cdots, \ell$ we have

$$\begin{aligned} w(s,t-jb) &\le \frac{1}{p_2^k}\prod_{i=1}^{k}(p_4 - \mu Q(s-ia,t-jb))w(s-ka,t-jb) \\ &\le \left[\frac{1}{p_2^k}\prod_{i=1}^{k}(p_4 - \mu Q(s-ia,t-jb))\right] w(s-ka,t-\ell b) \end{aligned}$$

and so

$$\begin{aligned} &w^\ell(s,t) \\ &\le \prod_{j=1}^{\ell}\left(\frac{p_4}{p_3}\right)^j w(s,t-jb) \\ &\le \prod_{j=1}^{\ell}\left\{\left(\frac{p_4}{p_3}\right)^j \left[\frac{1}{p_2^k}\prod_{i=1}^{k}(p_4 - \mu Q(s-ia,t-jb))\right] w(s-ka,t-\ell b)\right\} \\ &= \frac{1}{p_2^{k\ell}}\left(\frac{p_4}{p_3}\right)^{\ell(\ell+1)/2}\left[\prod_{j=1}^{\ell}\prod_{i=1}^{k}(p_4-\mu Q(s-ia,t-jb))\right] w^\ell(s-ka,t-\ell b), \end{aligned}$$

i.e.

$$w(s,t) \le \left[\frac{1}{p_2^{k\ell}}\left(\frac{p_4}{p_3}\right)^{\ell(\ell+1)/2}\prod_{j=1}^{\ell}\prod_{i=1}^{k}(p_4 - \mu Q(s-ia,t-jb))\right]^{1/\ell} \times w(s-ka,t-\ell b).$$

Similarly, we have

$$w(s,t) \le \left(\frac{1}{p_3}\right)^\ell \prod_{j=1}^{\ell}(p_4 - \mu Q(s,t-jb))w(s,t-\ell b)$$

and

$$\begin{aligned} &w^k(s,t) \\ &\le \prod_{i=1}^{k}\left(\frac{p_4}{p_2}\right)^i w(s-ia,t) \end{aligned}$$

$$\leq \frac{1}{p_3^{\ell k}}\left(\frac{p_4}{p_2}\right)^{k(k+1)/2}\left[\prod_{i=1}^{k}\prod_{j=1}^{\ell}(p_4-\mu Q(s-ia,t-jb))\right]w^k(s-ka,t-\ell b),$$

i.e.

$$w(s,t) \leq \left[\frac{1}{p_3^{\ell k}}\left(\frac{p_4}{p_2}\right)^{k(k+1)/2}\prod_{i=1}^{k}\prod_{j=1}^{\ell}(p_4-\mu Q(s-ia,t-jb))\right]^{1/k}$$
$$\times\ w(s-ka,t-\ell b).$$

The rest of the proof is similar to that of Theorem 1.20.2. ∎

Corollary 1.20.11. Assume that $k,\ \ell > 0$, and

$$\liminf_{s,t\to\infty}\frac{1}{k\ell}\sum_{j=1}^{\ell}\sum_{i=1}^{k}Q(s-ia,t-jb) > p_2^{-k}\left(\frac{p_4}{p_3}\right)^{(\ell+1)/2}\frac{k^k}{(k+1)^{k+1}} \tag{1.20.39}$$

or

$$\liminf_{s,t\to\infty}\frac{1}{\ell k}\sum_{i=1}^{k}\sum_{j=1}^{\ell}Q(s-ia,t-jb) > p_3^{-\ell}\left(\frac{p_4}{p_2}\right)^{(k+1)/2}\frac{\ell^\ell}{(\ell+1)^{\ell+1}}. \tag{1.20.40}$$

Then, every solution of (E_{61}) is oscillatory.

Theorem 1.20.12. Assume that $k,\ \ell > 0$

$$\liminf_{s,t\to\infty}Q(s,t) = q > 0 \tag{1.20.41}$$

and

$$\limsup_{s,t\to\infty}Q(s,t) > p_4-\frac{p_1+p_2+p_3}{p_4}q > 0. \tag{1.20.42}$$

Then, every solution of (E_{61}) is oscillatory.

Proof. Suppose to the contrary $z(s,t)$ is an eventually positive solution. Let $w(s,t)$ be defined as in Lemma 1.20.1. Then, from (1.20.3) we have

$$p_1w(s+a,t+b)+p_2w(s+a,t)+p_3w(s,t+b)-p_4w(s,t)$$
$$+Q(s,t)w(s-a,t-b) \leq 0. \tag{1.20.43}$$

From (1.20.41) for any $\varepsilon > 0$, we have

$$Q(s,t) > q-\varepsilon \quad \text{for} \quad s\geq S,\ t\geq T.$$

Thus, from (1.20.43) and the above inequality, we obtain

$$w(s,t) \geq \frac{1}{p_4}(q-\varepsilon)w(s-a,t-b)$$

$$w(s,t) \;\geq\; \frac{1}{p_4}(q-\varepsilon)w(s-a,t)$$

and

$$w(s,t) \;\geq\; \frac{1}{p_4}(q-\varepsilon)w(s,t-b).$$

Substituting above inequalities in (1.20.43), we get

$$\left(\frac{p_1+p_2+p_3}{p_4}(q-\varepsilon) - p_4 + Q(s,t)\right) w(s,t) \;<\; 0,$$

which implies

$$\limsup_{s,t\to\infty} Q(s,t) \;\leq\; p_4 - \frac{p_1+p_2+p_3}{p_4}q.$$

But, this contradicts (1.20.42). ∎

Theorem 1.20.13. Assume that there exist $S \geq s_0$, $T \geq t_0$ such that if $k>0$, $\ell = 0$

$$\sup_{\lambda\in E, s\geq S, t\geq T} \lambda \prod_{i=1}^{k}(p_4 - \lambda Q(s-ia,t)) \;<\; p_2^k. \tag{1.20.44}$$

Then, every solution of (E_{61}) is oscillatory.

Proof. From (1.20.3), we have

$$p_1 w(s+a,t+b) + p_2 w(s+a,t) + p_3 w(s,t+b) - p_4 w(s,t)$$
$$+Q(s,t)w(s-ka,t) \;\leq\; 0.$$

Let $\mu \in \mathcal{S}$. Then,

$$\begin{aligned} w(s,t) \;&\leq\; \frac{1}{p_2}(p_4 - \mu Q(s-a,t))w(s-a,t) \\ &\leq\; \left(\frac{1}{p_2}\right)^k \prod_{i=1}^{k}(p_4 - \mu Q(s-ia,t))w(s-ka,t). \end{aligned}$$

The rest of the proof is similar to that of Theorem 1.20.2. ∎

Corollary 1.20.14. Assume that $k>0$, $\ell = 0$, and

$$\liminf_{s,t\to\infty} Q(s,t) \;>\; \frac{p_4^{k+1}k^k}{p_2^k(k+1)^{k+1}}. \tag{1.20.45}$$

Then, every solution of (E_{61}) is oscillatory.

Theorem 1.20.15. Assume that there exist $S \geq s_0$, $T \geq t_0$ such that if $\ell > 0$, $k = 0$

$$\sup_{\lambda \in E, s \geq S, t \geq T} \lambda \prod_{j=1}^{\ell} (p_4 - \lambda Q(s, t - jb)) \ < \ p_3^{\ell}. \tag{1.20.46}$$

Then, every solution of (E_{61}) is oscillatory.

Proof. From (1.20.3), we have

$$p_1 w(s+a, t+b) + p_2 w(s+a, t) + p_3 w(s, t+b) - p_4 w(s,t) + Q(s,t) w(s, t - \ell b) \ \leq \ 0.$$

Let $\mu \in \mathcal{S}$. Then,

$$\begin{aligned} w(s,t) \ &\leq \ \frac{1}{p_3}(p_4 - \mu Q(s, t-b)) w(s, t-b) \\ &\leq \ \left(\frac{1}{p_3}\right)^{\ell} \prod_{j=1}^{\ell} (p_4 - \mu Q(s, t - jb)) w(s, t - \ell b). \end{aligned}$$

The rest of the proof is similar to that of Theorem 1.20.2. ∎

Corollary 1.20.16. Assume that $\ell > 0$, $k = 0$, and

$$\liminf_{s,t \to \infty} Q(s,t) \ > \ \frac{p_4^{\ell+1} \ell^{\ell}}{p_3^{\ell} (\ell+1)^{\ell+1}}. \tag{1.20.47}$$

Then, every solution of (E_{61}) is oscillatory.

Theorem 1.20.17. Assume that $k = \ell = 0$, and

$$\limsup_{s,t \to \infty} Q(s,t) \ > \ p_4. \tag{1.20.48}$$

Then, every solution of (E_{61}) is oscillatory.

Proof. Suppose $z(s,t)$ is an eventually positive solution. Let $w(s,t)$ be defined as in Lemma 1.20.1. Then, from (1.20.3) we have

$$p_1 w(s+a, t+b) + p_2 w(s+a, t) + p_3 w(s, t+b) - p_4 w(s,t) + Q(s,t) w(s,t) \ \leq \ 0.$$

Hence, it follows that

$$(-p_4 + Q(s,t))\, w(s,t) \ < \ 0,$$

which implies

$$\limsup_{s,t\to\infty} Q(s,t) \;\leq\; p_4.$$

But, this contradicts (1.20.48). ■

Example 1.20.1. Consider the partial difference equation with continuous variables

$$z\left(s+\frac{1}{2},t-1\right)+z\left(s+\frac{1}{2},t\right)+e^2 z(s,t-1)-z(s,t)+2ez(s-1,t+2) \;=\; 0. \tag{E_{62}}$$

It is easy to see that (E_{62}) satisfies all the conditions of Corollary 1.20.7, so every solution of this equation is oscillatory. In fact, $z(s,t)=(-e)^{2s+t}$ is such a solution.

Chapter 2
Oscillation of Functional Differential Equations

2.1. Introduction

The purpose of this chapter is to present some recent results pertaining to the oscillation of n–th order functional differential equations with deviating arguments, and functional differential equations of neutral type. We shall mainly deal with integral criteria guaranteeing oscillation. While several results of this chapter were originally formulated for more complicated and/or more general differential equations, we discuss here a simplified version which makes the oscillation theory of functional differential equations transparent. We remark here that because of the large number of oscillation theorems presented in this chapter it was impossible to prove all these results. Instead we selected the proofs of only those results which we thought would best illustrate the various strategies and ideas involved.

The plan of this chapter is as follows: In Section 2.2 we introduce the terms oscillation, nonoscillation and almost oscillation of solutions of functional differential equations. Here several lemmas are also collected which are repeatedly used throughout this chapter. In Section 2.3 we shall obtain sufficient conditions as well as necessary and sufficient conditions for the oscillation of various types of ordinary differential equations which have been studied extensively in the recent years. In Section 2.4 we shall list a total of 28 results which as in Section 2.3 provide sufficient conditions, and necessary and sufficient conditions for the oscillation of all solutions of some special functional differential equations. One of the basic techniques in *Oscillation Theory* is to acquire criteria by comparing the given differential equation with others of the same order and/or of lower order whose oscillatory behavior is known in advance. In Sections 2.5 – 2.7 we shall present several such results. Sections 2.8 – 2.13 respectively are devoted to the study of the oscillatory behavior of nth order equations with middle term of order $(n-1)$, forced equations, forced equations with middle term of order $(n-1)$, superlinear forced equations, sublinear forced equations, and perturbed functional equations. In Sections 2.14 and 2.15 we shall establish the oscillation of functional differential equations of neutral type by comparing respectively with equations of nonneutral type, and with equations of the same form. In Section 2.16 we shall present easily verifiable sufficient conditions for the oscillation of neutral differential equations of mixed type. Sections 2.17 – 2.19 study higher order functional differential equations involving quasi–derivatives. In Section 2.17 we shall establish oscillation criteria by comparing such equations with second order ordinary linear and first order delay differential equations. Section 2.18 uses the results of Section 2.17 to examine such equations of neutral and damped type. In Section 2.19 we shall consider forced such equations with advanced arguments, and present oscillation criteria which do not hold for the corresponding ordinary differential equations. Finally, in Section 2.20 we shall provide sufficient conditions which guarantee the oscillation of all bounded solutions of certain higher order linear systems of delay differential equations.

2.2. Definitions, Notations and Preliminaries

As in Chapter 1, in what follows $\mathbb{R}$ denotes the real line, and $\mathbb{R}_0$ the interval $[0,\infty)$. Given continuous functions $F : \mathbb{R}_0 \times \mathbb{R}^{n+2} \to \mathbb{R}$, g, $\tau : \mathbb{R}_0 \to \mathbb{R}$, $\lim_{t\to\infty} g(t) = \infty$ and $\lim_{t\to\infty} \tau(t) = \infty$, we shall begin our study of oscillatory behavior of solutions of several particular cases of the following general nth order functional differential equation

$$x^{(n)}(t) + \delta F\left(t, x(t), x[g(t)], x'(t), \cdots, x^{(n-1)}(t), x^{(n)}[\tau(t)]\right) = 0, \quad (E_1)$$

where $\delta = \pm 1$.

By a *solution* of (E_1) we mean a function $x(t)$, $t \in [t_x, \infty) \subset \mathbb{R}_0$ which is n times continuously differentiable and satisfies (E_1) on the interval $[t_x, \infty)$. The number $t_x \geq 0$ depends on the particular solution $x(t)$ under consideration. Let S denote the set of all solutions of (E_1). A function $x \in S$ is said to be *oscillatory* if it has an unbounded set of zeros in $[t_x, \infty)$. If all functions $x \in S$ are oscillatory, then the equation (E_1) is said to be oscillatory. We say that a property P holds *eventually or for all large* t if there exists a $T \geq 0$ such that P holds for all $t \geq T$.

Now let $x(t)$, $t \in [t_0, \infty) \subset \mathbb{R}_0$ be an n–times continuously differentiable function satisfying the following property: there exists an integer m, $0 \leq m \leq n$ which is odd if n is even, and even if n is odd such that for $t \geq t_0$

$$\begin{aligned} x(t)x^{(i)}(t) &> 0, \quad i = 0, 1, \cdots, m \\ (-1)^{n+i}x(t)x^{(i)}(t) &< 0, \quad i = m+1, m+2, \cdots, n. \end{aligned}$$

Such a function will be referred to as a *function of degree* m for $t \geq t_0$.

The equation (E_1) is said to be *almost oscillatory* if

(i) for $\delta = 1$ and n even, every solution of (E_1) is oscillatory,

(ii) for $\delta = 1$ and n odd, every solution $x(t)$ of (E_1) is either oscillatory, or strongly decreasing in the sense that $|x^{(i)}(t)| \to 0$ monotonically as $t \to \infty$, $i = 0, 1, \cdots, n-1$,

(iii) for $\delta = -1$ and n even, every solution $x(t)$ of (E_1) is oscillatory, strongly decreasing, or else strongly increasing in the sense that

$$|x^{(i)}(t)| \to \infty \quad \text{monotonically as } t \to \infty, \quad i = 0, 1, \cdots, n-1, \tag{2.2.1}$$

(iv) for $\delta = -1$ and n odd, every solution $x(t)$ of (E_1) is either oscillatory or strongly increasing.

The equation (E_1) is said to have *Property A* if it is almost oscillatory and (2.2.1) is replaced by

$$\left| \frac{x(t)}{t^{n-1}} \right| \to \infty \quad \text{as } t \to \infty.$$

The following fundamental lemma is due to Kiguradge [160].

Lemma 2.2.1. Let $x(t)$ be a function such that it and each of its derivative up to order $(n-1)$ inclusive is absolutely continuous and of constant sign in an interval (t_0, ∞). If $x^{(n)}(t)$ is of constant sign and not identically zero on any interval of the form $[t_1, \infty)$ for some $t_1 \geq t_0$, then there exist a $t_x \geq t_0$ and an integer m, $0 \leq m \leq n$ with $n+m$ even for $x^{(n)}(t) \geq 0$, or $n+m$ odd for $x^{(n)}(t) \leq 0$, and such that for every $t \geq t_x$,

$$m > 0 \quad \text{implies} \quad x^{(k)}(t) > 0, \quad k = 0, 1, \cdots, m-1$$

and

$$m \leq n-1 \quad \text{implies} \quad (-1)^{m+k} x^{(k)}(t) > 0, \quad k = m, m+1, \cdots, n-1.$$

As a consequence of Lemma 2.2.1, we have

Lemma 2.2.2. If the function $x(t)$ is as in Lemma 2.2.1 and

$$x^{(n-1)}(t) x^{(n)}(t) \leq 0 \quad \text{for all } t \geq t_x,$$

then for every λ, $0 < \lambda < 1$, there exists a constant $M > 0$ such that

$$|x[\lambda t]| \geq M t^{n-1} |x^{(n-1)}(t)| \quad \text{for all large } t.$$

Other useful version of Lemma 2.2.2 is due to Philos [243], which we will now state.

Lemma 2.2.3. Let $x(t)$ be as in Lemma 2.2.2. If $\lim_{t\to\infty} x(t) \neq 0$, then for every λ, $0 < \lambda < 1$

$$x(t) \geq \frac{\lambda}{(n-1)!} t^{n-1} x^{(n-1)}(t) \quad \text{for all large } t.$$

Let $x(t)$ be as in Lemma 2.2.1. Then, it is easy to verify that for $t, s \in \mathbb{R}_0$

$$x^{(i)}(t) = \sum_{j=i}^{k-1} \frac{(t-s)^{j-i}}{(j-i)!} x^{(j)}(s) + \int_s^t \frac{(t-u)^{k-i-1}}{(k-i-1)!} x^{(k)}(u) du, \quad 0 \leq i < k \leq n.$$

This relation is known as *Taylor's formula with integral remainder.*

The following two lemmas are given in [282].

Lemma 2.2.4. Let $x(t)$ be a bounded n–times differentiable function on an interval $[t_0, \infty)$ with

$$x(t) > 0 \quad \text{and} \quad (-1)^n x^{(n)}(t) \geq 0 \quad \text{for} \quad t \geq t_0.$$

Then, there exists a $t_x \geq t_0$ such that

$$(-1)^k x^{(k)}(t) \geq 0 \quad \text{for every} \quad t \geq t_x, \quad k = 1, 2, \cdots, n$$

and

$$x(\xi) \geq \frac{(-1)^{n-1} x^{(n-1)}(\eta)}{(n-1)!} (\eta - \xi)^{n-1} \quad \text{for every } \xi, \eta \text{ with } t_x \leq \xi \leq \eta.$$

Lemma 2.2.5. If $x(t)$ is a function as in Lemma 2.2.4, then there exists a $t_x \geq t_0$ such that for every $t \geq t_x$

$$x(t) \geq (-1)^n \int_t^\infty \frac{(\xi - t)^{n-1}}{(n-1)!} x^{(n)}(\xi) d\xi.$$

The following two lemmas are proved by Staikos and Sficas [277].

Lemma 2.2.6. If $x(t)$ is an n–times differentiable function on $[t_0, \infty)$ with $x^{(k)}(t)$, $k = 0, 1, \cdots, n$ of constant sign on $[t_0, \infty)$, then for any $i = 0, 1, \cdots, n-2$ with $\lim_{t\to\infty} x^{(i)}(t) = c$, $c \in \mathbb{R}$, it follows that $\lim_{t\to\infty} x^{(i+1)}(t) = 0$.

Lemma 2.2.7. Consider the linear differential equation

$$z'(t) - \frac{\mu}{t} z(t) + \frac{h(t)}{t} = 0, \tag{2.2.2}$$

where μ is a positive integer, and h is a continuous function on $[T,\infty)$, $T \geq t_0 > 0$. If $\lim_{t\to\infty} |h(t)| = \infty$ and $x(t)$ is a solution of (2.2.2) with $x(T) = 0$, then $\lim_{t\to\infty} x(t) = \pm\infty$.

As in Mahfoud [211], we denote by

$$\mathbb{R}_{t_0} = (-\infty, t_0) \cup (t_0, \infty) \quad \text{for any } t_0 > 0$$

and consider the spaces

$$C(\mathbb{R}) = \{f : \mathbb{R} \to \mathbb{R} \mid f \text{ is continuous and } xf(x) > 0 \text{ for } x \neq 0\}$$

and

$$C_B(\mathbb{R}_{t_0}) = \{f \in C(\mathbb{R}) \,:\, f \text{ is of bounded variation on any interval } [a,b] \subset \mathbb{R}_{t_0}\}.$$

Lemma 2.2.8. Suppose $t_0 > 0$ and $f \in C(\mathbb{R})$. Then, $f \in C_B(\mathbb{R}_{t_0})$ if and only if $f(x) = G(x)H(x)$ for all $x : \mathbb{R}_{t_0}$, where $G : \mathbb{R}_{t_0} \to \mathbb{R}_+$ is nondecreasing on $(-\infty, -t_0)$ and nonincreasing on (t_0, ∞), and $H : \mathbb{R}_{t_0} \to \mathbb{R}$ and nondecreasing on $\mathbb{R}_{t_0}$.

Definition 2.2.1. We call G in Lemma 2.2.8 a *positive component* of f, H a *nondecreasing component* of f, and the ordered pair (G, H) a pair of *components* of f.

The following lemma is due to Koplatadze and Chanturia [169].

Lemma 2.2.9. Let a, $\tau : [t_0, \infty) \to \mathbb{R}$ be continuous functions, $a(t) \geq 0$ eventually, $\tau(t) \leq t$, $\tau'(t) \geq 0$ for $t \geq t_0$, and $\lim_{t\to\infty} \tau(t) = \infty$. If

$$\liminf_{t\to\infty} \int_{\tau(t)}^{t} a(s)ds > \frac{1}{e},$$

then

(i) the inequality

$$x'(t) + a(t)x[\tau(t)] \leq 0 \quad \text{eventually}$$

has no eventually positive solution,

(ii) the inequality

$$x'(t) + a(t)x[\tau(t)] \geq 0 \quad \text{eventually}$$

has no eventually negative solution,

(iii) the equation

$$x'(t) + a(t)x[\tau(t)] = 0$$

is oscillatory.

For inequalities with advanced arguments, the following lemma extends a result of Onose [234].

Lemma 2.2.10. Let $a, \tau : [t_0, \infty) \to \mathbb{R}$ be continuous functions, $a(t) \geq 0$ eventually, $\tau(t) \geq t$, and $\tau'(t) \geq 0$ for $t \geq t_0$. If

$$\liminf_{t\to\infty} \int_t^{\tau(t)} a(s)ds > \frac{1}{e},$$

then

(i) the inequality

$$x'(t) - a(t)x[\tau(t)] \geq 0 \quad \text{eventually}$$

has no eventually positive solution,

(ii) the inequality

$$x'(t) - a(t)x[\tau(t)] \leq 0 \quad \text{eventually}$$

has no eventually negative solution,

(iii) the equation

$$x'(t) - a(t)x[\tau(t)] = 0$$

is oscillatory.

The following lemma has been extracted from Kusano [177] and Ladas and Stavroulakis [198].

Lemma 2.2.11. Suppose that a and h are positive constants, and let

$$a^{1/n}\left(\frac{h}{n}\right)e > 1.$$

Then,

(i) for n odd, the inequality

$$x^{(n)}(t) - ax[t+h] \geq 0$$

has no eventually positive solution,

(ii) for n odd, the inequality

$$x^{(n)}(t) + ax[t-h] \leq 0$$

has no eventually positive solution,

(iii) for n even, the inequality

$$x^{(n)}(t) - ax[t-h] \ \geq \ 0$$

has no eventually positive bounded solution,

(iv) for n even, the inequality

$$x^{(n)}(t) - ax[t+h] \ \geq \ 0$$

has no eventually positive solution, i.e. this inequality has no solution $x(t)$ with $x^{(i)}(t) > 0$ eventually, $i = 0, 1, \cdots, n$.

The following lemma is obtained from Corollaries 4.2 and 4.3 in [165].

Lemma 2.2.12. Suppose that $q : [t_0, \infty) \to \mathbb{R}$ is continuous, $q(t) \geq 0$ eventually, and σ is a positive real number. Then, the following hold:

(i) If

$$\limsup_{t\to\infty} \int_t^{t+\sigma} \frac{(s-t)^i (t-s+\sigma)^{n-i-1}}{i!\ (n-i-1)!} q(s)ds \ > \ 1$$

holds for some $i = 0, 1, \cdots, n-1$, then the inequality

$$x^{(n)}(t) \ \geq \ q(t)x[t+\sigma]$$

has no eventually positive solution $x(t)$ which satisfies

$$x^{(j)}(t) \ > \ 0 \quad \text{eventually}, \quad j = 0, 1, \cdots, n.$$

(ii) If

$$\limsup_{t\to\infty} \int_{t-\sigma}^{t} \frac{(t-s)^i (s-t+\sigma)^{n-i-1}}{i!\ (n-i-1)!} q(s)ds \ > \ 1$$

holds for some $i = 0, 1, \cdots, n-1$, then the inequality

$$(-1)^n y^{(n)}(t) \ \geq \ q(t)y[t-\sigma]$$

has no eventually positive solution $y(t)$ which satisfies

$$(-1)^j y^{(j)}(t) \ > \ 0 \quad \text{eventually}, \quad j = 0, 1, \cdots, n.$$

As in Kitamura [165], for a function $g : \mathbb{R}_0 \to \mathbb{R}$ we put

$$\begin{aligned}
\sigma(t) &= \min\{t,\ g(t)\} \\
\mathcal{A}_g &= \{t \in \mathbb{R}_0 : g(t) > t\} \\
\mathcal{R}_g &= \{t \in \mathbb{R}_0 : g(t) < t\} \\
r(t) &= \max\{\min\{s,\ g(s)\} : 0 \leq s \leq t\} \\
\rho(t) &= \min\{\max\{s,\ g(s)\} : s \geq t\}.
\end{aligned}$$

Note that the following inequalities hold

$$g(s) \leq r(t) \quad \text{for} \quad r(t) < s < t$$

and

$$g(s) \geq \rho(t) \quad \text{for} \quad t < s < \rho(t).$$

These notations will be used frequently in later sections.

We shall also need the following elementary result.

Lemma 2.2.13 [131]. Let A and B be nonnegative. If $\lambda > 1$, then

$$A^\lambda - \lambda AB^{\lambda-1} + (\lambda - 1)B^\lambda \geq 0$$

and if $0 < \lambda < 1$, then

$$A^\lambda - \lambda AB^{\lambda-1} - (1 - \lambda)B^\lambda \leq 0.$$

In the above inequalities equality hold if and only if $A = B$.

2.3. Ordinary Differential Equations

Here we shall list many existing results for the oscillation of the following ordinary differential equations which are interesting special cases of (E_1), and have been studied extensively

$$x^{(n)}(t) + q(t)x^\lambda(t) = 0 \qquad (E_2)$$

$$x^{(n)}(t) + q(t)|x(t)|^\gamma \text{sgn}\, x(t) = 0 \qquad (E_3)$$

and

$$x^{(n)}(t) + q(t)f(x(t)) = 0, \qquad (E_4)$$

where $q : [t_0, \infty) \to [0, \infty)$ and $f : \mathbb{R} \to \mathbb{R}$ are continuous, $q(t) \not\equiv 0$ on any ray of the form $[t^*, \infty)$ for some $t^* \geq t_0 \geq 0$, $xf(x) > 0$ for $x \neq 0$, $f'(x) \geq 0$ for $|x| \geq k > 0$, $\lambda = r/s$, where r and s are positive odd integers, and γ is a positive real number.

In 1963, Licko and Švec [205] established the following interesting criterion for almost oscillation of (E_2).

Theorem 2.3.1. The equation (E_2) is almost oscillatory if and only if

$$\int^\infty s^{(n-1)\lambda} q(s)ds = \infty, \quad 0 < \lambda < 1 \qquad (2.3.1)$$

$$\int^\infty s^{n-1} q(s)ds = \infty, \quad \lambda > 1. \qquad (2.3.2)$$

The sufficiency part of Theorem 2.3.1 for n even and $\lambda > 1$ was given for the first time by Kiguradze [160, Theorem 5] in 1962. The case n even and $0 < \lambda < 1$ in Theorem 2.3.1 extends a result of Belohorec [33] who considered second order equations, and the case n even and $\lambda > 1$ in Theorem 2.3.1 generalizes a result obtained by Atkinson [30] for second order equations.

For $\lambda = 1$ many papers have appeared for the almost oscillation of the equation (E_2). In extending a result of Fite [56], Mikusinski [216] showed that the condition

$$\int^{\infty} s^{n-1-\epsilon} q(s)ds = \infty \quad \text{for some } \epsilon, \ 0 < \epsilon < n-1 \tag{2.3.3}$$

is sufficient for (E_2) with $\lambda = 1$ and n even to be oscillatory. When n is odd, he claimed that (E_2) with $\lambda = 1$ is almost oscillatory provided

$$\int^{\infty} s^{n-1} q(s)ds = \infty. \tag{2.3.4}$$

However, the proof given in [216] for this case is incorrect. A simple counterexample (see also Ananeva and Balaganskii [29], Kartsatos [150] and Ševelo [258]) is provided by the equation

$$x^{(3)}(t) + \frac{3}{8t^3}x(t) = 0, \quad t > 0.$$

Here condition (2.3.4) is satisfied, but $x(t) = t^{2/3}$ is a nonoscillatory solution which does not converge to zero as $t \to \infty$. Minkusinski's conclusion for n odd is true if (2.3.4) is replaced by (2.3.3). Next, Ananeva and Balaganskii [29] proved that if

$$\int^{\infty} s^{n-2} q(s)ds = \infty, \tag{2.3.5}$$

then the equation (E_2) with $\lambda = 1$ is almost oscillatory.

Thus, the interesting case occurs when $0 < \epsilon < 1$. For this case Kartsatos [141] studied more general equations of type (E_5) (below), and when specialized to equation (E_3) his result reads as follows:

Theorem 2.3.2. The equation (E_3) is almost oscillatory if

$$\int^{\infty} s^{n-1} q(s)ds = \infty, \quad \gamma > 1 \tag{2.3.6}$$

$$\int^{\infty} s^{n-1-\epsilon} q(s)ds = \infty, \quad \gamma = 1 \quad \text{for some } \epsilon > 0 \tag{2.3.7}$$

$$\int^{\infty} s^{(n-1)\gamma} q(s)ds = \infty, \quad 0 < \gamma < 1. \tag{2.3.8}$$

One of the first oscillation results for the more general differential equation

$$x^{(n)}(t) + H(t, x(t)) = 0, \quad n \text{ is even} \tag{E_5}$$

where $H : [t_0, \infty) \times \mathbb{R} \to \mathbb{R}$ is continuous, $xH(t,x) > 0$ for $x \neq 0$, $t \geq t_0$ and $H(t,x)$ is nondecreasing in x is due to Kiguradze [160], and it is concerned with the superlinear case.

Theorem 2.3.3. Let $H(t,x) = F(t,x^2)x$, where $F(t,y)$ is defined on $[t_0, \infty) \times \mathbb{R}$ and nonnegative and increasing in y. Moreover, assume that there exists a function $g : \mathbb{R}_+ \to \mathbb{R}_+$ such that $g'(x) \geq 0$, and

$$F(t,x^2) \geq F(t,c^2)g(x) \quad \text{for every } x \text{ in } \mathbb{R}_+ \text{ and any } c > 0$$

with $x > c$. Furthermore, suppose that for every $\epsilon > 0$

$$\int_{\epsilon}^{\infty} \frac{du}{ug(u)} < \infty \quad \text{and} \quad \int^{\infty} s^{n-1} F(s,c^2)ds = \infty.$$

Then, the equation (E_5) is almost oscillatory.

Kartsatos [142] showed that the conclusion of Theorem 2.3.3 remains valid for the equation (E_4). In fact, he proved the following:

Theorem 2.3.4. If condition (2.3.4) holds, and

$$\int_{\pm\epsilon}^{\pm\infty} \frac{du}{f(u)} < \infty \quad \text{for some } \epsilon > 0, \tag{2.3.9}$$

then the equation (E_4) is almost oscillatory.

Kamenev [139] extended Theorem 2.3.4 and established the following criterion:

Theorem 2.3.5. Suppose that there exists a nondecreasing continuously differentiable function $\phi : \mathbb{R}_+ \to \mathbb{R}_+$ such that

$$\int^{\pm\infty} \frac{du}{[\phi(|u|^{1/(n-1)})f(u)]} < \infty \tag{2.3.10}$$

and

$$\int^{\infty} s^{n-1} \frac{q(s)}{\phi(s)} ds = \infty, \tag{2.3.11}$$

then the equation (E_4) with $n \geq 2$ even is oscillatory.

In 1981, Kreith and Kusano [174] considered equation (E_5) and introduced new definitions for superlinearity and sublinearity of $H(t,x)$ which unify the corresponding definitions found in the literature, and obtained characterizations for the oscillation of the equation (E_5) with such super–sublinearity so as to cover all the known criteria presented above.

Definition 2.3.1. The function $H(t,x)$ is called *superlinear* if

$$\int^{\infty} \frac{H(\phi(u),c)}{H(\phi(u),u)}du < \infty \quad \text{and} \quad \int^{\infty} \frac{H(\phi(u),-c)}{H(\phi(u),-u)}du < \infty \tag{2.3.12}$$

for some constant $c > 0$ and every strictly increasing continuous function $\phi : \mathbb{R}_+ \to \mathbb{R}_+$ such that $\phi(u) \to \infty$ as $u \to \infty$.

Definition 2.3.2. The function $H(t,x)$ is called *sublinear* if

$$\int_{+0} \frac{H(\psi(u),c\eta(u))}{H(\psi(u),u\eta(u))}du < \infty \quad \text{and} \quad \int_{+0} \frac{H(\psi(u),-c\eta(u))}{H(\psi(u),-u\eta(u))}du < \infty \tag{2.3.13}$$

for some constant $c > 0$ and every pair of strictly decreasing continuous functions ψ, $\eta : (0,\theta) \to \mathbb{R}_+$, $\theta > 0$ such that $\psi(u) \to \infty$ and $\eta(u) \to \infty$ as $u \to 0$.

Theorem 2.3.6. Let n be even and $H(t,x)$ be superlinear. Then, the equation (E_5) is oscillatory if and only if

$$\int^{\infty} s^{n-1}|H(s,c)|ds = \infty \quad \text{for every } c \neq 0. \tag{2.3.14}$$

Proof. Let $x(t)$ be a nonoscillatory solution of the equation (E_5), say $x(t) > 0$ for $t \geq t_0 \geq 0$. Then, there exists a $T > t_0$ such that for $t \geq T$

$$\begin{aligned} x(t) &\geq \int_T^t \frac{(s-T)^{n-1}}{(n-1)!}H(s,x(s))ds + \frac{(t-T)^{n-1}}{(n-1)!}\int_t^{\infty} H(s,x(s))ds \\ &\geq \int_T^t \frac{(s-T)^{n-1}}{(n-1)!}H(s,x(s))ds = Y(t), \quad \text{(say)}. \end{aligned} \tag{2.3.15}$$

Let $u = Y(t)$. Since

$$\frac{du}{dt} = \frac{(t-T)^{n-1}}{(n-1)!}H(t,x(t)) > 0, \qquad t > T$$

$u = Y(t)$ has the inverse function, which we denote by $t = y(u)$. It is clear that $y(u) \to \infty$ as $u \to \infty$. Noting that

$$H(t,x(t)) \geq H(t,Y(t))$$

and using the fact that $H(t,x)$ is nondecreasing in x and (2.3.14), we have

$$\begin{aligned}\int_{t_1}^{t_2} \frac{(s-T)^{n-1}}{(n-1)!} H(s,c)ds &\leq \int_{t_1}^{t_2} \frac{(s-T)^{n-1}}{(n-1)!} H(s,x(s)) \frac{H(s,c)}{H(s,Y(s))} ds \\ &= \int_{Y(t_1)}^{Y(t_2)} \frac{H(y(u),c)}{H(y(u),u)} du \end{aligned} \tag{2.3.16}$$

for any $t_2 > t_1 > T$ and some constant $c > 0$. Letting $t_2 \to \infty$ in (2.3.16) and using (2.3.12), we have for some $c > 0$

$$\int_{t_1}^{\infty} \frac{(s-T)^{n-1}}{(n-1)!} H(s,c)ds \leq \int_{Y(t_1)}^{\infty} \frac{H(y(u),c)}{H(y(u),u)} du < \infty,$$

which contradicts (2.3.14).

Conversely, if (2.3.14) is not satisfied for some $c \neq 0$, then we can solve the integral equation

$$x(t) = \frac{c}{2} + \int_t^{\infty} \frac{(s-t)^{n-1}}{(n-1)!} H(s,x(s))ds \tag{2.3.17}$$

by the method of successive approximations, or prove the existence of a solution by the Schauder–Tychonoff fixed point theorem. The solution of (2.3.17) is clearly a nonoscillatory solution of the equation (E_5). ■

Theorem 2.3.7. Let n be even and $H(t,x)$ be sublinear. Then, the equation (E_5) is oscillatory if and only if

$$\int^{\infty} |H(s, cs^{n-1})|ds = \infty \quad \text{for every } c \neq 0. \tag{2.3.18}$$

Proof. Let $x(t)$ be a nonoscillatory solution of the equation (E_5) say $x(t) > 0$ for $t \geq t_0 \geq 0$. Then, there exists a $T > t_0$ such that (2.3.15) holds for $t \geq T$. Hence, we have

$$x(t) \geq \frac{(t-T)^{n-1}}{(n-1)!} \int_t^{\infty} H(s,x(s))ds \geq \frac{t^{n-1}}{n!} \int_t^{\infty} H(s,x(s))ds \tag{2.3.19}$$

for $t \geq T_1$ provided $T_1 > T$ is sufficiently large. Put

$$u = \Psi(t) = \frac{1}{n!} \int_t^{\infty} H(s,x(s))ds$$

and denote its inverse function $t = \psi(u)$. Since we must have $x^{(n-1)}(t) \to 0$ as $t \to \infty$, it is obvious that $\psi(u) \to \infty$ as $u \to 0$, and

$$\frac{du}{dt} = -\frac{1}{n!} H(t,x(t)).$$

In view of (2.3.19) and the fact that $H(t,x)$ is nondecreasing in x, we have

$$H(t,x(t)) \geq H(t,t^{n-1}\Psi(t)) \quad \text{for} \quad t \geq T_1$$

and

$$\begin{aligned}\frac{1}{n!}\int_{t_1}^{t_2} H(s,cs^{n-1})ds &\leq \int_{t_1}^{t_2} \frac{H(s,x(s))}{n!}\frac{H(s,cs^{n-1})}{H(s,s^{n-1}\Psi(s))}ds \\ &= \int_{\Psi(t_2)}^{\Psi(t_1)} \frac{H(\psi(u),c\psi^{n-1}(u))}{H(\psi(u),u\psi^{n-1}(u))}du \qquad (2.3.20)\end{aligned}$$

for any $t_2 > t_1 > T_1$ and some $c > 0$. Letting $t_2 \to \infty$ and using (2.3.13), we obtain from (2.3.20) that for some $c > 0$

$$\int_{t_1}^{\infty} H(s,cs^{n-1})ds \leq n!\int_{+0}^{\Psi(t_1)} \frac{H(\psi(u),c\psi^{n-1}(u))}{H(\psi(u),u\psi^{n-1}(u))}du < \infty,$$

which contradicts (2.3.18).

If (2.3.18) is violated for some $c > 0$ (respectively $c < 0$), then the equation (E_5) has a nonoscillatory solution $x(t)$ satisfying

$$\lim_{t\to\infty} \frac{x(t)}{t^{n-1}} = \text{constant} > 0 \quad (\text{respectively} < 0),$$

which is obtained as a solution to the integral equation

$$x(t) = \frac{c}{2}t^{n-1} + \int_T^t \frac{(t-s)^{n-2}}{(n-2)!}\int_s^{\infty} H(u,x(u))duds,$$

where $T > 0$ is chosen suitably large. ∎

In [43] Chanturia considered the equation (E_2) with $\lambda = 1$ and n even, and obtained the following oscillation criterion which improves those presented in [141,216].

Theorem 2.3.8. If

$$\liminf_{t\to\infty} t^{n-1}\int_t^{\infty} q(s)ds > \frac{M_n}{n-1},$$

where M_n denotes the maximum of

$$P_n(x) = x(1-x)\cdots(n-1-x)$$

on $[0,1]$, then the equation (E_2) with $\lambda = 1$ and n even is oscillatory.

Recently, for a special case of the equation (E_2), namely

$$x^{(n)}(t) + \frac{C}{t^{\alpha}}x(t) = 0, \quad n \text{ is even} \qquad (E_6)$$

where C and α are constants and $C > 0$, Tang and Shen [286] obtained the following necessary and sufficient conditions for the oscillation of all solutions.

Theorem 2.3.9. The equation (E_6) is oscillatory if and only if $\alpha < n$, or $\alpha = n$ and $C > M_n$, where M_n is defined in Theorem 2.3.8.

In the same paper [286], Tang and Shen also obtained the following sufficient condition for the equation (E_2) with $\lambda = 1$ and n even, to be oscillatory.

Theorem 2.3.10. Suppose that there exists a continuous function $\varphi : [0, \infty) \to [0, 1]$ such that

$$\int_{t_0}^{\infty} (1 - \varphi(s))Q(s) \exp\left[2\int_{t_0}^{u} (\varphi(u)Q(u))^{1/2} du\right] ds = \infty, \qquad (2.3.21)$$

where

$$Q(t) = \begin{cases} \dfrac{1}{(n-3)!}\displaystyle\int_t^{\infty} (s-t)^{n-3} q(s)ds, & n > 2 \\ q(t), & n = 2. \end{cases}$$

Then, the equation (E_2) with $\lambda = 1$ and n even is oscillatory.

Proof. We shall give the proof only for the case $n > 2$. In fact, the case $n = 2$ can be considered similarly. Let $x(t)$ be a nonoscillatory solution of the equation (E_2), say $x(t) > 0$ for $t \geq t_0 \geq 0$. Then, by a result proved in [43], there exists a $t_1 \geq t_0$ such that

$$x(t) > 0, \quad (-1)^{i+1} x^{(i)}(t) > 0, \quad i = 1, 2, \cdots, n-1 \quad \text{and} \quad x^{(n)}(t) \leq 0 \quad \text{for } t \geq t_1. \qquad (2.3.22)$$

Integrating (E_2), $(n-2)$ times from $t \geq t_1$ to u, using (2.3.22) and letting $u \to \infty$, we get

$$x''(t) \leq -\frac{1}{(n-3)!}\int_t^{\infty} (s-t)^{n-3} q(s)x(s)ds \leq -Q(t)x(t), \quad t \geq t_1. \qquad (2.3.23)$$

Set $w(t) = x'(t)/x(t)$. Then, we obtain

$$w'(t) + w^2(t) + Q(t) \leq 0 \quad \text{for} \quad t \geq t_1. \qquad (2.3.24)$$

For the function $\varphi(t)$ by using the mean value inequality and (2.3.24), we find

$$w'(t) + 2(\varphi(t)Q(t))^{1/2} w(t) + (1 - \varphi(t))Q(t) \leq 0$$

and so

$$\frac{d}{dt}\left[w(t)\exp\left(2\int_{t_1}^{t}(\varphi(s)Q(s))^{1/2}ds\right)\right]$$

$$+(1-\varphi(t))Q(t)\exp\left(2\int_{t_1}^{t}(\varphi(s)Q(s))^{1/2}ds\right) \le 0, \quad t \ge t_1.$$

Integrating the above inequality from t_1 to ∞, we get

$$\int_{t_1}^{\infty}(1-\varphi(s))Q(s)\exp\left(2\int_{t_1}^{s}(\varphi(u)Q(u))^{1/2}du\right)ds \le w(t_1) < \infty,$$

which contradicts (2.3.21). ∎

2.4. Functional Differential Equations

Here we shall consider the following functional differential equations which are interesting particular cases of (E_1)

$$x^{(n)}(t) + \delta q(t) f(x[g(t)]) = 0 \qquad (E_7;\delta)$$

and

$$x^{(n)}(t) + \delta H(t, x[g(t)]) = 0, \qquad (E_8;\delta)$$

where $\delta = \pm 1$, g, $q : [t_0, \infty) \to \mathbb{R}$, $f : \mathbb{R} \to \mathbb{R}$ and $H : [t_0, \infty) \times \mathbb{R} \to \mathbb{R}$ are continuous, $q(t) \ge 0$ eventually, $xf(x) > 0$ and $xH(t,x) > 0$ for $x \ne 0$ and $\lim_{t\to\infty} g(t) = \infty$.

In the literature there are many known oscillation criteria for equations which are more general than $(E_7;\delta)$ and $(E_8;\delta)$, and when these results are specialized to above equations, they become almost the same. So, we will give only one significant result and refer to the others which are repeated.

The following theorem is due to Kusano and Onose [183], and it extends Theorem 3.1.5 of Kamenev [139].

Theorem 2.4.1. Let $f'(x) \ge 0$ for $x \ne 0$, $g(t) \le t$, and $g'(t) \ge 0$ for $t \ge t_0$. In addition, assume that there exists a function $\phi(y)$ such that

$$\phi \in C^1[[0,\infty), \mathbb{R}], \quad \phi(y) > 0, \quad \phi'(y) \ge 0$$

$$\int_{\pm\epsilon}^{\pm\infty} \frac{dy}{f(y)\phi(y^{1/(n-1)})} < \infty \quad \text{for every} \quad \epsilon > 0$$

and

$$\int^{\infty} \frac{(g(s))^{n-1} q(s)}{\phi(g(s))} ds = \infty.$$

Then, the equation $(E_7;1)$ is almost oscillatory.

Next, Sficas and Staikos [263] obtained the following interesting result.

Theorem 2.4.2. Let $g(t) \leq t$ and $g'(t) \geq 0$ for $t \geq t_0$, and consider the functions q_1, q_2, ϕ, ψ subject to the following conditions:

(i) q_1, q_2 are nonnegative and locally integrable on $[t_0, \infty)$,

(ii) ϕ, ψ are defined at least on $\mathbb{R}_+$ and such that for any $y \neq 0$

$$y\phi(y) > 0 \quad \text{and} \quad y\psi(y) > 0,$$

(iii) the function $y\psi(y)$ is nondecreasing for $y > 0$, nonincreasing for $y < 0$ and such that

$$\int^{\pm\infty} \frac{dy}{y\psi(y)} < \infty,$$

(iv) the function $\dfrac{\phi(y)}{y\psi(y)}$ is nonincreasing for $y > 0$ and nondecreasing for $y < 0$, and

(v) for every μ sufficiently large

$$\int^{\infty} q_1(s)\frac{\phi[\mu g^{n-1}(s)]}{\psi[\mu g^{n-1}(s)]}ds = \infty \quad \text{and} \quad \int^{\infty} q_2(s)\frac{\phi[-\mu g^{n-1}(s)]}{\psi[-\mu g^{n-1}(s)]}ds = \infty.$$

If for any $t \geq t_0$

$$q_1(t)\phi(y) \leq H(t,y) \quad \text{for } y > 0 \quad \text{and} \quad H(t,y) \leq q_2(t)\phi(y) \quad \text{for } y < 0,$$

then the equation $(E_8; 1)$ is almost oscillatory.

Many other oscillatory criteria for the above equations are known in the literature. Here, we refer to the works of Kusano and Onose [177,182–184], Ladas [187], Marusiak [213], Onose [225–234], Ševelo and Vareh [259,260], Staikos and Sficas [277–281], Grammatikopoulos, Sficas and Staikos [124], and the references cited therein.

The following results are due to Grace and Lalli [90,91,93].

Theorem 2.4.3. Let n be even, $0 < \sigma(t) = \min\{t, g(t)\}$, $\sigma'(t) > 0$ for $t \geq t_0$, $\lim_{t\to\infty} \sigma(t) = \infty$, $f'(x) \geq 0$ for $x \neq 0$, and

$$\int^{\pm\infty} \frac{du}{f(u)} < \infty.$$

Then, each of the following conditions ensures the oscillation of the equation $(E_7; 1)$:

(I) $$\int^{\infty} \sigma'(s)\sigma^{n-2}(s)\left(\int_s^{\infty} q(u)du\right) ds = \infty,$$

(II) there exists a differentiable function $z : [t_0, \infty) \to \mathbb{R}_+$ such that $\int^{\infty} z(s)q(s)ds = \infty$, and either

(i) $z'(t) \geq 0$ and $\beta(t) = \left(\dfrac{z'(t)}{\sigma^{n-2}(t)\sigma'(t)}\right)' \leq 0$ for $t \geq t_0$, or

(ii) $z'(t) \geq 0$ and $\displaystyle\int_{t_0}^{t} |\beta(s)|ds < \infty$.

Theorem 2.4.4. Let n be even, $0 < \sigma(t) = \min\{t, g(t)\}$, $\sigma'(t) \geq 0$ for $t \geq t_0$, $\lim_{t\to\infty} \sigma(t) = \infty$,

$$-f(-xy) \geq f(xy) \;\geq\; Kf(x)f(y) \quad \text{for} \quad xy \neq 0,$$

f is sublinear, i.e. $\int_{\pm 0} du/f(u) < \infty$, and $f'(x) \geq 0$ for $x \neq 0$. If

$$\int^{\infty} f(\sigma^{n-1}(s))q(s)ds \;=\; \infty,$$

then the equation $(E_7; 1)$ is oscillatory.

Theorem 2.4.5. Let n be even, $0 < \sigma(t) = \min\{t, 2g(t)\}$, $\sigma'(t) > 0$ for $t \geq t_0$, $\lim_{t\to\infty} \sigma(t) = \infty$, and $f'(x) \geq \alpha > 0$ for $x \neq 0$. Suppose further that there exists a differentiable function $z : [t_0, \infty) \to \mathbb{R}_+$ such that

$$\limsup_{t\to\infty} \int_{t_0}^{t} \left[z(s)q(s) - \frac{(z'(s))^2}{2\alpha M_{1/2}\sigma^{n-2}(s)\sigma'(s)z(s)}\right] ds \;=\; \infty, \tag{2.4.1}$$

where $M_{1/2} = \dfrac{2^{2-2n}}{(n-1)!}$ ($M_{1/2}$ is as in Lemma 2.2.2 for $\lambda = 1/2$). Then, the equation $(E_7; 1)$ is oscillatory.

We note that the condition (2.4.1) can be written as

$$\limsup_{t\to\infty} \int_{t_0}^{t} z(s)q(s)ds \;=\; \infty$$

and

$$\lim_{t\to\infty} \int_{t_0}^{t} \frac{(z'(s))^2}{\sigma^{n-2}(s)\sigma'(s)z(s)} ds \;<\; \infty.$$

In [94], Grace and Lalli improved Theorem 2.4.5. In addition they extended Ohriska's result [223]. Their result can be stated as follows:

Theorem 2.4.6. Let n be even, $g(t) \leq t$ for $t \geq t_0$, and $f(x) = x$. If

(i) $\limsup_{t\to\infty} g^{n-1}(t)\int_t^\infty q(s)ds > (n-1)!$ and $g'(t) \geq 0$ for $t \geq t_0$, or

(ii) $\limsup_{t\to\infty} t^{n-1}\int_{\gamma(t)}^\infty q(s)ds > (n-1)!$, where $\gamma(t) = \sup\{s \geq t_0 : g(s) \leq t\}$ for $t \geq t_0$,

then the equation $(E_7; 1)$ is oscillatory.

Next, Grace and Lalli [92,94] obtained the following:

Theorem 2.4.7. Let n be even, $0 < \sigma(t) = \min\{t, g(t)\}$, $\sigma'(t) > 0$ for $t \geq t_0$, $\lim_{t\to\infty}\sigma(t) = \infty$, and $f'(x) \geq \alpha > 0$ for $x \neq 0$. If there exists a function $z \in C^1[[t_0,\infty), \mathbb{R}_+]$ such that

$$\limsup_{t\to\infty}\frac{1}{t^{m-1}}\int_{t_0}^t\left[(t-s)^{m-1}z(s)q(s)\right.$$
$$\left.-\frac{(t-s)^{m-3}[z'(s)(t-s)-(m-1)z(s)]^2}{2M_{1/2}\alpha z(s)\sigma^{n-2}(s)\sigma'(s)}\right]ds \;=\; \infty, \qquad (2.4.2)$$

where $m \geq 3$ and $M_{1/2}$ is as in Theorem 2.4.5, then the equation $(E_7; 1)$ is oscillatory.

Once again condition (2.4.2) can be written as

$$\limsup_{t\to\infty}\frac{1}{t^{m-1}}\int_{t_0}^t (t-s)^{m-1}z(s)q(s)ds \;=\; \infty$$

and

$$\lim_{t\to\infty}\frac{1}{t^{m-1}}\int_{t_0}^t (t-s)^{m-3}\frac{[z'(s)(t-s)-(m-1)z(s)]^2}{z(s)\sigma^{n-2}(s)\sigma'(s)}ds \;<\; \infty.$$

Before we state further results we need to introduce the following:

$$A(t) \;=\; \exp\left(-2a\int_{t_0}^t \sigma^{n-2}(s)\sigma'(s)ds\right),$$

where $a = (1/2)M_{1/2}\alpha$, $M_{1/2}$ and α are defined in Theorem 2.4.7, and let

$$\Psi(t) \;=\; \int_{t_0}^t A(s)ds.$$

Also, we define a sequence of functions $C_k(t)$, $k = 0, 1, \cdots$ as follows:

$$C_0(t) \;=\; \int_t^\infty q(s)ds \quad \text{assuming it exists}$$

$$C_1(t) = \int_t^\infty \sigma^{n-2}(s)\sigma'(s)C_0^2(s)ds$$

and

$$C_{k+1}(t) = \int_t^\infty \sigma^{n-2}(s)\sigma'(s)[C_0(s)+aC_k(s)]^2ds, \qquad k=1,2,\cdots.$$

Theorem 2.4.8. Let n be even, $0<\sigma(t)=\min\{t,g(t)\}$, $\sigma'(t)\geq 0$ for $t\geq t_0$, $\lim_{t\to\infty}\sigma(t)=\infty$, and $f'(x)\geq\alpha>0$ for $x\neq 0$. Suppose that there exists a positive integer m such that $C_k(t)$ is defined for $k=0,1,\cdots,m-1$ but $C_m(t)$ does not exist, then the equation $(E_7;1)$ is oscillatory.

Theorem 2.4.9. Let n be even, $0<\sigma(t)=\min\{t,g(t)\}$, $\sigma'(t)\geq 0$ for $t\geq t_0$, $\lim_{t\to\infty}\sigma(t)=\infty$, $f'(x)\geq\alpha>0$ for $x\neq 0$, and

$$\int^{\pm\infty}\frac{du}{f(u)} < \infty.$$

Suppose that there exists a positive integer m such that $C_k(t)$ is defined for $k=0,1,\cdots,m-1$, and

$$\int^\infty \sigma^{n-2}(s)\sigma'(s)\left[C_0(s)+aC_m(s)\right]ds = \infty.$$

Then, the equation $(E_7;1)$ is oscillatory.

Theorem 2.4.10. Let n be even, $0<\sigma(t)=\min\{t,g(t)\}$, $\sigma'(t)>0$ for $t\geq t_0$, $\lim_{t\to\infty}\sigma(t)=\infty$, and $f'(x)\geq\alpha>0$ for $x\neq 0$. If

$$\limsup_{t\to\infty}\left[\left(\psi(t)+\left[(1+\psi(t))\int_{t_0}^t\frac{ds}{\sigma^{n-1}(s)\sigma'(s)}\right]^{1/2}\right)^{-1}\left(\int_{t_0}^t C_0(s)ds\right)\right]=\infty,$$

then the equation $(E_7;1)$ is oscillatory.

In 1982, Ivanov, Kitamura, Kusano and Shevelo [135] further studied $(E_7;\delta)$ and established the following results:

Theorem 2.4.11. Let $n\geq 3$ be odd, and there exists a positive constant M such that $f(x)$ is monotonically increasing for $|x|\geq M$,

$$\int_{\pm M}^{\pm\infty}\frac{du}{f(u)} < \infty, \qquad \int^\infty s^{n-1}q(s)ds = \infty,$$

$$\int_{A_g}(g(s)-s)^{n-1}q(s)ds = \infty \quad\text{and}\quad \int_{A_g}s^{n-2}(g(s)-s)\,q(s)ds = \infty.$$

Then, the equation $(E_7; -1)$ is oscillatory.

Theorem 2.4.12. Let n be even, and there exist positive numbers M and m such that $f(u)$ is monotonically increasing for $|u| \geq M$ and $|u| \leq m$, and

$$\int_{\pm M}^{\pm\infty} \frac{du}{f(u)} < \infty \quad \text{and} \quad \int_{\pm 0}^{\pm m} \frac{du}{f(u)} < \infty.$$

If

$$\int_{\mathcal{R}_g} (s - g(s))^{n-1} q(s) ds = \infty,$$

$$\int_{\mathcal{A}_g} (g(s) - s)^{n-1} q(s) ds = \infty \quad \text{and} \quad \int_{\mathcal{A}_g} s^{n-2} (g(s) - s) q(s) ds = \infty,$$

then the equation $(E_7; -1)$ is oscillatory.

Theorem 2.4.13. Let n be even, and there exists a positive constant M such that $f(u)$ is monotonically increasing for $|u| \geq M$, and

$$\int_{\pm M}^{\pm\infty} \frac{du}{f(u)} < \infty.$$

If

$$\int^{\infty} s^{n-1} q(s) ds = \infty \quad \text{and} \quad \int_{\mathcal{A}_g} s^{n-2} (g(s) - s) q(s) ds = \infty,$$

then the equation $(E_7; 1)$ is oscillatory.

Theorem 2.4.14. Let $n \geq 3$ be odd, and there exist positive numbers M and m such that $f(x)$ is monotonically increasing for $|u| \geq M$ and $|u| \leq m$, and

$$\int_{\pm M}^{\pm\infty} \frac{du}{f(u)} < \infty \quad \text{and} \quad \int_{0}^{\pm m} \frac{du}{f(u)} < \infty.$$

If

$$\int_{\mathcal{A}_g} s^{n-2} (g(s) - s) q(s) ds = \infty \quad \text{and} \quad \int_{\mathcal{R}_g} (s - g(s))^{n-1} q(s) ds = \infty,$$

then the equation $(E_7; 1)$ is oscillatory.

The following interesting result is due to Kusano [178].

Theorem 2.4.15. Let n be even, $0 < \sigma(t) = \min\{t, g(t)\}$, $\lim_{t\to\infty} \sigma(t) = \infty$, $g'(t) \geq 0$ for $t \geq t_0$ and of mixed type, and suppose that there is a constant $\epsilon > 0$ such that

$$\int^{\infty} (\sigma(s))^{n-1-\epsilon} q(s)ds = \infty.$$

Suppose moreover that there exist two sequences $\{t_k\}$ and $\{\tau_k\}$ such that $t_k \in \mathcal{A}_g$, $t_k \to \infty$ as $k \to \infty$ and $\tau_k \in \mathcal{R}_g$, $\tau_k \to \infty$ as $k \to \infty$. If

$$\int_{t_k}^{g(t_k)} (g(s) - g(t_k))^{n-1} q(s)ds \geq (n-1)!$$

and

$$\int_{g(\tau_k)}^{\tau_k} (g(\tau_k) - g(s))^{n-1} q(s)ds \geq (n-1)!$$

for all $k = 1, 2, \cdots$, then the equation $(E_7; -1)$ is oscillatory.

Kitamura [165] gave complete criteria for the oscillation of $(E_7; \delta)$, $\delta = \pm 1$. His results require the condition

$$\inf\left\{\frac{f(\xi x)}{f(\xi)} \;:\; \xi \neq 0\right\} > 0 \quad \text{for any} \quad x > 0 \tag{2.4.3}$$

and defines

$$\begin{aligned} w[f](x) &= \operatorname{sgn} x \cdot \inf\left\{\frac{f(\xi|x|)}{f(\xi)} \;:\; \xi x \neq 0\right\} \quad \text{if} \quad x \neq 0 \\ &= 0 \qquad \text{if} \quad x = 0. \end{aligned}$$

Theorem 2.4.16. Let n be even.

(i) Suppose that

$$\int^{\pm\infty} \frac{du}{f(u)} < \infty.$$

Then, the condition

$$\int^{\infty} \sigma^{n-1}(s)q(s)ds = \infty$$

is sufficient for the equation $(E_7; 1)$ to be oscillatory. If in addition

$$\liminf_{t\to\infty} \frac{g(t)}{t} > 0,$$

then the condition

$$\int^{\infty} s^{n-1}q(s)ds = \infty$$

is a necessary and sufficient condition for the equation $(E_7; 1)$ to be oscillatory.

(ii) Suppose that condition (2.4.3) holds, and

$$\int_{\pm 0}^{\pm\infty} \frac{du}{w[f](u)} < \infty.$$

Then, the condition

$$\int^{\infty} q(s) \left| f\left(\pm\sigma^{n-1}(s)\right)\right| ds = \infty$$

is sufficient for the equation $(E_7; 1)$ to be oscillatory. If in addition

$$\limsup_{t\to\infty} \frac{g(t)}{t} < \infty,$$

then the condition

$$\int^{\infty} q(s) \left| f\left(\pm g^{n-1}(s)\right)\right| ds = \infty$$

is necessary and sufficient for the equation $(E_7; 1)$ to be oscillatory.

(iii) Suppose that

$$\frac{f(x)}{x} \geq k \quad \text{for} \quad x \neq 0,$$

where k is a positive constant. If

$$\int^{\infty} \sigma^{n-1}(s) g^{-\epsilon}(s) q(s) ds = \infty \quad \text{for some} \quad \epsilon > 0,$$

then the equation $(E_7; 1)$ is oscillatory.

Theorem 2.4.17. Let n be odd.

(i) Suppose that

$$\int^{\pm\infty} \frac{du}{f(u)} < \infty.$$

Then, the condition

$$\int^{\infty} \sigma^{n-1}(s) q(s) ds = \infty$$

is sufficient for the equation $(E_7; 1)$ to be almost oscillatory. If in addition

$$\liminf_{t\to\infty} \frac{g(t)}{t} > 0,$$

then the condition

$$\int^{\infty} s^{n-1} q(s) ds = \infty$$

is a necessary and sufficient condition for the equation $(E_7; 1)$ to be almost oscillatory.

(ii) Suppose that condition (2.4.3) holds, and

$$\int_{\pm 0} \frac{du}{w[f](u)} < \infty.$$

Then, the conditions

$$\int^{\infty} s^{n-1} q(s) ds = \infty \quad \text{and} \quad \int^{\infty} q(s) \left| f\left(\pm g(s) \sigma^{n-2}(s)\right) \right| ds = \infty$$

are sufficient for the equation $(E_7; 1)$ to be almost oscillatory. If in addition

$$\limsup_{t \to \infty} \frac{g(t)}{t} < \infty,$$

then the conditions

$$\int^{\infty} s^{n-1} q(s) ds = \infty \quad \text{and} \quad \int^{\infty} q(s) \left| f\left(\pm g^{n-1}(s)\right) \right| ds = \infty$$

are necessary and sufficient for the equation $(E_7; 1)$ to be almost oscillatory.

(iii) Suppose that

$$\int_{\pm 0}^{\pm \infty} \frac{du}{f(u)} < \infty.$$

Then, the conditions

$$\int^{\infty} g(s) \sigma^{n-2}(s) q(s) ds = \infty \quad \text{and} \quad \int_{\mathcal{R}_g} (s - g(s))^{n-1} q(s) ds = \infty$$

are sufficient for the equation $(E_7; 1)$ to be oscillatory. If in addition

$$0 < \liminf_{t \to \infty} \frac{g(t)}{t} \le \limsup_{t \to \infty} \frac{g(t)}{t} < 1,$$

then the condition

$$\int^{\infty} s^{n-1} q(s) ds = \infty$$

is a necessary and sufficient condition for the equation $(E_7; 1)$ to be oscillatory.

(iv) Suppose that condition (2.4.3) holds, and

$$\int_{\pm 0}^{\pm\infty} \frac{du}{w[f](u)} < \infty.$$

Then, the conditions

$$\int^{\infty} q(s) \left| f\left(\pm g(s)\sigma^{n-2}(s)\right) \right| ds = \infty$$

and

$$\int_{\mathcal{R}_g} (s - g(s))^{n-1} q(s) ds = \infty \quad \text{or} \quad \int_{\mathcal{R}_g} q(s) \left| f\left(\pm(s - g(s)\right)^{n-1} \right| ds = \infty$$

are sufficient for the equation $(E_7; 1)$ to be oscillatory. If in addition

$$\limsup_{t\to\infty} \frac{g(t)}{t} > 1,$$

then the condition

$$\int^{\infty} q(s) \left| f\left(\pm g^{n-1}(s)\right) \right| ds = \infty$$

is necessary and sufficient for the equation $(E_7; 1)$ to be oscillatory.

(v) Suppose that

$$\frac{f(x)}{x} \geq k > 0 \quad \text{for} \quad x \neq 0.$$

If

$$\int^{\infty} \sigma^{n-2}(s) g^{1-\epsilon}(s) q(s) ds = \infty \quad \text{for some} \quad \epsilon > 0$$

and

$$\int^{\infty} s^{n-1} q(s) ds = \infty,$$

then the equation $(E_7; 1)$ is almost oscillatory. If in addition

$$\limsup_{t\to\infty} \int_{\tau(t)}^{t} \frac{(s - \tau(t))^{n-i-1}}{(n-i-1)!} \frac{(\tau(t) - \tau(s))^{i}}{i!} q(s) ds > \frac{1}{k}$$

for some $i = 0, 1, \cdots, n-1$, then the equation $(E_7; 1)$ is oscillatory.

Theorem 2.4.18. Let n be even.

(i) Suppose that

$$\int^{\pm\infty} \frac{du}{f(u)} < \infty.$$

Then, the equation $(E_7;-1)$ is almost oscillatory if

$$\int^{\infty} s\sigma^{n-2}(s)q(s)ds = \infty$$

and

$$\int^{\infty} q(s)\left|f\left(cg^{n-1}(s)\right)\right|ds = \infty \quad \text{for any} \quad c \neq 0.$$

Suppose moreover that

$$\liminf_{t\to\infty} \frac{g(t)}{t} > 0.$$

Then, the equation $(E_7;-1)$ is almost oscillatory if and only if

$$\int^{\infty} s^{n-1}q(s)ds = \infty$$

and

$$\int^{\infty} q(s)\left|f\left(cg^{n-1}(s)\right)\right|ds = \infty \quad \text{for any} \quad c \neq 0.$$

(ii) Suppose that condition (2.4.3) holds, and

$$\int_{\pm 0} \frac{du}{w[f](u)} < \infty.$$

Then, the equation $(E_7;-1)$ is almost oscillatory if

$$\int^{\infty} s^{n-1}q(s)ds = \infty \quad \text{and} \quad \int^{\infty} q(s)\left|f\left(\pm g(s)\sigma^{n-2}(s)\right)\right|ds = \infty.$$

Suppose moreover that

$$\limsup_{t\to\infty} \frac{g(t)}{t} < \infty.$$

Then, the equation $(E_7;-1)$ is almost oscillatory if and only if

$$\int^{\infty} s^{n-1}q(s)ds = \infty \quad \text{and} \quad \int^{\infty} q(s)\left|f\left(\pm g^{n-1}(s)\right)\right|ds = \infty.$$

(iii) Suppose that

$$\int_{\pm 0}^{\pm\infty} \frac{du}{f(u)} < \infty.$$

Then, the equation $(E_7;-1)$ is oscillatory if

$$\int^{\infty} sg(s)\sigma^{n-3}(s)q(s)ds = \infty \quad \text{for} \quad n > 2$$

$$\int_{\mathcal{R}_g} (s-g(s))^{n-1}\, q(s)ds \;=\; \infty \quad \text{and} \quad \int_{\mathcal{A}_g} (g(s)-s)^{n-1}\, q(s)ds \;=\; \infty.$$

(iv) Suppose that condition (2.4.3) holds, and

$$\int_{\pm 0}^{\pm\infty} \frac{du}{w[f](u)} \;<\; \infty.$$

Then, the equation $(E_7; -1)$ is oscillatory if

$$\int^{\infty} q(s) \left| f\left(\pm s g(s)\sigma^{n-3}(s)\right)\right| ds \;=\; \infty \quad \text{for} \quad n > 2$$

$$\int_{\mathcal{R}_g} (s-g(s))^{n-1} q(s)ds \;=\; \infty \quad \text{or} \quad \int_{\mathcal{R}_g} q(s) \left| f\left(\pm(s-g(s))^{n-1}\right)\right| ds \;=\; \infty$$

and

$$\int_{\mathcal{A}_g} (g(s)-s)^{n-1} q(s)ds \;=\; \infty \quad \text{or} \quad \int_{\mathcal{A}_g} q(s) \left| f\left(\pm(g(s)-s)^{n-1}\right)\right| ds \;=\; \infty.$$

(v) Suppose that

$$\frac{f(x)}{x} \;\geq\; k \;>\; 0 \quad \text{for} \quad x \neq 0.$$

If

$$\int^{\infty} s\sigma^{n-3}(s) g^{1-\epsilon}(s) q(s)ds \;=\; \infty \quad \text{for some} \quad \epsilon > 0,$$

$$\int^{\infty} s^{n-1} q(s)ds \;=\; \infty \quad \text{and} \quad \int^{\infty} g^{n-1}(s) q(s)ds \;=\; \infty,$$

then the equation $(E_7; -1)$ is almost oscillatory. If in addition

$$\limsup_{t\to\infty} \int_{\tau(t)}^{t} \frac{(s-\tau(t))^{n-i-1}}{(n-i-1)!} \frac{(\tau(t)-g(s))^{i}}{i!} q(s)ds \;>\; \frac{1}{k}$$

for some $i = 0, 1, \cdots, n-1$, and

$$\limsup_{t\to\infty} \int_{t}^{\rho(t)} \frac{(g(s)-\rho(t))^{j}}{j!} \frac{(\rho(t)-s)^{n-j-1}}{(n-j-1)!} q(s)ds \;>\; \frac{1}{k}$$

for some $j = 0, 1, \cdots, n-1$, then the equation $(E_7; -1)$ is oscillatory.

Theorem 2.4.19. Let n be odd.

(i) Suppose that

$$\int^{\pm\infty} \frac{du}{f(u)} \;<\; \infty.$$

Then, the equation $(E_7; -1)$ is almost oscillatory if

$$\int^{\infty} s\sigma^{n-2}(s)q(s)ds = \infty$$

and

$$\int^{\infty} q(s)\left|f\left(cg^{n-1}(s)\right)\right| ds = \infty \quad \text{for any} \quad c \neq 0.$$

Suppose moreover that

$$\liminf_{t\to\infty} \frac{g(t)}{t} > 0.$$

Then the equation $(E_7; -1)$ is almost oscillatory if and only if

$$\int^{\infty} s^{n-1}q(s)ds = \infty$$

and

$$\int^{\infty} q(s)\left|f\left(cg^{n-1}(s)\right)\right| ds = \infty \quad \text{for any} \quad c \neq 0.$$

(ii) Suppose that condition (2.4.3) holds, and

$$\int_{\pm 0} \frac{du}{w[f](u)} < \infty.$$

Then, the equation $(E_7; -1)$ is almost oscillatory if

$$\int^{\infty} q(s)\left|f\left(\pm\sigma^{n-1}(s)\right)\right| ds = \infty.$$

Suppose moreover that

$$\limsup_{t\to\infty} \frac{g(t)}{t} < \infty.$$

Then, the equation $(E_7; -1)$ is almost oscillatory if and only if

$$\int^{\infty} q(s)\left|f\left(\pm g^{n-1}(s)\right)\right| ds = \infty.$$

(iii) Suppose that

$$\int^{\pm\infty} \frac{du}{f(u)} < \infty.$$

Then, the equation $(E_7; -1)$ is oscillatory if

$$\int^{\infty} s\sigma^{n-2}(s)q(s)ds = \infty \quad \text{and} \quad \int_{A_g} (g(s) - s)^{n-1} q(s)ds = \infty.$$

Suppose moreover that

$$\liminf_{t\to\infty} \frac{g(t)}{t} > 1.$$

Then, the equation $(E_7; -1)$ is oscillatory if and only if

$$\int^{\infty} s^{n-1} q(s) ds = \infty.$$

(iv) Suppose that

$$\int_{\pm 0}^{\pm\infty} \frac{du}{w[f](u)} < \infty.$$

Then, the equation $(E_7; -1)$ is oscillatory if

$$\int^{\infty} q(s) \left| f\left(\pm s\sigma^{n-2}(s)\right)\right| ds = \infty$$

and

$$\int_{A_g} (g(s) - s)^{n-1} q(s) ds = \infty \text{ or } \int_{A_g} q(s) \left| f\left(\pm (g(s) - s)^{n-1}\right)\right| ds = \infty.$$

Suppose moreover that

$$1 < \liminf_{t\to\infty} \frac{g(t)}{t} \leq \limsup_{t\to\infty} \frac{g(t)}{t} < \infty.$$

Then, the equation $(E_7; -1)$ is oscillatory if and only if

$$\int^{\infty} q(s) \left| f\left(\pm g^{n-1}(s)\right)\right| ds = \infty.$$

(v) Suppose that

$$\frac{f(x)}{x} \geq k > 0 \quad \text{for} \quad x \neq 0.$$

If

$$\int^{\infty} s\sigma^{n-2}(s) g^{-\epsilon}(s) q(s) ds = \infty \quad \text{for some} \quad \epsilon > 0$$

and

$$\int^{\infty} g^{n-1}(s) q(s) ds = \infty,$$

then the equation $(E_7; -1)$ is almost oscillatory. If in addition

$$\limsup_{t\to\infty} \int_t^{\rho(t)} \frac{(g(s) - \rho(t))^j}{j!} \frac{(\rho(t) - s)^{n-j-1}}{(n-j-1)!} q(s) ds > \frac{1}{k}$$

for some $j = 0, 1, \cdots, n-1$, then the equation $(E_7; -1)$ is oscillatory.

Now, we shall consider $(E_7;\delta)$, $\delta = \pm 1$ when $g(t)$ is an advanced argument, and present Werbowski's [302] results which are for the case $f(x) = x$.

In the following results we shall assume that the function g is quickly increasing, i.e.

$$g(t) - g(s) \ \geq \ t - s \quad \text{for} \quad t \geq s \geq t_0.$$

Theorem 2.4.20. Let n be even. If

$$\liminf_{t\to\infty} \int_t^{g(t)} \int_s^\infty \frac{(v-s)^{n-2}}{(n-2)!} q(v)dvds \ > \ \frac{1}{e},$$

then the equation $(E_7;1)$ is oscillatory.

We note that the equation

$$x''(t) + \frac{1}{4t^2}x(t) \ = \ 0, \quad t > 0$$

has a nonoscillatory solution $x(t) = \sqrt{t}$, while the advanced differential equation

$$x''(t) + \frac{1}{4t^2}x[t^2] \ = \ 0, \quad t > 0$$

is oscillatory by Theorem 2.4.20. Clearly the advanced argument generates oscillation.

Theorem 2.4.21. Let n be odd, the condition of Theorem 2.4.20 hold, and let $r : \mathbb{R}_+ \to \mathbb{R}_+$ be a continuous function such that $r(t) \leq t \leq g[r(t)]$. If for some $\nu = 1, 2, \cdots, n-1$

$$\liminf_{t\to\infty} \int_t^{g[r(t)]} \int_{r(v)}^{v} \frac{(v-s)^{n-\nu-1}}{(n-\nu-1)!} \frac{(g(s) - g[r(v)])^{\nu-1}}{(\nu-1)!} q(s)dsdv \ > \ \frac{1}{e}, \tag{2.4.4}$$

then $(E_7;-1)$ is oscillatory.

Theorem 2.4.22. Let $n > 2$ be even. If (2.4.4),

$$\int^\infty s^{n-1}q(s)ds \ = \ \infty$$

and

$$\liminf_{t\to\infty} \int_t^{g(t)} \int_s^\infty (v-s)^{n-3} \left(g(v) - g(s)\right) q(v)dvds \ > \ \frac{c_n}{e},$$

where $c_n = (n-2-((-1)^n+1)/2)!$, then every solution $x(t)$ of $(E_7;-1)$ is either oscillatory or $\lim_{t\to\infty} x^{(k)}(t) = 0$ monotonically, $k = 0, 1, \cdots, n-1$.

When the function g is slowly increasing, i.e.

$$t - s > g(t) - g(s) \geq 0 \quad \text{for} \quad t > s \geq t_0,$$

Werbowski [302] obtained the following:

Theorem 2.4.23. The equation $(E_7; 1)$ is almost oscillatory if

$$\liminf_{t\to\infty} \int_t^{g(t)} \int_s^\infty \frac{(g(v) - g(s))^{n-2}}{(n-2)!} q(v)dvds > \frac{1}{e}.$$

Theorem 2.4.24. If $\displaystyle\int^\infty sq(s)ds = \infty$ for $n = 2$, and when $n > 2$

$$\liminf_{t\to\infty} \int_t^{g(t)} \int_s^\infty (v - s)(g(v) - g(s))^{n-3} q(v)dvds > \frac{c_n}{e},$$

where c_n is defined in Theorem 2.4.22, and (2.4.4) is satisfied, then every solution $x(t)$ of $(E_7; -1)$ for n odd is oscillatory, while for n even it is either oscillatory or $x^{(k)}(t) \to 0$ monotonically as $t \to \infty$, $k = 0, 1, \cdots, n-1$.

Finally, when g is nondecreasing, Werbowski [302] obtained the following results:

Theorem 2.4.25. Let n be odd. If (2.4.4), and

$$\liminf_{t\to\infty} \int_t^{g(t)} \frac{s}{g(s)} \int_s^\infty \frac{(v - s)^{n-2}}{(n-2)!} q(v)dvds > \frac{1}{e},$$

then $(E_7; -1)$ is oscillatory.

Theorem 2.4.26. Let $n > 2$ be even. If (2.4.4),

$$\int^\infty s^{n-1} q(s)ds = \infty$$

and

$$\liminf_{t\to\infty} \int_t^{g(t)} \frac{s}{g(s)} \int_s^\infty (v - s)^{n-3}(g(v) - s)q(v)dvds > \frac{c_n}{e},$$

where c_n is as in Theorem 2.4.22, then the conclusion of Theorem 2.4.22 holds.

Now let n be even, and

$$\int_{t_0}^\infty q(s)ds < \infty.$$

We define a sequence $\{c_k(t)\}$, $k = 0, 1, \cdots$, as follows

$$c_0(t) = \int_t^\infty q(s)ds$$

and

$$c_k(t) = c_0(t) + \int_t^\infty a(s)c_{k-1}^2(s)ds, \quad k = 1, 2, \cdots$$

where $a(t) = (1/2)M_{1/2}\alpha\sigma^{n-2}(t)\sigma'(t)$, and $M_{1/2}$ and α are defined in Theorem 2.4.5.

The following result is extracted from Grace [81].

Theorem 2.4.27. Let n be even, $0 < \sigma(t) = \min\{t, g(t)\}$, $\sigma'(t) \geq 0$ for $t \geq t_0$, $\lim_{t\to\infty} \sigma(t) = \infty$, and $f'(x) \geq \alpha > 0$ for $x \neq 0$. Then, $(E_7; 1)$ is oscillatory if either one of the following holds:

(I) There exists an integer $m > 0$ such that $c_k(t)$, $k = 0, 1, \cdots, m-1$ are defined, and

$$\lim_{t\to\infty} \int_{t_0}^t a(s)c_{m-1}^2(s)ds = \infty.$$

(II) $c_k(t)$ is defined for $k = 0, 1, \cdots$ but there is a $T^* \geq t_0$ such that

$$\lim_{k\to\infty} c_k(T^*) = \infty.$$

(III) There exists a k such that $c_k(t)$ is defined, and

$$\limsup_{t\to\infty} c_k(t) \left(\int_{t_0}^t a(s)ds \right) > 1.$$

Finally, we shall prove the following result for the oscillation of $(E_7; 1)$ when n is even, and

$$A(t) = \int_t^\infty q(s)ds \geq 0, \quad t \geq t_0.$$

Theorem 2.4.28. Let n be even, $0 < \sigma(t) = \min\{t, 2g(t)\}$, $\sigma'(t) > 0$ for $t \geq t_0$, $\lim_{t\to\infty} \sigma(t) = \infty$, and $f(x) \operatorname{sgn} x \geq |x|^\gamma$ for $x \neq 0$ and γ is a positive real number. If for all positive constants c_1, c_2, θ, $0 < \theta < 1$

(i) $$\int^\infty \sigma^{n-2}(s)\sigma'(s) \left[A(s) + \gamma c_1 \int_s^\infty A^2(\tau)\sigma^{n-2}(\tau)\sigma'(\tau)d\tau \right] ds = \infty$$ when $\gamma > 1$,

(ii) $$\limsup_{t\to\infty} \frac{\sigma^{n-1}(t)}{(n-1)!} \left[A(t) + \theta \int_t^\infty \frac{\sigma^{n-2}(s)}{(n-2)!} A^2(s)ds \right] > 1$$ when $\gamma = 1$,

$$\text{(iii)}\ \limsup_{t\to\infty} \frac{\sigma^{n-1}(t)}{(n-1)!}(A(t))^{1/\gamma}\left[1+\frac{\theta\gamma}{(n-2)!}\frac{1}{c_2A(t)}\int_t^\infty \sigma^{n-2}(s)\frac{\sigma'(s)}{2}\right.$$

$$\left.\times\ A^{1+1/\gamma}(s)ds\right] > c_2 \ \text{when}\ A(t)>0 \ \text{and}\ 0<\gamma<1,$$

then the equation $(E_7;1)$ is oscillatory.

Proof. Let $x(t)$ be an eventually positive solution of $(E_7;1)$, say $x(t)>0$ for $t \ge t_0 \ge 0$. By Lemma 2.2.1 there exists a $t_1 \ge t_0$ such that

$$x^{(n-1)}(t) > 0 \ \text{and}\ x'(t) > 0 \ \text{for}\ t \ge t_1 \tag{2.4.5}$$

and

$$x^{(n)}(t) + q(t)x^{\gamma}[g(t)] \le 0 \ \text{for}\ t \ge t_1.$$

Now there exist a $t_2 \ge t_1$, and constants $K_i > 0$, $i = 1,2$ such that $\sigma(t) \ge 2t_1$ for $t \ge t_2$

$$x\left[\frac{\sigma(t)}{2}\right] \ge K_1 \ \text{and}\ x^{(n-1)}(t) \le K_2 \ \text{for}\ t \ge t_2 \tag{2.4.6}$$

and

$$x^{(n)}(t) + q(t)x^{\gamma}\left[\frac{\sigma(t)}{2}\right] \le 0 \ \text{for}\ t \ge t_2. \tag{2.4.7}$$

Integrating (2.4.7) from t to $u \ge t \ge t_0$ and letting $u \to \infty$, we get

$$x^{(n-1)}(t) \ge x^{\gamma}\left[\frac{\sigma(t)}{2}\right]\left[A(t)+\gamma\int_t^\infty x^{-\gamma-1}\left[\frac{\sigma(s)}{2}\right]x'\left[\frac{\sigma(s)}{2}\right]\right.$$

$$\left.\times\ \frac{\sigma'(s)}{2}x^{(n-1)}(s)ds\right]. \tag{2.4.8}$$

We need to consider the following three cases:

(i) Let $\gamma > 1$. Then, by Lemma 2.2.3 there exist a $t_3 \ge t_2$, and θ_1, $\theta_2 > 0$, $0 < \theta_1$, $\theta_2 < 1$ such that for $t \ge t_3$

$$x'\left[\frac{\sigma(t)}{2}\right] \ge \frac{\theta_1}{(n-2)!}\sigma^{n-2}(t)x^{(n-1)}(t) \tag{2.4.9}$$

and

$$x\left[\frac{\sigma(t)}{2}\right] \ge \frac{\theta_2}{(n-1)!}\sigma^{n-1}(t)x^{(n-1)}(t). \tag{2.4.10}$$

Using (2.4.9) in (2.4.8), we obtain

$$x'\left[\frac{\sigma(t)}{2}\right] \ge \frac{\theta_1\sigma^{n-2}(t)}{(n-2)!}x^{\gamma}\left[\frac{\sigma(t)}{2}\right]\left[A(t)+\gamma\int_t^\infty \frac{\theta_1}{(n-2)!}\sigma^{n-2}(s)\frac{\sigma'(s)}{2}\right.$$

$$\times A^2(s)x^{\gamma-1}\left[\frac{\sigma(s)}{2}\right]ds\Bigg]$$

or

$$\frac{x'\left[\frac{\sigma(t)}{2}\right]\frac{\sigma'(t)}{2}}{x^\gamma\left[\frac{\sigma(t)}{2}\right]} \geq \frac{\theta_1}{(n-2)!}\sigma^{n-2}(t)\frac{\sigma'(t)}{2}\left[A(t)+\gamma K_1^{\gamma-1}\frac{\theta_1}{(n-2)!}\right.$$

$$\left.\times\int_t^\infty \sigma^{n-2}(s)\frac{\sigma'(s)}{2}A^2(s)ds\right].$$

Integrating this inequality from t_3 to $t \geq t_3$ we get the desired contradiction.

(ii) Let $\gamma = 1$. From (2.4.8), we have

$$x^{(n-1)}(t) \geq A(t)x\left[\frac{\sigma(t)}{2}\right]. \tag{2.4.11}$$

Using (2.4.9) – (2.4.11) in (2.4.8), we obtain

$$1 \geq \frac{\theta_2}{(n-1)!}\sigma^{n-1}(t)\left[A(t)+\frac{\theta_1}{(n-2)!}\int_t^\infty \sigma^{n-2}(s)\frac{\sigma'(s)}{2}A^2(s)ds\right].$$

Taking lim sup on both sides of the above inequality as $t \to \infty$, we get the desired contradiction.

(iii) Let $0 < \gamma < 1$. From (2.4.6) and (2.4.8), we have

$$K_2 \geq x^{(n-1)}(t) \geq A(t)x^\gamma\left[\frac{\sigma(t)}{2}\right] \quad \text{for} \quad t \geq t_2 \tag{2.4.12}$$

or

$$x\left[\frac{\sigma(t)}{2}\right] \leq K_2^{1/\gamma}(A(t))^{-1/\gamma} \tag{2.4.13}$$

and

$$x^{\gamma-1}\left[\frac{\sigma(t)}{2}\right] \geq K_2^{(\gamma-1)/\gamma}(A(t))^{(1-\gamma)/\gamma}. \tag{2.4.14}$$

Also, from (2.4.8) – (2.4.10), we find

$$x^{(n-1)}(t)x^{1-\gamma}\left[\frac{\sigma(t)}{2}\right]$$

$$\geq A(t)x\left[\frac{\sigma(t)}{2}\right]+\gamma x\left[\frac{\sigma(t)}{2}\right]\int_t^\infty x^{-\gamma-1}\left[\frac{\sigma(s)}{2}\right]x'\left[\frac{\sigma(s)}{2}\right]\frac{\sigma'(s)}{2}x^{(n-1)}(s)ds$$

$$\geq \frac{\theta_2 A(t)}{(n-1)!}\sigma^{n-1}(t)x^{(n-1)}(t)+\frac{\gamma\theta_2}{(n-1)!}\sigma^{n-1}(t)x^{(n-1)}(t)$$

$$\times\int_t^\infty A^2(s)\frac{\theta_1}{(n-2)!}\sigma^{n-2}(s)\frac{\sigma'(s)}{2}x^{\gamma-1}\left[\frac{\sigma(s)}{2}\right]ds$$

or

$$x^{1-\gamma}\left[\frac{\sigma(t)}{2}\right] \geq \frac{\theta_2}{(n-1)!}\sigma^{n-1}(t)\left[A(t)+\frac{\gamma\theta_1}{k\,(n-2)!}\int_t^\infty A^2(s)\sigma^{n-2}(s)\right.$$
$$\left.\times\ \sigma'(s)A^{1+1/\gamma}(s)ds\right],$$

where $k = K_2^{(1-\gamma)/\gamma}$. Using (2.4.13) in the above inequality, we obtain

$$k(A(t))^{(\gamma-1)/\gamma} \geq \frac{\theta_2}{(n-1)!}\sigma^{n-1}(t)\left[A(t)+\frac{\gamma}{k}\frac{\theta_1}{(n-2)!}\int_t^\infty \sigma^{n-2}(s)\frac{\sigma'(s)}{2}\right.$$
$$\left.\times\ A^{1+1/\gamma}(s)ds\right]$$

or

$$k \geq \frac{\theta_2}{(n-1)!}\sigma^{n-1}(t)(A(t))^{1/\gamma}\left[1+\frac{\gamma}{kA(t)}\frac{\theta_1}{(n-2)!}\int_t^\infty \sigma^{n-2}(s)\frac{\sigma'(s)}{2}\right.$$
$$\left.\times\ A^{1+1/\gamma}(s)ds\right].$$

Taking $\limsup$ on both sides of the above inequality as $t \to \infty$ we obtain a contradiction to (iii). ∎

2.5. Comparison of Equations of the Same Form

An important technique in *Oscillation Theory* is to acquire comparison results for the oscillatory and asymptotic behavior of solutions of differential equations with and without deviating arguments. We trace this idea back to 1971, when Kartsatos [145] obtained results of this nature for ordinary differential equations. We begin with the following lemma which is essentially due to Kartsatos [148], however the present version is taken from Foster and Grimmer [60].

Lemma 2.5.1. Consider

$$x^{(n)}(t) + H(t, x(t)) = 0 \tag{E_9}$$

and

$$z^{(n)}(t) + H(t, z(t)) \leq 0, \tag{2.5.1}$$

where $H : [t_0, \infty) \times \mathbb{R} \to \mathbb{R}$, $t_0 \geq 0$ is continuous, increasing in its second variable, and such that $uH(t,u) > 0$ for $u \neq 0$. Let $z(t)$ be an eventually positive solution of the inequality (2.5.1). Then, for some

t_0, $z(t)$ has degree m for $t \geq t_0$, where m is an integer such that $0 \leq m \leq n-1$. Moreover, m is odd if n is even and m is even if n is odd. If $1 \leq m \leq n-1$ and if x_0 is such that $0 < x_0 \leq z(t_0)$, then there exists a solution $x(t)$ of the equation (E_9) with $x(t_0) = x_0$ satisfying for $t \geq t_0$

$$0 < x^{(k)}(t) \leq z^{(k)}(t), \quad k = 0, 1, \cdots, m$$

$$0 > (-1)^{n+k} x^{(k)}(t) \geq (-1)^{n+k} z^{(k)}(t), \quad k = m+1, m+2, \cdots, n.$$

If $m = 0$ where n is odd and x_∞ satisfies $0 < x_\infty < z(\infty)$, then there exists a solution $x(t)$ of (E_9) with $\lim_{t\to\infty} x(t) = x_\infty$ such that

$$0 < (-1)^k x^{(k)}(t) \leq (-1)^k z^{(k)}(t), \quad k = 0, 1, \cdots, n, \quad t \geq t_0.$$

This lemma shows that the existence of an eventually positive solution of the inequality (2.5.1) implies the same for the equation (E_9).

A special case of Theorem 2.1 of Kartsatos [148] is the following:

Theorem 2.5.2. Let $H_i(t, u)$, $i = 1, 2$ be continuous on $[t_0, \infty) \times \mathbb{R}$, $t_0 \geq 0$, increasing in the second variable, and such that

$$\begin{aligned} &uH_i(t, u) > 0 \quad \text{for} \quad u \neq 0, \quad i = 1, 2, \quad \text{and} \\ &H_1(t, u) \operatorname{sgn} u \leq H_2(t, u) \operatorname{sgn} u \end{aligned} \tag{2.5.2}$$

and the equation

$$x^{(n)}(t) + H_1(t, x(t)) = 0, \quad n \text{ is even}$$

is oscillatory. Then, the equation

$$x^{(n)}(t) + H_2(t, x(t)) = 0, \quad n \text{ is even}$$

is also oscillatory.

Several results of the similar nature, but for more general equations with deviating arguments, appeared later in the works of Kusano and Naito [180], Philos, Sficas and Staikos [249], Grace and Lalli [105] and many others.

The following results are taken from [180], and are concerned with the comparison of equations of the form

$$x^{(n)}(t) + \delta H_1(t, x[g_1(t)]) = 0 \tag{$E_{10}; \delta$}$$

and

$$y^{(n)}(t) + \delta H_2(t, y[g_2(t)]) = 0, \tag{$E_{11}; \delta$}$$

where $\delta = \pm 1$, $H_i : [t_0, \infty) \times \mathbb{R} \to \mathbb{R}$, $i = 1, 2$ are continuous, $t_0 \geq 0$, $g_i : [t_0, \infty) \to \mathbb{R}$, $i = 1, 2$ are continuous, and $\lim_{t\to\infty} g_i(t) = \infty$, $i = 1, 2$.

Theorem 2.5.3. Suppose that in addition to the condition (2.5.2),

$$g_1(t) \leq g_2(t), \quad t \geq t_0 \tag{2.5.3}$$

and

$$H_1(t,x) \quad \text{is nondecreasing in } x \text{ for each } t \geq t_0. \tag{2.5.4}$$

If the equation $(E_{11};\delta)$ has Property A, then the equation $(E_{10};\delta)$ has Property A.

Theorem 2.5.3 is equivalent to the following:

Theorem 2.5.4. Suppose that conditions (2.5.2) – (2.5.4) hold.

(i) If the equation $(E_{11};1)$ has a nonoscillatory solution $y(t)$ satisfying

$$\liminf_{t\to\infty} |y(t)| > 0, \tag{2.5.5}$$

then the equation $(E_{10};1)$ has a nonoscillatory solution $x(t)$ satisfying

$$\liminf_{t\to\infty} |x(t)| > 0. \tag{2.5.6}$$

(ii) If the equation $(E_{11};-1)$ has a nonoscillatory solution $y(t)$ satisfying (2.5.5), and

$$\limsup_{t\to\infty} |y^{(n-1)}(t)| < \infty, \tag{2.5.7}$$

then the equation $(E_{10};-1)$ has a nonoscillatory solution $x(t)$ satisfying (2.5.6), and

$$\limsup_{t\to\infty} |x^{(n-1)}(t)| < \infty. \tag{2.5.8}$$

A more general comparison theorem is the following:

Theorem 2.5.5. Suppose that conditions (2.5.2) – (2.5.4) hold.

(i) If there exists a nonoscillatory function $y(t)$ satisfying $\liminf_{t\to\infty} |y(t)| > 0$, and the inequality

$$\{y^{(n)}(t) + H_2(t, y[g_2(t)])\} \operatorname{sgn} y(t) \leq 0 \tag{2.5.9}$$

holds eventually, then the equation $(E_{10};1)$ has a nonoscillatory solution $x(t)$ satisfying $\liminf_{t\to\infty} |x(t)| > 0$.

(ii) If there exists a nonoscillatory function $y(t)$ satisfying

$$\liminf_{t\to\infty} |y(t)| > 0, \quad \limsup_{t\to\infty} |y^{(n-1)}(t)| < \infty, \tag{2.5.10}$$

and the inequality

$$\{y^{(n)}(t) - H_2(t, y[g_2(t)])\} \operatorname{sgn} y(t) \geq 0 \tag{2.5.11}$$

holds eventually, then the equation $(E_{10};-1)$ has a nonoscillatory solution $x(t)$ satisfying

$$\liminf_{t\to\infty} |x(t)| \;>\; 0, \qquad \limsup_{t\to\infty} |x^{(n-1)}(t)| \;<\; \infty. \tag{2.5.12}$$

For the particular case $H_1 = H_2$ and $g_1 = g_2$, Kusano and Naito [180] obtained the following:

Corollary 2.5.6. (i) The equation $(E_{10};1)$ has a solution $x(t)$ such that $\liminf_{t\to\infty} |x(t)| > 0$ if and only if there exists a function $y(t)$ satisfying the inequality (2.5.9) eventually, and $\liminf_{t\to\infty} |y(t)| > 0$.
(ii) The equation $(E_{10};-1)$ has a solution $x(t)$ such that

$$\liminf_{t\to\infty} |x(t)| \;>\; 0 \quad \text{and} \quad \limsup_{t\to\infty} |x^{(n-1)}(t)| \;<\; \infty$$

if and only if there exists a function $y(t)$ satisfying the inequality (2.5.11) eventually,

$$\liminf_{t\to\infty} |y(t)| \;>\; 0 \quad \text{and} \quad \limsup_{t\to\infty} |y^{(n-1)}(t)| \;<\; \infty.$$

Also, Philos [241] established the following criterion:

Theorem 2.5.7. If $y(t)$ is a positive and strictly decreasing solution of the integral inequality

$$y(t) \;\geq\; \int_t^\infty \frac{(s-t)^{n-1}}{(n-1)!} H(s, y[g(s)])ds,$$

where $H : [t_0,\infty) \times \mathbb{R} \to \mathbb{R}$ and $g : [t_0,\infty) \to \mathbb{R}$ are continuous, $t_0 \geq 0$, $xH(t,x) > 0$ for $x \neq 0$, $t \geq t_0$, $H(t,x)$ is increasing in the second variable, $g(t) < t$ and $\lim_{t\to\infty} g(t) = \infty$, then there exists a positive solution $x(t)$ of the equation

$$(-1)^n x^{(n)}(t) \;=\; H(t, x[g(t)])$$

such that $x(t) \leq y(t)$ eventually, and satisfy

$$\lim_{t\to\infty} x^{(i)}(t) \;=\; 0 \quad \text{monotonically}, \quad i = 0, 1, \cdots, n-1.$$

An immediate corollary of Theorem 2.5.7 is the following:

Corollary 2.5.8. If $y(t)$ is a positive bounded solution of

$$(-1)^n y^{(n)}(t) \;\geq\; H(t, y[g(t)]),$$

then the conclusion of Theorem 2.5.7 holds.

Next, we shall discuss comparison results in which oscillatory and asymptotic properties of solutions of differential equations with deviating arguments are inherited from the same properties of solutions of ordinary equations. These type of results have been addressed by Mahfoud [210] who considered the equation

$$x^{(n)}(t) + \delta H(t, x[g(t)]) = 0, \qquad (E_{12}; \delta)$$

where $n \geq 2$, $\delta = \pm 1$, $H : [t_0, \infty) \times \mathbb{R} \to \mathbb{R}$, $g : [t_0, \infty) \to \mathbb{R}$, $t_0 \geq 0$ are continuous, $g(t) \leq t$ for $t \geq t_0$, $\lim_{t\to\infty} g(t) = \infty$, $xH(t,x) > 0$ for $x \neq 0$, and $H(t,x)$ is nondecreasing in x.

Theorem 2.5.9. Suppose g is continuously differentiable on $[t_0, \infty)$, $t_0 \geq 0$, and $g'(t) > 0$ for $t \geq t_0$.

(i) If for n even the equation $(E_{12}; 1)$ has a nonoscillatory solution, then the equation

$$\frac{d^n y(s)}{ds^n} + \frac{1}{g'(g^{-1}(s))} H(g^{-1}(s), y(s)) = 0, \qquad (E_{13})$$

has a nonoscillatory solution, where g^{-1} is the inverse function of g.

(ii) If for n odd the equation $(E_{12}; 1)$ has an unbounded nonoscillatory solution, so does (E_{13}).

Next, he obtained the following result which extends Lovelady's Theorems 6 and 15 in [209], and essentially provides a comparison between $(E_{12}; \delta)$ and the delay equation

$$x^{(n)}(t) + \delta H(t, x[h(t)]) = 0, \qquad (E_{14}; \delta)$$

where $h : [t_0, \infty) \to \mathbb{R}$ is continuous, $h(t) \leq t$ and $\lim_{t\to\infty} h(t) = \infty$.

Theorem 2.5.10. Suppose $g(t) \geq h(t)$ for $t \geq t_0$.

(i) If for n even the equation $(E_{12}; 1)$ has a nonoscillatory solution, so does the equation $(E_{14}; 1)$.

(ii) If for n odd the equation $(E_{12}; 1)$ has an unbounded nonoscillatory solution, so does the equation $(E_{14}; 1)$.

The following result directly follows from Theorems 2.5.9 and 2.5.10.

Theorem 2.5.11. Suppose that $g(t) \geq h(t)$, h is continuously differentiable on $[t_0, \infty)$, and $h'(t) > 0$ for $t \geq t_0$.

(i) If for n even, the equation

$$\frac{d^n y(s)}{ds^n} + \frac{1}{h'(h^{-1}(s))} H(h^{-1}(s), y(s)) = 0 \qquad (E_{15})$$

is oscillatory, so is the equation $(E_{12};1)$.
(ii) If for n odd, the equation (E_{15}) has no bounded nonoscillatory solutions, neither does the equation $(E_{12};1)$.

Finally, Mahfoud [212] presented the following criterion which concerns $(E_{12};1)$ with bounded delays.

Theorem 2.5.12. Suppose $\tau(t) = t - g(t)$ is bounded. Then,

(i) for n even, the equation $(E_{12};1)$ is oscillatory if and only if the equation $(E_{12};1)$ with $g(t) = t$ is oscillatory,
(ii) for n odd, every nonoscillatory solution of the equation $(E_{12};1)$ is bounded if and only if every nonoscillatory solution of the equation $(E_{12};1)$ with $g(t) = t$ is bounded.

Next, Philos [240], and Kusano and Naito [180] established more general comparison results. The following criterion due to them generalizes Theorems 2.5.9 – 2.5.11.

Theorem 2.5.13. Suppose that g and h are continuously differentiable $g'(t) > 0$, $h'(t) > 0$, $h(t) \leq g(t)$ for $t \geq t_0$. If the equation

$$z^{(n)}(t) + \delta \frac{g'(t)}{h'(h^{-1}(g(t)))} H(h^{-1}(g(t)), z[g(t)]) = 0 \qquad (E_{16};\delta)$$

has Property A, then the equation $(E_{14};\delta)$ has Property A.

A special case of this result is the following:

Corollary 2.5.14. Suppose that h is a continuously differentiable function, and $h'(t) > 0$ for $t \geq t_0$. If the equation

$$z^{(n)}(t) + \frac{\delta}{h'(h^{-1}(t))} H(h^{-1}(t), z(t)) = 0$$

has Property A, then so does the delay equations $(E_{14};\delta)$.

The following two generalizations of Theorem 2.5.12 are also adapted from [180].

Theorem 2.5.15. Let g be a continuously differentiable function, and $g'(t) > 0$ for $t \geq t_0$. Then, for every constant $c \geq 0$, $(E_{12};\delta)$ has Property A if and only if

$$y^{(n)}(t) + \delta H(t, y[g(t) - c]) = 0$$

has Property A.

Theorem 2.5.16. Let g be a continuously differentiable function on $[t_0, \infty)$, $g'(t) > 0$, $\lim_{t\to\infty} g(t) = \infty$, and g_i, $i = 1, 2$ are continuous functions on $[t_0, \infty)$ such that $\lim_{t\to\infty} g_i(t) = \infty$, $i = 1, 2$ and $|g_1(t) - g(t)|$ and $|g_2(t) - g(t)|$ are bounded. Then, the equation

$$x^{(n)}(t) + \delta H(t, x[g_1(t)]) = 0$$

has Property A if and only if the equation

$$y^{(n)}(t) + \delta H(t, y[g_2(t)]) = 0$$

has Property A, where H is as in $(E_{12}; \delta)$.

For odd order delay equations

$$x^{(n)}(t) + q_1(t)x[g_1(t)] = 0 \tag{E_{17}}$$

and

$$y^{(n)}(t) + q_2(t)y[g_2(t)] = 0, \tag{E_{18}}$$

where q_i, $g_i : [t_0, \infty) \to \mathbb{R}$, $i = 1, 2$ are continuous, $q_i(t) \geq 0$ eventually, $i = 1, 2$, Gopalsamy et. al. [63] proved the following:

Theorem 2.5.17. Assume that $q_1(t) \leq q_2(t)$, $g_2(t) \leq g_1(t) < t$, and

$$\int^{\infty} g_2^{n-2}(s)q_2(s)ds = \infty, \qquad n \geq 3.$$

Then, every solution of the equation (E_{17}) is oscillatory implies the same for the equation (E_{18}).

2.6. Comparison of Equations with Others of Lower Order

In separate papers Lovelady [207,208] has related oscillation of solutions of certain linear differential equations of odd order greater than three and even order greater than four to oscillation of associated second order equations. A unified version of these results is the following:

Theorem 2.6.1. Suppose $q : [t_0, \infty) \to [0, \infty)$ is continuous and eventually positive for $t_0 \geq 0$, and

$$\int^{\infty} s^{n-3}q(s)ds < \infty \quad \text{where} \quad n > 2. \tag{2.6.1}$$

Suppose also that the equation

$$w''(t) + Q(t)w(t) = 0 \tag{E_{19}}$$

is oscillatory, where

$$Q(t) = \int_t^\infty \frac{(s-t)^{n-3}}{(n-3)!} q(s)ds, \qquad t > 0.$$

Then, every solution of the equation

$$x^{(n)}(t) + q(t)x(t) = 0, \qquad t > 0$$

is oscillatory if n is even, while if n is odd and $x(t)$ is nonoscillatory, then $x^{(i)}(t) \to 0$ monotonically, $i = 1, 2, \cdots, n-1$ and $x(t) \to \gamma$ (finite real number) as $t \to \infty$.

In [296] Trench has unified the proof of Lovelady's results to nonlinear equations of arbitrary order greater than 3. In fact, he adopted Lovelady's methods to study the oscillation of the equation

$$x^{(n)}(t) + H(t, x(t)) = 0, \qquad t \geq 0, \quad n \geq 3 \tag{E_{20}}$$

where $f : [t_0, \infty) \times \mathbb{R} \to \mathbb{R}$ is continuous, and satisfies

$$\frac{H(t,x)}{x} \geq q(t) \geq 0, \qquad x \neq 0 \tag{2.6.2}$$

where $q : [t_0, \infty) \to \mathbb{R}_+$ and satisfies (2.6.1), and proved the following:

Theorem 2.6.2. Suppose that (2.6.2) holds, and the equation (E_{19}) is oscillatory.

(i) If n is even, then the equation (E_{20}) is oscillatory.

(ii) If n is odd and $x(t)$ is a nonoscillatory solution of the equation (E_{20}), then $x^{(i)}(t) \to 0$ monotonically, $i = 1, 2, \cdots, n-1$ as $t \to \infty$, and $\lim_{t\to\infty} x(t) = \gamma$ (finite real number). In this case, $\gamma = 0$ if

$$\int^\infty s^{n-1} q(s)ds = \infty.$$

Next, Philos [243] considered the linear delay equation

$$x^{(n)}(t) + \delta q(t) x[g(t)] = 0, \tag{$E_{21};\delta$}$$

where $\delta = \pm 1$, g, $q : [t_0, \infty) \to \mathbb{R}$, $q(t) \geq 0$ eventually, $g(t) < t$ and $\lim_{t\to\infty} g(t) = \infty$, and proved the following results.

Theorem 2.6.3. Let $n > 1$, and let θ, $0 < \theta < 1$ be such that

(i) every nonoscillatory solution $y(t)$ of the first order equation

$$y'(t) + \frac{\theta}{(n-1)!} g^{n-1}(t) q(t) y[g(t)] = 0 \tag{E_{22}}$$

satisfies $\lim_{t\to\infty} y(t) \neq 0$, or
(ii) the equation (E_{22}) is oscillatory.

Then, the equation $(E_{21}; 1)$ is almost oscillatory.

Theorem 2.6.4. Let $n > 2$, and let θ, $0 < \theta < 1$ be such that

(i) every bounded solution of the second order equation

$$z''(t) - \frac{\theta}{(n-2)!} g^{n-2}(t) q(t) z[g(t)] = 0 \qquad (E_{23})$$

is oscillatory, or

(ii) every nonoscillatory solution $z(t)$ of the equation (E_{23}) satisfies $\lim_{t\to\infty} z(t) \neq 0$.

Then, we have

(a) for n even, every solution $x(t)$ of the equation $(E_{21}; -1)$ is oscillatory or $x^{(i)}(t) \to 0$ monotonically as $t \to \infty$, $i = 0, 1, \cdots, n-1$, or $\lim_{t\to\infty} |x^{(i)}(t)| \to \infty$, $i = 0, 1, \cdots, n-2$,
(b) for n odd, every solution $x(t)$ of the equation $(E_{21}; -1)$ is oscillatory, or $\lim_{t\to\infty} |x^{(i)}(t)| \to \infty$, $i = 0, 1, \cdots, n-2$.

If in addition

$$\int^{\infty} g^{n-1}(s) q(s) ds = \infty, \qquad (2.6.3)$$

then the equation $(E_{21}; -1)$ is almost oscillatory.

Theorems 2.6.3 and 2.6.4 can be extended rather easily to more general equations of the form

$$x^{(n)}(t) + \delta q(t) f(x[g(t)]) = 0, \qquad (E_{24}; \delta)$$

where $\delta = \pm 1$, g and q are as in $(E_{21}; \delta)$, $f : \mathbb{R} \to \mathbb{R}$ is continuous, $xf(x) > 0$, $f'(x) \geq 0$ for $x \neq 0$, and

$$-f(-xy) \geq f(xy) \geq f(x)f(y) \quad \text{for} \quad xy \neq 0.$$

However, in this case equations (E_{22}) and (E_{23}) are respectively replaced by

$$y'(t) + q(t) f\left(\frac{\theta}{(n-1)!}\right) f\left(g^{n-1}(t)\right) f(y[g(t)]) = 0$$

and

$$z''(t) - q(t) f\left(\frac{\theta}{(n-2)!}\right) f\left(g^{n-2}(t)\right) f(z[g(t)]) = 0.$$

As an application of Philos' results in [243] to $(E_{24}; \delta)$ the following corollaries are immediate.

Corollary 2.6.5. Let $n > 1$, and

(i)

$$\frac{f(x)}{x} \geq k > 0 \quad \text{for} \quad x \neq 0, \tag{2.6.4}$$

and suppose that

$$\limsup_{t\to\infty} \int_{g(t)}^{t} g^{n-1}(s)q(s)ds > \frac{(n-1)!}{k}$$

or

$$\liminf_{t\to\infty} \int_{g(t)}^{t} g^{n-1}(s)q(s)ds > \frac{(n-1)!}{ke},$$

(ii)

$$\int_{\pm 0} \frac{du}{f(u)} < \infty, \tag{2.6.5}$$

and

$$\int^{\infty} f\left(g^{n-1}(s)\right) q(s)ds = \infty.$$

Then, the equation $(E_{24};1)$ is almost oscillatory.

Corollary 2.6.6. Let $n > 2$, condition (2.6.3) holds, and suppose that

(i) condition (2.6.4) holds, and either

$$\limsup_{t\to\infty} \int_{g(t)}^{t} g^{n-2}(s)(g(t) - g(s))q(s)ds > \frac{(n-2)!}{k}$$

or

$$\limsup_{t\to\infty} \int_{g(t)}^{t} g^{n-2}(s)(s - g(t))q(s)ds > \frac{(n-2)!}{k},$$

(ii) condition (2.6.5) holds, and

$$\int^{\infty} (s - g(s)) f\left(g^{n-2}(s)\right) q(s)ds = \infty.$$

Then, the equation $(E_{24};-1)$ is almost oscillatory.

2.7. Further Comparison Results

For n even and the function f not necessarily nondecreasing, Mahfoud [211] compared $(E_{24};1)$ with a equation of the same type of the form

$$y^{(n)}(t) + Q(t)F(y[h(t)]) = 0, \quad n \text{ is even} \tag{E_{25}}$$

where $Q, h : [t_0, \infty) \to \mathbb{R}$, $F : \mathbb{R} \to \mathbb{R}$ are continuous, $Q(t) \geq 0$ eventually, $\lim_{t\to\infty} h(t) = \infty$, and $xF(x) > 0$ for $x \neq 0$. His result is contained in the following theorem.

Theorem 2.7.1. Let $Q(t) \leq q(t)$, $h(t) \leq g(t)$ for $t \geq t_0$, $F(x)$ sgn $x \leq f(x)$ sgn x, and let G and H be the pair of components of F with H being the nondecreasing. If the equation

$$x^{(n)}(t) + Q(t)G\left(\pm cg^{n-1}(t)\right)H(x[h(t)]) \;=\; 0, \qquad n \text{ even}$$

is oscillatory for every $c \neq 0$, then the equation $(E_{24}; 1)$ with n even is also oscillatory.

The following criterion due to Grace and Lalli [105] extends Mahfoud's result to more general equations.

Theorem 2.7.2. Let G and H be a pair of components of f with H being the nondecreasing one. Moreover, assume that $h(t) \leq g(t)$ for $t \geq t_0$, and for every $c \neq 0$

$$q(t)G\left(\pm cg^{n-1}(t)\right) \;\geq\; Q(t), \qquad t \text{ large}$$

and $F'(x) \geq 0$ for $x \neq 0$, and $H(x)$ sgn $x \geq F(x)$ sgn x. If the equation (E_{25}) is oscillatory, then the equation $(E_{24}; 1)$ with n even is also oscillatory.

In 1995, Grace [83] considered the equation $(E_{24}; 1)$ with n even, g any deviating argument, and f not necessarily a monotonic function, and compared its oscillation with the oscillation of a second order linear ordinary differential equation. In the following, we shall state and prove his result.

Theorem 2.7.3. Suppose $f \in C(\mathbb{R}_{t_0})$, $t_0 > 0$ and let G and H be a pair of continuous components with H nondecreasing, and

$$H(x) \text{ sgn } x \geq x^{\gamma} \quad \text{for} \quad x \neq 0 \quad \text{where} \quad \gamma \text{ is a positive constant.}$$

Moreover, assume that there exists an increasing differentiable function $\sigma : [t_0, \infty) \to \mathbb{R}_+$ such that

$$\sigma(t) \;\leq\; \min\{t, g(t)\} \quad \text{and} \quad \sigma(t) \to \infty \quad \text{as} \quad t \to \infty.$$

If for every $c \geq 1$, the linear equation

$$(a(t)y'(t))' + \frac{1}{2\,(n-2)!}G\left(cg^{n-1}(t)\right)q(t)Q(t)y(t) \;=\; 0 \qquad (E_{26})$$

is oscillatory, where $a(t) = (\sigma^{n-2}(t)\sigma'(t))^{-1}$, and

$$Q(t) = \begin{cases} \alpha_1, \text{ any positive constant} \quad \text{if } \gamma > 1 \\ \alpha_2, \text{ any constant, } 0 < \alpha_2 \leq 1 \quad \text{if } \gamma = 1 \\ \alpha_3\sigma^{(\gamma-1)(n-1)}(t), \ \alpha_3 \text{ is any constant, } 0 < \alpha_3 < 1 \text{ if } 0 < \gamma < 1 \end{cases}$$

then the equation $(E_{24}; 1)$ with n even is oscillatory.

Proof. Let $x(t)$ be a nonoscillatory solution of the equation $(E_{24}; 1)$, say $x(t) > 0$ and $x[g(t)] > 0$ for $t \geq t_0 > 0$. By Lemma 2.2.1 there exists a $t_1 \geq t_0$ such that

$$x'(t) > 0 \quad \text{and} \quad x^{(n-1)}(t) > 0 \quad \text{for } t \geq t_1.$$

Since $x(t)$ is an increasing function and $x^{(n-1)}(t)$ is a decreasing function for $t \geq t_1$, there exist positive constants k_1 and k_2 such that for $t \geq t_1$

$$x[\sigma(t)] \geq k_1 \tag{2.7.1}$$

and

$$x^{(n-1)}(t) \leq k_2. \tag{2.7.2}$$

By successive integration of (2.7.2), we conclude that there exist a $t_2 \geq t_1$ and a constant $k^* \geq 1$ such that

$$x[g(t)] \leq k^* g^{n-1}(t) \quad \text{and} \quad x[\sigma(t)] \leq k^* \sigma^{n-1}(t), \quad t \geq t_2. \tag{2.7.3}$$

Furthermore, let us consider an arbitrary constant α such that $\alpha > 1$. Then, by applying Lemma 2.2.2, we conclude that there exists a large $t_3 \geq 2t_2$ such that

$$x'\left[\frac{\sigma(t)}{2}\right] \geq \frac{\sigma^{n-2}(t)}{\alpha\,(n-2)!}\, x^{(n-1)}(t) \quad \text{for } t \geq t_3. \tag{2.7.4}$$

Now, we define a function $w(t)$ as follows

$$w(t) = -\frac{x^{(n-1)}(t)}{x\left[\frac{\sigma(t)}{2}\right]} \quad \text{for } t \geq t_3.$$

Then, for $t \geq t_3$ we have

$$\begin{aligned} w'(t) &= q(t)\frac{f(x[g(t)])}{x\left[\frac{\sigma(t)}{2}\right]} + \frac{x^{(n-1)}(t)x'\left[\frac{\sigma(t)}{2}\right]\frac{\sigma'(t)}{2}}{x^2\left[\frac{\sigma(t)}{2}\right]} \\ &= F(t)q(t) + \frac{1}{P(t)}w^2(t), \end{aligned} \tag{2.7.5}$$

where

$$F(t) = \frac{f(x[g(t)])}{x\left[\frac{\sigma(t)}{2}\right]} \quad \text{and} \quad P(t) = \frac{x^{(n-1}(t)}{x'\left[\frac{\sigma(t)}{2}\right]\frac{\sigma'(t)}{2}}. \tag{2.7.6}$$

The Ricatti equation (2.7.5) has a solution on $[t_3, \infty)$. But, from a well-known result [285] it is equivalent to the nonoscillation of the linear equation

$$(P(t)u'(t))' + q(t)F(t)u(t) \;=\; 0. \tag{2.7.7}$$

Using (2.7.3) and (2.7.4) in (2.7.6), we get

$$P(t) \;=\; \frac{2x^{(n-1)}(t)}{x'\left[\frac{\sigma(t)}{2}\right]\frac{\sigma'(t)}{2}} \;\le\; \frac{2\alpha\,(n-2)!}{\sigma'(t)\sigma^{n-2}(t)} \;=\; (2\alpha\,(n-2)!)a(t), \quad \text{for} \quad t \ge t_3 \tag{2.7.8}$$

and

$$\begin{aligned} F(t) \;&=\; \frac{G(x[g(t)])H(x[g(t)])}{x\left[\frac{\sigma(t)}{2}\right]} \;\ge\; \frac{G\left(k^* g^{n-1}(t)\right) x^{\gamma}[\sigma(t)]}{x\left[\frac{\sigma(t)}{2}\right]} \\ &\ge\; G\left(k^* g^{n-1}(t)\right) x^{\gamma-1}[\sigma(t)] \left(\frac{x[\sigma(t)]}{x\left[\frac{\sigma(t)}{2}\right]}\right) \\ &\ge\; G\left(k^* g^{n-1}(t)\right) x^{\gamma-1}[\sigma(t)]. \end{aligned} \tag{2.7.9}$$

Now there are three cases to consider:

Case 1 $\gamma > 1$. From (2.7.1), it follows that

$$x^{\gamma-1}[\sigma(t)] \;\ge\; k^{\gamma-1} \quad \text{for} \quad t \ge t_3$$

and hence (2.7.9) becomes

$$F(t) \;\ge\; k_1^{\gamma-1} G\left(k^* g^{n-1}(t)\right) \quad \text{for} \quad t \ge t_3.$$

Case 2 $\gamma = 1$. In this case

$$F(t) \;\ge\; G\left(k^* g^{n-1}(t)\right) \quad \text{for} \quad t \ge t_3.$$

Case 3 $0 < \gamma < 1$. From (2.7.2), we have

$$x^{\gamma-1}[\sigma(t)] \;\ge\; \left(k^* \sigma^{n-1}(t)\right)^{\gamma-1} \quad \text{for} \quad t \ge t_3$$

and hence (2.7.9) becomes

$$F(t) \;\ge\; (k^*)^{\gamma-1} \left(\sigma^{n-1}(t)\right)^{\gamma-1} G\left(k^* g^{n-1}(t)\right) \quad \text{for} \quad t \ge t_3.$$

Now an application of the well-known Picone–Sturm comparison theorem (see [74]) to equation (2.7.7) yields the nonoscillation of the linear equation

$$(a(t)y'(t))' + \frac{1}{2\alpha\,(n-2)!} G\left(k^* g^{n-1}(t)\right) q(t) Q^*(t) y(t) \;=\; 0,$$

where

$$Q^*(t) = \begin{cases} k_1^{\gamma-1} & \text{if } \gamma > 1 \\ 1 \quad \text{if } \gamma = 1 \\ k^{*(\gamma-1)}\sigma^{(\gamma-1)(n-1)}(t) & \text{if } 0 < \gamma < 1. \end{cases}$$

This contradicts the hypothesis that the equation (E_{26}) is oscillatory. ∎

Next, Grace [74] considered the equation

$$x^{(n)}(t) + f(t, x[t-\tau], x'[t-\sigma]) = 0, \quad n \text{ is even} \qquad (E_{27})$$

where $f : [t_0, \infty) \times \mathbb{R}^2 \to \mathbb{R}$, $t_0 \geq 0$ is continuous, τ and σ are nonnegative constants with $\sigma \geq \tau$, and studied the oscillatory and asymptotic behavior of this equation by comparing it with certain differential equations of the same or lower order whose oscillatory behavior is known. The asuumption on f, namely

$$\begin{cases} \text{there exist a continuous function } q : [t_0, \infty) \to \mathbb{R}_+ \\ \text{and positive numbers } \lambda \text{ and } \mu \text{ such that} \\ f(t, x, y) \operatorname{sgn} x \geq q(t)|x|^\lambda |y|^\mu \quad \text{for } t \geq t_0 \text{ and } xy \neq 0 \end{cases} \qquad (2.7.10)$$

allows the obtained results in [74] to be applicable to equations of the form

$$x^{(n)}(t) + q(t)|x[t-\tau]|^\lambda |x'[t-\sigma]|^\mu \operatorname{sgn} x[t-\tau] = 0, \quad n \text{ is even.}$$

The following is the list of results established by Grace for the equation (E_{27}).

Theorem 2.7.4. Let condition (2.7.10) hold. If for every $c > 0$ the equation

$$x^{(n)}(t) + cq(t)|x[t-\tau]|^\lambda \operatorname{sgn} x[t-\tau] = 0 \qquad (E_{28})$$

is oscillatory, and for every $c^* > 0$ all bounded solutions of the equation

$$y^{(n-1)}(t) + c^* q(t)|y[t-\sigma]|^\mu \operatorname{sgn} y[t-\sigma] = 0 \qquad (E_{29})$$

are oscillatory, then the equation (E_{27}) is oscillatory.

Theorem 2.7.5. Let condition (2.7.10) hold. If for every $c > 0$ the equation (E_{28}) is oscillatory, and for every θ, $0 < \theta < 1$ all bounded solutions of the equation

$$y^{(n-1)}(t) + (\theta t^\lambda q(t))|y[t-\tau]|^{\lambda+\mu} \operatorname{sgn} y[t-\tau] = 0 \qquad (E_{30})$$

are oscillatory, then the equation (E_{27}) is oscillatory.

Theorem 2.7.6. Let condition (2.7.10) hold. If for every $c > 0$ the equation (E_{28}) is oscillatory, and for every $c^* > 0$ every solution $y(t)$

of the equation (E_{29}) is either oscillatory, or $y^{(i)}(t) \to 0$ monotonically as $t \to \infty$, $i = 0, 1, \cdots, n-1$, then every solution $x(t)$ of the equation (E_{27}) is either oscillatory, or $x^{(i)}(t) \to 0$ monotonically as $t \to \infty$, $i = 1, 2, \cdots, n-1$.

Theorem 2.7.7. Let condition (2.7.10) hold. If for every θ, θ_i, $0 < \theta$, $\theta_i < 1$, $i = 3, 5, \cdots, n-1$ all bounded solutions of the equation (E_{30}) are oscillatory, and the equations

$$z''(t) + \theta_i K_i t^{\alpha_i} q(t)|z[t-\sigma]|^{\lambda+\mu} \operatorname{sgn} z[t-\sigma] \;=\; 0, \quad i = 3, 5, \cdots, n-1$$

are oscillatory, where

$$K_i \;=\; \frac{1}{(n-i)!}\left(\frac{1}{(i-1)!}\right)^{\lambda}\left(\frac{1}{(i-2)!}\right)^{\mu}, \quad i = 3, 5, \cdots, n-1$$

and

$$\alpha_i \;=\; n - i - 1 + \lambda(i-1) + \mu(i-2), \quad i = 3, 5, \cdots, n-1$$

then the equation (E_{27}) is oscillatory.

Theorem 2.7.8. Let $\tau > 0$ and condition (2.7.10) holds. If for every θ, θ_1, $0 < \theta$, $\theta_1 < 1$, every bounded solution of the equation (E_{30}) is oscillatory, and either

(a) the equation

$$w'(t) + \theta_1 K t^{\alpha} q(t)|w[t-\tau]|^{\lambda+\mu} \operatorname{sgn} w[t-\tau] \;=\; 0 \qquad (E_{31})$$

is oscillatory, where

$$K \;=\; \left(\frac{1}{(n-1)!}\right)^{\lambda}\left(\frac{1}{(n-2)!}\right)^{\mu} \quad \text{and} \quad \alpha \;=\; (n-1)\lambda + (n-2)\mu$$

(b) every nonoscillatory solution $w(t)$ of the equation (E_{31}) satisfies $\lim_{t\to\infty} w(t) \neq 0$,

then the equation (E_{27}) is oscillatory.

Recently, Agarwal and Grace [4] have extended the above results to more general equations of the form

$$x^{(n)}(t) + \delta f\left(t, x[g(t)], \frac{d}{dt}x[h(t)]\right) \;=\; 0, \qquad (E_{32};\delta)$$

where $\delta = \pm 1$, $f : [t_0, \infty) \times \mathbb{R}^2 \to \mathbb{R}$ and g, $h : [t_0, \infty) \to \mathbb{R}$ are continuous, f satisfies condition (2.7.10), $0 < h(t) \le t$, $h'(t) > 0$ for $t \ge t_0$, $\lim_{t\to\infty} g(t) = \infty$, and $\lim_{t\to\infty} h(t) = \infty$. For n even we state and prove these results in the following:

Theorem 2.7.9. Let n be even, condition (2.7.10) holds, and $h(t) \le g(t) \le t$ for $t \ge t_0$. If for every θ_i, $0 < \theta_i < 1$, $i = 1, 2$ the equations

$$y'(t) + \left(\frac{\theta_1}{(n-1)^{\lambda}((n-2)!)^{\gamma}}\right)(h(t))^{(n-2)\gamma}H(t)q(t)|y[h(t)]|^{\gamma}\operatorname{sgn} y[h(t)] = 0 \tag{E_{33}}$$

and

$$z'(t) + \left(\frac{\theta_2}{(2^{n-2}(n-2)!)^{\gamma}}\right)(t-h(t))^{(n-2)\gamma}H(t)q(t)\left|z\left[\frac{t+h(t)}{2}\right]\right|^{\gamma} \times \operatorname{sgn} z\left[\frac{t+h(t)}{2}\right] = 0 \tag{E_{34}}$$

are oscillatory, where $\gamma = \lambda + \mu \le 1$ and $H(t) = h^{\lambda}(t)(h'(t))^{\mu}$, then the equation $(E_{32}; 1)$ is oscillatory.

Proof. Let $x(t)$ be a nonoscillatory solution of $(E_{32}; 1)$, say $x(t) > 0$ for $t \ge t_0 \ge 0$. By Lemma 2.2.1 there exists a $t_1 \ge t_0$ such that $x^{(n-1)}(t) > 0$ and $x'(t) > 0$ for $t \ge t_1$. We distinguish the following two cases:

(I) $x^{(n)}(t) \le 0$, $x^{(n-1)}(t) > 0, \cdots, x''(t) > 0$ and $x'(t) > 0$ for $t \ge t_1$, and

(II) $x^{(n)}(t) \le 0$, $x^{(n-1)}(t) > 0, \cdots, x''(t) < 0$ and $x'(t) > 0$ for $t \ge t_1$.

Assume (I) holds. By Lemma 2.2.3 there exist $t_2 \ge t_1$ and b_i, $0 < b_i < 1$, $i = 1, 2$ such that for $t \ge t_2$

$$x[g(t)] \ \ge\ x[h(t)] \ \ge\ \frac{b_1}{(n-1)!}(h(t))^{n-1}x^{(n-1)}[h(t)] \tag{2.7.11}$$

and

$$\frac{d}{dt}x[h(t)] = x'[h(t)]h'(t) \ge \frac{b_2}{(n-2)!}(h(t))^{n-2}h'(t)x^{(n-1)}[h(t)]. \tag{2.7.12}$$

Using (2.7.10) – (2.7.12) in the equation $(E_{32}; 1)$, we get

$$x^{(n)}(t) + \left(\frac{b_1}{(n-1)!}\right)^{\lambda}\left(\frac{b_2}{(n-2)!}\right)^{\mu}(h(t))^{(n-2)\gamma}H(t)q(t)\left(x^{(n-1)}[h(t)]\right)^{\gamma} \le 0, \quad t \ge t_2.$$

Setting $w(t) = x^{(n-1)}(t)$, $t \ge t_2$ to obtain

$$w'(t) + \left[\frac{b_1^{\lambda}b_2^{\mu}}{((n-1)!)^{\lambda}((n-2)!)^{\mu}}\right](h(t))^{(n-2)\gamma}H(t)q(t)(w[h(t)])^{\gamma} \le 0, \quad t \ge t_2. \tag{2.7.13}$$

Integrating (2.7.13) from t to $T \geq t \geq t_2$ and letting $T \to \infty$, we find

$$w(t) \geq \left[\frac{b_1^\lambda b_2^\mu}{(n-1)^\lambda((n-2)!)^\mu}\right] \int_t^\infty (h(s))^{(n-2)\gamma} H(s) q(s)(w[h(s)])^\gamma ds.$$

The function $w(t) = x^{(n-1)}(t)$ is clearly strictly decreasing for $t \geq t_2$. Hence by Theorem 2.5.7 there exists a positive solution $y(t)$ of (E_{33}) with $y(t) \to 0$ as $t \to \infty$. We note that this contradicts the assumptions of our theorem.

Assume (II) holds. By Lemma 2.2.3 there exists a $T_1 \geq t_1$ and a constant b, $0 < b < 1$ such that

$$x[g(t)] \geq x[h(t)] \geq bh(t)x'[h(t)] \quad \text{for} \quad t \geq T_1. \tag{2.7.14}$$

Using (2.7.10) and (2.7.14) in $(E_{32}; 1)$ and setting $v(t) = x'(t)$ for $t \geq T_1$, we get

$$v^{(n-1)}(t) + b^\lambda H(t) q(t)(v[h(t)])^\gamma \leq 0 \quad \text{for} \quad t \geq T_1. \tag{2.7.15}$$

It is clear that the function $v(t)$ satisfies

$$(-1)^i v^{(i)}(t) > 0, \quad 0 \leq i \leq n-1 \quad \text{and} \quad t \geq T_1. \tag{2.7.16}$$

Now by Lemma 2.2.4 there exists a $T \geq T_1$ such that

$$v[h(t)] \geq \left[\frac{(t-h(t))^{n-2}}{2^{n-2}(n-2)!}\right] v^{(n-2)} \left[\frac{t+h(t)}{2}\right] \quad \text{for} \quad T \leq h(t) \leq \frac{t+h(t)}{2}. \tag{2.7.17}$$

Thus, (2.7.15) takes the form

$$u'(t) + \left[\frac{b^\lambda}{(2^{n-2}(n-2)!)^\gamma}\right] (t-h(t))^{(n-2)\gamma} H(t) q(t) \left(u\left[\frac{t+h(t)}{2}\right]\right)^\gamma \leq 0, \quad t \geq T \tag{2.7.18}$$

where $u(t) = v^{(n-2)}(t)$, $t \geq T$. The rest of the proof is similar to that of case (I). ∎

The following corollary in view of the results established in [169] and [199] is immediate.

Corollary 2.7.10. Let n be even, condition (2.7.10) holds, and $h(t) \leq g(t) \leq t$ for $t \geq t_0$. If $\gamma = \lambda + \mu = 1$, then

$$\liminf_{t\to\infty} \int_{h(t)}^t (h(s))^{n-2} H(s) q(s) ds > \frac{(n-1)^\lambda (n-2)!}{e}$$

and

$$\liminf_{t\to\infty} \int_{(t+h(t))/2}^{t} (s-h(s))^{n-2} H(s)q(s)ds > \frac{2^{n-2}(n-2)!}{e}$$

and, if $\gamma = \lambda + \mu < 1$, then

$$\int^{\infty} (h(s))^{(n-2)\gamma} H(s)q(s)ds = \infty$$

and

$$\int^{\infty} (s-h(s))^{(n-2)\gamma} H(s)q(s)ds = \infty$$

imply that the equation $(E_{32}; 1)$ is oscillatory.

In Theorem 2.7.9 if we let $B = \min\{\theta_1, \theta_2\}$, and

$$Q(t) = \min\left\{ \frac{(h(t))^{(n-2)\gamma}}{(n-1)^{\lambda}}, \left(\frac{t-h(t)}{2}\right)^{(n-2)\gamma} \right\},$$

then we have the following oscillation criterion:

Corollary 2.7.11. Let n be even, condition (2.7.10) holds, and $h(t) \le g(t) \le t$ for $t \ge t_0$. If for every B, $0 < B < 1$ the equation

$$w'(t) + \left[\frac{B}{((n-2)!)^{\gamma}}\right] Q(t)H(t)q(t) \left| w\left[\frac{t+h(t)}{2}\right]\right|^{\gamma} \operatorname{sgn} w\left[\frac{t+h(t)}{2}\right] = 0$$

is oscillatory, where $\gamma = \lambda + \mu \le 1$, then the equation $(E_{32}; 1)$ is oscillatory.

Theorem 2.7.9 and Corollaries 2.7.10 and 2.7.11 are applicable only when $\gamma = \lambda + \mu \le 1$. Our next result deals with the case $\lambda \le 1$ and $\mu \le 1$.

Theorem 2.7.12. Let n be even, condition (2.7.10) holds, $g(t) \le t$, and $g'(t) \ge 0$ for $t \ge t_0$. If for every positive constants k_i, $i = 1, 2$ the equations

$$y'(t) + \left(\frac{k_1}{((n-1)!)^{\lambda}}\right) (h'(t))^{\mu} (g(t))^{(n-1)\lambda} q(t) |y[g(t)]|^{\lambda} \operatorname{sgn} y[g(t)] = 0 \tag{E_{35}}$$

and

$$z'(t) + \left(\frac{k_2}{(2^{n-2}(n-2)!)^{\mu}}\right) (t-h(t))^{(n-1)\mu} (h'(t))^{\mu} q(t) \left| z\left[\frac{t+h(t)}{2}\right]\right|^{\mu} \times \operatorname{sgn} z\left[\frac{t+h(t)}{2}\right] = 0 \tag{E_{36}}$$

are oscillatory, where $\lambda \leq 1$ and $\mu \leq 1$, then the equation $(E_{32}; 1)$ is oscillatory.

Proof. Let $x(t)$ be a nonoscillatory solution of $(E_{32}; 1)$, say $x(t) > 0$ for $t \geq t_0 \geq 0$. As in Theorem 2.7.9 we have the cases (I) and (II) for $t \geq t_1$.

Assume (I) holds. Then, there exist a $t_2 \geq t_1$ and positive constants b and C such that

$$\frac{d}{dt} x[h(t)] \geq Ch'(t) \quad \text{for} \quad t \geq t_2 \tag{2.7.19}$$

and

$$x[g(t)] \geq \frac{b}{(n-1)!}(g(t))^{n-1} x^{(n-1)}[g(t)] \quad \text{for} \quad t \geq t_2. \tag{2.7.20}$$

Using (2.7.10), (2.7.19) and (2.7.20) in $(E_{32}; 1)$, we get

$$w'(t) + \left[\frac{C^{\mu} b^{\lambda}}{((n-1)!)^{\lambda}}\right] (h'(t))^{\mu} (g(t))^{(n-1)\lambda} q(t) (w[g(t)])^{\lambda} \leq 0, \quad t \geq t_2$$

where $w(t) = x^{(n-1)}(t)$, $t \geq t_2$. Now proceeding as in Theorem 2.7.9(I), we obtain the desired contradiction.

Assume (II) holds. Then, there exist a $T \geq t_1$ and a positive constant C_1 such that (2.7.17) holds, and

$$x[g(t)] \geq C_1 \quad \text{for} \quad t \geq T. \tag{2.7.21}$$

Thus, (2.7.18) takes the form

$$u'(t) + \left[\frac{C_1^{\lambda}}{(2^{n-2}(n-2)!)^{\mu}}\right] (t - h(t))^{(n-2)\mu} (h'(t))^{\mu} q(t) \left(u\left[\frac{t + h(t)}{2}\right]\right)^{\mu} \leq 0, \quad t \geq T.$$

The rest of the proof is similar to that of Theorem 2.7.9(II). ∎

In the next result λ and μ are assumed to be arbitrary constants.

Theorem 2.7.13. Let n be even, condition (2.7.10) holds, $g(t) \leq t$, and $g'(t) \geq 0$ for $t \geq t_0$. If for every positive constants k_i, $i = 1, 2$ the equation

$$y^{(n)}(t) + k_1 (h'(t))^{\mu} q(t) |y[g(t)]|^{\lambda} \operatorname{sgn} y[g(t)] = 0 \tag{E_{37}}$$

is oscillatory, and every bounded solution of the equation

$$z^{(n-1)}(t) + k_2 (h'(t))^{\mu} q(t) |z[h(t)]|^{\mu} \operatorname{sgn} z[h(t)] = 0 \tag{E_{38}}$$

is oscillatory, then the equation $(E_{32};1)$ is oscillatory.

Proof. Let $x(t)$ be a nonoscillatory solution of $(E_{32};1)$, say $x(t)>0$ for $t\ge t_0\ge 0$. As in Theorem 2.7.9 cases (I) and (II) for $t\ge t_1\ge t_0$ hold.

In the case (I), (2.7.19) holds for $t\ge t_2\ge t_1$. Thus, the equation $(E_{32};1)$ leads to

$$x^{(n)}(t)+C^{\lambda}(h'(t))^{\mu}q(t)(x[g(t)])^{\lambda}\ \le\ 0 \quad \text{for } t\ge t_2.$$

However, then by Lemma 2.5.1 the equation

$$x^{(n)}(t)+C^{\lambda}(h'(t))^{\mu}q(t)(x[g(t)])^{\lambda}\ =\ 0$$

has a positive solution, which is a contradiction.

If (II) holds, then (2.7.21) is satisfied for $t\ge T\ge t_1$, and hence we have

$$v^{(n-1)}(t)+(C_1h'(t))^{\mu}q(t)(v[h(t)])^{\mu}\ \le\ 0 \quad \text{for} \quad t\ge T, \tag{2.7.22}$$

where $v(t)=x'(t)$ and (2.7.16) holds for $t\ge T$. Integrating (2.7.22), $(n-1)$–times from t to $T^*\ge T$, using (2.7.16) and letting $T^*\to\infty$, we obtain

$$v(t)\ \ge\ C_1^{\mu}\int_t^{\infty}\frac{(s-t)^{n-2}}{(n-2)!}(h'(s))^{\mu}q(s)(v[h(s)])^{\mu}ds.$$

The function $v(t)$ is positive and strictly decreasing on $[T,\infty)$. Thus, Theorem 2.5.7 ensures the existence of a positive solution z of the equation (E_{38}) with $\lim_{t\to\infty}z(t)=0$, which is a contradiction. ■

Corollary 2.7.14. Let n be even, condition (2.7.10) holds, and $h(t)\le g(t)\le t$ for $t\ge t_0$. Moreover, assume that equation (E_{34}) is oscillatory for every constant θ_2, $0<\theta_2<1$, or every bounded solution of equation (E_{38}) is oscillatory for every $k_2>0$. Then, all bounded solutions of $(E_{32};1)$ are oscillatory.

The following example illustrates the importance of the above results.

Example 2.7.1. Consider the differential equation

$$x^{(n)}(t)+q(t)\left|x\left[\frac{t}{2}\right]\right|^{\lambda}\left|\frac{d}{dt}x\left[\frac{t}{2}\right]\right|^{\mu}\operatorname{sgn} x\left[\frac{t}{2}\right]\ =\ 0, \quad t\ge 1 \tag{E_{39}}$$

where $q:[1,\infty)\to\mathbb{R}_+$ is continuous, and λ and μ are positive constants. We consider the following cases:

(i) Let $\lambda + \mu = \gamma \le 1$. From Theorem 2.7.9 the equation (E_{39}) is oscillatory if for every θ_1 and $\theta_2 > 0$ the equations

$$y'(t) + \left[\frac{\theta_1}{(n-1)^\lambda \left(2^{n-1}(n-2)!\right)^\gamma}\right] t^{(n-2)\gamma+\lambda} q(t) \left|y\left[\frac{t}{2}\right]\right|^\gamma \operatorname{sgn} y\left[\frac{t}{2}\right] = 0$$

and

$$z'(t) + \left[\frac{\theta_2}{\left(2^{2n-3}(n-2)!\right)^\gamma}\right] t^{(n-2)\gamma+\lambda} q(t) \left|z\left[\frac{3t}{4}\right]\right|^\gamma \operatorname{sgn} z\left[\frac{3t}{4}\right] = 0$$

are oscillatory. Also, (E_{39}) is oscillatory by Corollary 2.7.10, if we take $q(t) = t^{-(n-2)\gamma-\lambda-k}$, where k is a constant, $0 < k < 1$ when $\gamma = 1$, or $k = 1$ when $\gamma < 1$.

(ii) Let $\lambda \le 1$ and $\mu \le 1$. Then by Theorem 2.7.12 the equation (E_{39}) is oscillatory if for every positive constants k_1 and k_2 the equations

$$y'(t) + \left[\frac{k_1}{2^\mu \left(2^{n-1}(n-1)!\right)^\lambda}\right] t^{(n-1)\lambda} q(t) \left|y\left[\frac{t}{2}\right]\right|^\lambda \operatorname{sgn} y\left[\frac{t}{2}\right] = 0$$

and

$$z'(t) + \left[\frac{k_2}{\left(2^{2n-3}(n-2)!\right)^\mu}\right] t^{(n-2)\mu} q(t) \left|z\left[\frac{3t}{4}\right]\right|^\mu \operatorname{sgn} z\left[\frac{3t}{4}\right] = 0$$

are oscillatory. We also note that (E_{39}) is oscillatory if we take $q(t) = t^{n-2-k}$, $0 < k < 1$ when $\lambda = \mu = 1$, and $q(t) = \min\left\{t^{(n-1)\lambda}, t^{(n-2)\mu}\right\}/t$, $t > 1$ when $\lambda < 1$ and $\mu < 1$.

(iii) For any $\lambda > 0$ and $\mu \le 1$ we can apply Theorem 2.7.13 to conclude that (E_{39}) is oscillatory if for every positive constants k_1 and k_2 the equations

$$y^{(n)}(t) + \left(\frac{k_1}{2^\mu}\right) q(t) \left|y\left[\frac{t}{2}\right]\right|^\lambda \operatorname{sgn} y\left[\frac{t}{2}\right] = 0$$

and

$$z^{(n-1)}(t) + \left(\frac{k_2}{2^\mu}\right) q(t) \left|z\left[\frac{t}{2}\right]\right|^\mu \operatorname{sgn} z\left[\frac{t}{2}\right] = 0$$

are oscillatory. We can also select the function $q(t)$ depending on λ and μ and apply results of [91,150,161,166] to establish the oscillatory behavior of (E_{39}).

Now we shall state and prove results for the case when n is odd.

Theorem 2.7.15. Let n be odd, condition (2.7.10) holds, $g'(t) \ge 0$, and $g(t) \ge t$ for $t \ge t_0$. If for every positive constants k_i, $i = 1, 2, 3$ the equations

$$y'(t) + \left(\frac{k_1}{\left(2^{n-2}(n-2)!\right)^\mu}\right) \left(h'(t)(t-h(t))^{n-2}\right)^\mu q(t) \left|y\left[\frac{t+h(t)}{2}\right]\right|^\mu$$

$$\times \ \operatorname{sgn} y\left[\frac{t+h(t)}{2}\right] = 0 \quad (E_{40})$$

and

$$z'(t) + \left(\frac{k_2}{(2\,(n-2)!)^{\mu}}\right)\left[h'(t)(t-h(t))h^{n-2}(t)\right]^{\mu} q(t)\left|z\left[\frac{t+h(t)}{2}\right]\right|^{\mu}$$

$$\times \ \operatorname{sgn} z\left[\frac{t+h(t)}{2}\right] = 0 \quad (E_{41})$$

are oscillatory, and the differential inequality

$$w'(t) - \frac{k_3}{(2^{n-1}(n-1)!)^{\lambda}}\left(h'(t)h^{n-2}(t)\right)^{\mu}(g(t)-t)^{(n-1)\lambda}q(t)\left|w\left[\frac{t+g(t)}{2}\right]\right|^{\lambda}$$

$$\times \ \operatorname{sgn} w\left[\frac{t+g(t)}{2}\right] \ \geq \ 0 \quad (2.7.23)$$

has no eventually positive solution, then the equation $(E_{32};-1)$ is oscillatory.

Proof. Let $x(t)$ be a nonoscillatory solution of $(E_{32};-1)$, say $x(t)>0$ for $t\geq t_0\geq 0$. By Lemma 2.2.1, there exists a $t_1\geq t_0$ such that $x'(t)>0$ for $t\geq t_1$. Now we need to consider the following three cases:

(i) $x^{(n)}(t)\geq 0,\ x^{(n-1)}(t)<0,\ x^{(n-2)}(t)>0,\cdots,x''(t)<0,\ t\geq t_1,$

(ii) $x^{(n)}(t)\geq 0,\ x^{(n-1)}(t)<0,\cdots,x''(t)>0,\ t\geq t_1,$

(iii) $x^{(i)}(t)>0,\ 0\leq i\leq n,\ t\geq t_1.$

Assume (i) holds. Since $x(t)$ is an increasing function for $t\geq t_1$, there exist a $t_2\geq t_1$ and a positive constant k such that

$$x[g(t)] \ \geq \ k \quad \text{for} \quad t\geq t_2. \quad (2.7.24)$$

Using condition (2.7.10) and (2.7.24) in $(E_{32};-1)$, we get

$$x^{(n)}(t) \ \geq \ k^{\lambda}q(t)\left(\frac{d}{dt}x[h(t)]\right)^{\mu} \quad \text{for} \quad t\geq t_2. \quad (2.7.25)$$

Setting $v(t)=x'(t),\ t\geq t_2$ to obtain

$$v^{(n-1)}(t) \ \geq \ k^{\lambda}q(t)(h'(t))^{\mu}(v[h(t)])^{\mu} \quad \text{for} \quad t\geq t_2. \quad (2.7.26)$$

By Lemma 1 in [282], there exists a $t_3\geq t_2$ such that

$$\begin{aligned} v[h(t)] \ &\geq \ \frac{(t-h(t))^{n-2}}{2^{n-2}(n-2)!}\left(-v^{(n-2)}\left[\frac{t+h(t)}{2}\right]\right) \\ &= \ \frac{(t-h(t))^{n-2}}{2^{n-2}(n-2)!}\,z\left[\frac{t+h(t)}{2}\right] \quad \text{for} \quad t\geq t_3, \end{aligned} \quad (2.7.27)$$

where $z(t) = -v^{(n-2)}(t) > 0$ for $t \geq t_1$. Thus, inequality (2.7.26) can be written as

$$z'(t) + \left[\frac{k^{\lambda}}{(2^{n-2}(n-2)!)^{\mu}}\right]\left((t-h(t))^{n-2}h'(t)\right)^{\mu} q(t)\left(z\left[\frac{t+h(t)}{2}\right]\right)^{\mu} \leq 0 \quad \text{for} \quad t \geq t_3.$$

Now proceeding as in Theorem 2.7.9(I) we find the desired contradiction.

Assume (ii) holds. As in (i) we obtain (2.7.24) for $t \geq t_2$. By Lemma 2.2.3 there exist a $T_1 \geq t_2$ and a positive constant b, $0 < b < 1$ such that

$$\frac{d}{dt}x[h(t)] = x'[h(t)]h'(t) \geq \left(\frac{b}{(n-2)!}\right)(h(t))^{n-2}h'(t)x^{(n-2)}[h(t)], \quad t \geq T_1.$$

Thus,

$$u''(t) \geq \left[\frac{k^{\lambda}b^{\mu}}{((n-2)!)^{\mu}}\right]\left((h(t))^{n-2}h'(t)\right)^{\mu} q(t)(u[h(t)])^{\mu}, \quad t \geq T_1 \qquad (2.7.28)$$

where $u(t) = x^{(n-2)}(t) > 0$ for $t \geq T_1$. Next, by Lemma 2.2.4 there exists a $T_2 \geq T_1$ such that

$$u[h(t)] \geq \frac{(t-h(t))}{2}\left(-u'\left[\frac{t+h(t)}{2}\right]\right) \quad \text{for} \quad t \geq T_2$$

and consequently (2.7.28) leads to

$$w'(t) + \left[\frac{k^{\lambda}b^{\mu}}{(2\,(n-2)!)^{\mu}}\right]\left((h(t))^{n-2}h'(t)(t-h(t))\right)^{\mu} q(t)\left(w\left[\frac{t+h(t)}{2}\right]\right)^{\mu} \leq 0, \quad t \geq T_2$$

where $w(t) = -u'(t) > 0$ for $t \geq T_2$. The rest of the proof is similar to that of Theorem 2.7.9(I).

Finally, assume (iii) holds. Then, there exist a $T^* \geq t_1$ and a positive constant B such that

$$\frac{d}{dt}x[h(t)] \geq B(h(t))^{n-2}h'(t) \quad \text{for} \quad t \geq T^*$$

and hence equation $(E_{32}; -1)$ gives

$$x^{(n)}(t) \geq B^{\mu}\left(h(t))^{n-2}h'(t)\right)^{\mu} q(t)(x[g(t)])^{\lambda} \quad \text{for} \quad t \geq T^*. \qquad (2.7.29)$$

Let $T \geq T^*$. Then, Taylor's formula for the case (iii) leads to

$$x(u) \geq \frac{(u-v)^{n-1}}{(n-1)!}x^{(n-1)}(v) \quad \text{for} \quad u \geq v \geq T.$$

Putting $u = g(t)$ and $v = (t + g(t))/2$ in the above inequality, we get

$$x[g(t)] \geq \frac{(g(t)-t)^{n-1}}{2^{n-1}(n-1)!} x^{(n-1)}\left[\frac{t+g(t)}{2}\right], \quad g(t) \geq \frac{t+g(t)}{2} \geq T. \quad (2.7.30)$$

Using (2.7.30) in (2.7.29), we obtain

$$W'(t) \geq \left[\frac{B^{\mu}}{(2^{n-1}(n-1)!)^{\lambda}}\right] ((h(t))^{n-2}h'(t))^{\mu} (g(t)-t)^{(n-1)\lambda}$$

$$\times\, q(t)\left(W\left[\frac{t+g(t)}{2}\right]\right)^{\lambda} \quad \text{for} \quad t \geq T,$$

where $W(t) = x^{(n-1)}(t)$ is a positive solution for $t \geq T$ of the above inequality, which is a contradiction to our assumption. ∎

If we let $k = \min\{k_1, k_2\}$, and

$$Q(t) = \min\left\{\left(\frac{h'(t)(t-h(t))^{n-2}}{2^{n-2}\,(n-2)!}\right)^{\mu}, \left(\frac{h'(t)(t-h(t))(h(t))^{n-2}}{2\,(n-2)!}\right)^{\mu}\right\}$$

then, Theorem 2.7.15 can be restated as follows:

Theorem 2.7.16. Let n be odd, condition (2.7.10) holds, $g'(t) \geq 0$, and $g(t) \geq t$ for $t \geq t_0$. If for every positive constants k and k_3, the equation

$$y'(t) + kQ(t)q(t)\left|y\left[\frac{t+h(t)}{2}\right]\right|^{\mu} \operatorname{sgn} y\left[\frac{t+h(t)}{2}\right] = 0$$

is oscillatory, and the inequality (2.7.23) has no eventually positive solution, then the equation $(E_{32}; -1)$ is oscillatory.

Theorem 2.7.17. Let n be odd, condition (2.7.10) holds, $g'(t) \geq 0$, and $g(t) \geq t$ for $t \geq t_0$. If for every positive constants k_1 and k_3, the equation (E_{40}) is oscillatory and the inequality (2.7.23) has no eventually positive solution, and

$$\int^{\infty} s(h'(s))^{\mu} q(s)\,ds = \infty, \quad (2.7.31)$$

then equation $(E_{32}; -1)$ is oscillatory.

Proof. Let $x(t)$ be a nonoscillatory solution of $(E_{32}; -1)$, say $x(t) > 0$ for $t \geq t_0 \geq 0$. We proceed as in Theorem 2.7.15 and only consider case (ii). Since $x''(t) > 0$, $x'(t) > 0$ for $t \geq t_1$ there exist a $T \geq t_1$ and positive constants d_1 and d_2 such that

$$x[g(t)] \geq d_1 \quad \text{and} \quad \frac{d}{dt}x[h(t)] \geq d_2 h'(t) \quad \text{for} \quad t \geq T. \quad (2.7.32)$$

We multiply the equation $(E_{32};-1)$ by t, use (2.7.10) and (2.7.32), and integrate from T to $t \geq T$, to obtain

$$tx^{(n-1)}(t) - \int_T^t x^{(n-1)}(s)ds \;\geq\; N + d_1^\lambda d_2^\mu \int_T^t s(h'(s))^\mu q(s)ds,$$

where N is a real number. Now by (2.7.31), we get

$$\lim_{t\to\infty}\left[tx^{(n-1)}(t) - x^{(n-2)}(t)\right] \;=\; \infty$$

and thus by Lemma 2.2.7, we find that $x^{(n-1)}(t) \to \infty$ as $t \to \infty$, which is a contradiction. ∎

Theorem 2.7.18. Let n be odd, condition (2.7.10) holds, $g'(t) \geq 0$, and $g(t) \geq t$ for $t \geq t_0$. If for every positive constants k_i, $i = 1,2,3$ all bounded solutions of the equations

$$y^{(n-1)}(t) - k_1\,(h'(t))^\mu\, q(t)|y[h(t)]|^\mu \;\mathrm{sgn}\; y[h(t)] \;=\; 0 \qquad (E_{42})$$

and

$$z''(t) - \left(\frac{k_2}{((n-2)!)^\mu}\right)\left(h^{n-2}(t)h'(t)\right)^\mu q(t)|z[h(t)]|^\mu \;\mathrm{sgn}\; z[h(t)] \;=\; 0$$

are oscillatory and the inequality (2.7.23) has no eventually positive solution, then the equation $(E_{32};-1)$ is oscillatory.

Proof. Let $x(t)$ be a nonoscillatory solution of $(E_{32};-1)$, say $x(t) > 0$ for $t \geq t_0 \geq 0$. As in Theorem 2.7.15 we have the cases (i) – (iii). The proof of the case (iii) is similar to that of Theorem 2.7.15(iii). Now we assume (i) holds. As in Theorem 2.7.15(i), we see that (2.7.26) holds for $t \geq t_2$. Integrating (2.7.26), $(n-1)$ times from t to $T \geq t \geq t_2$ and using the fact that $(-1)^i v^{(i)}(t) > 0$ for $t \geq t_2$ and $0 \leq i \leq n-1$, we find

$$v(t) \;\geq\; k^\lambda \int_t^\infty \frac{(s-t)^{n-2}}{(n-2)!}(h'(s))^\mu q(s)\,(v[h(s)])^\mu\, ds, \quad t \geq t_2.$$

Now we proceed as in Theorem 2.7.13(II) to obtain the desired contradiction. Finally, we assume that (ii) holds. As in the proof of Theorem 2.7.15(ii), we note that (2.7.28) holds for $t \geq T_1$. Once again, we integrate (2.7.28) from t to $T^* \geq T_1$ and let $T^* \to \infty$, to get

$$u(t) \;\geq\; \left[\frac{k^\lambda b^\mu}{((n-2)!)^\mu}\right]\int_t^\infty (s-t)\left((h(s))^{n-2}h'(s)\right)^\mu q(s)(u[h(s)])^\mu ds, \; t \geq T_1.$$

The rest of the proof is similar to that of Theorem 2.7.13(II). ∎

In view of the results established in [169] and their analogs to advanced equations, and the results for first order equations and/or inequalities presented in [165] the following corollary is immediate.

Corollary 2.7.19. Let n be odd, condition (2.7.10) holds, $g'(t) \geq 0$, and $g(t) \geq t$ for $t \geq t_0$. If for every positive constants k_1, k_2 and k_3,

$$\liminf_{t\to\infty} \int_{(t+h(t))/2}^{t} h'(s)(s-h(s))^{n-2}q(s)ds > \frac{2^{n-2}(n-2)!}{k_1 e} \quad \text{when} \quad \mu = 1$$

or

$$\int^{\infty} \left(h'(s)(s-h(s))^{n-2}\right)^{\mu} q(s)ds = \infty \quad \text{when} \quad \mu < 1,$$

$$\liminf_{t\to\infty} \int_{(t+h(t))/2}^{t} h'(s)(s-h(s))(h(s))^{n-2}q(s)ds > \frac{2(n-2)!}{k_2 e} \quad \text{when} \quad \mu = 1$$

or

$$\int^{\infty} \left(h'(s)(s-h(s))(h(s))^{n-2}\right)^{\mu} q(s)ds = \infty \quad \text{when} \quad \mu < 1,$$

and either

$$\liminf_{t\to\infty} \int_{t}^{(t+g(t))/2} \left(h'(s)(h(s))^{n-2}\right)^{\mu} (g(s)-s)^{n-1}q(s)ds > \frac{2^{n-1}(n-1)!}{k_3 e}$$

$$\text{when} \quad \lambda = 1 \quad \text{and} \quad \mu > 0,$$

or

$$\int^{\infty} \left(h'(s)(h(s))^{n-2}\right)^{\mu} (g(s)-s)^{(n-1)\lambda}ds = \infty \quad \text{when} \quad \lambda > 1 \quad \text{and} \quad \mu > 0,$$

then $(E_{32}; -1)$ is oscillatory.

Corollary 2.7.20. Let n be odd, condition (2.7.10) holds, and assume that the equation (E_{40}) is oscillatory for every constant $k_1 > 0$, or every bounded solution of the equation (E_{42}) is oscillatory for every constant $k_1 > 0$. Then, every bounded solution of the equation $(E_{32}; -1)$ is oscillatory.

Example 2.7.2. For the equation

$$x^{(n)}(t) - t^{2-n}\left|x\left[\frac{3t}{2}\right]\right|^{\lambda}\left|\frac{d}{dt}x\left[\frac{t}{2}\right]\right|^{\mu} \operatorname{sgn} x\left[\frac{3t}{2}\right] = 0, \quad t \geq 1 \qquad (E_{43})$$

where n is odd, and λ and μ are positive constants, all conditions of Corollary 2.7.19 are satisfied for $0 < \mu \leq 1$ and $\lambda \geq 1$, and hence (E_{43}) is oscillatory.

2.8. Equations with Middle Term of Order $(n-1)$

We shall consider the general equation

$$x^{(n)}(t) + p(t)|x^{(n-1)}(t)|^{\beta}x^{(n-1)}(t) + H(t, x[g(t)]) = 0, \quad n \text{ even} \quad (E_{44})$$

where $p : [t_0, \infty) \to \mathbb{R}_0$, $g : [t_0, \infty) \to \mathbb{R}$, $H : [t_0, \infty) \times \mathbb{R} \to \mathbb{R}$, $t_0 \geq 0$ are continuous, $xH(t, x) > 0$ for $x \neq 0$, $t \geq t_0$, H is nondecreasing in the second variable, $\lim_{t\to\infty} g(t) = \infty$, and $\beta \geq 0$. The following lemma is proved in Kartsatos and Onose [156] for $\beta = 0$, and extended for $\beta \neq 0$ by Grace and Lalli [91].

Lemma 2.8.1. Let $p(t) \leq m(t)$ where $m : [t_0, \infty) \to \mathbb{R}_+$ is continuous, and assume that for every $T \geq t_0$

$$\lim_{t\to\infty} \int_T^t \exp\left(-\int_T^u m(s)ds\right) du = \infty \quad \text{if } \beta = 0 \tag{2.8.1}$$

and

$$\left(1 + \int_T^t m(s)ds\right)^{-1/\beta} \notin \mathcal{L}(T, \infty) \quad \text{if } \beta \neq 0. \tag{2.8.2}$$

Then, if $x(t)$ is a nonoscillatory solution of the equation (E_{44}) we must have

$$x(t)x^{(n-1)}(t) > 0 \quad \text{eventually.}$$

The above lemma has also been established by Naito [219] for $\beta = 0$ by using Langehop's inequality. We further note that part of its proof for $n = 2$ with $\beta = 0$ goes back to Bobisud [36].

As an application of Lemma 2.8.1, Kartsatos [150], Grace and Lalli [91], and Kosmala [171] obtained the following oscillation criterion:

Theorem 2.8.2. Let conditions (2.8.1) and (2.8.2) hold. If the equation

$$x^{(n)}(t) + H(t, x[g(t)]) = 0, \quad n \text{ is even}$$

is oscillatory, then the equation (E_{44}) is oscillatory.

For $\beta = 0$ the equation (E_{44}) reduces to

$$x^{(n)}(t) + p(t)x^{(n-1)}(t) + H(t, x[g(t)]) = 0, \quad n \text{ even} \quad (E_{45})$$

for which Kosmala [171] assumed that $g(t) \leq t$,

$$\int^{\infty} p(s)ds > -\infty \tag{2.8.3}$$

and

$$H(t,u) = H_1(t,u) + H_2(t,u),$$

where $H_i : [t_0, \infty) \times \mathbb{R} \to \mathbb{R}$, $i = 1,2$ are continuous, increasing in the second variable, and such that $xH_i(t,x) > 0$ for $x \neq 0$, $i = 1,2$, $t \geq t_0$. Further, for all positive constants λ and μ there exists a $T(\lambda,\mu)$ such that

$$\lambda p(t) \leq H_1\left(t, \mu g^{n-2}(t)\right), \quad t \geq T \tag{2.8.4}$$

and the equation

$$x^{(n)}(t) + H_2(t, x[g(t)]) = 0 \quad \text{is oscillatory.} \tag{2.8.5}$$

Theorem 2.8.3. Let conditions (2.8.1), (2.8.3) – (2.8.5) hold. If for every positive constant k and $t_0 \geq 0$

$$\int_{t_0}^{\infty} sH(s, \pm k) = \pm\infty, \tag{2.8.6}$$

then the equation (E_{45}) is oscillatory.

Condition (2.8.6) is relaxed in the following results:

Theorem 2.8.4. Let conditions (2.8.1), (2.8.3) – (2.8.5) hold, and $(t^2p(t))' \geq 0$ eventually. If for every constant $k > 0$ and $t_0 \geq 0$

$$\int_{t_0}^{\infty} s^2 H(s, \pm k)ds = \pm\infty,$$

then the equation (E_{45}) is oscillatory.

Theorem 2.8.5. Let conditions (2.8.1), (2.8.3) – (2.8.5) hold. If for every constants M and $k > 0$

$$\lim_{t\to\infty} \int_{t_0}^{t} \exp\left(-\int_s^t p(u)du\right) \left[\int_{t_0}^{s} H(u, \pm k)du - M\right] ds = \pm\infty,$$

then the equation (E_{45}) is oscillatory.

Now in the case when condition (2.8.1) is violated we shall present criteria for the oscillation and asymptotic behavior of the equation (E_{45}).

Theorem 2.8.6. Let n be even, $p'(t) \leq 0$ for $t \geq t_0 \geq 0$, and for every constant $c > 0$

$$\int_{t_0}^{\infty} H(s,c)ds = \infty. \tag{2.8.7}$$

Then, every solution $x(t)$ of the equation (E_{45}) is oscillatory, or $x^{(i)}(t) \to 0$ monotonically as $t \to \infty$, $i = 0, 1, \cdots, n-2$.

Proof. Let $x(t)$ be an eventually positive solution of the equation (E_{45}), say $x(t) > 0$ and $x[g(t)] > 0$ for $t \geq t_0 \geq 0$. We claim that $x^{(n-1)}(t)$ is eventually of one sign. To prove it, assume that $x^{(n-1)}(t)$ is oscillatory. There exists a $t_1 \geq t_0$ such that $x^{(n-1)}(t_1) = 0$, and

$$x^{(n)}(t_1) \;=\; -H(t_1, x[g(t_1)]) \;<\; 0,$$

which implies that $x^{(n-1)}(t)$ cannot have another zero after it vanishes once. Thus, $x^{(n-1)}(t)$ is eventually of fixed sign.

Now, we consider the following two cases:

(I) $x^{(n-1)}(t) > 0$ eventually, and (II) $x^{(n-1)}(t) < 0$ eventually.

(I) Suppose $x^{(n-1)}(t) > 0$ eventually. Then,

$$x^{(n)}(t) + H(t, x[g(t)]) \;\leq\; 0 \quad \text{eventually.}$$

Since n is even, we have

$$x^{(n)}(t) \;\leq\; 0, \;\; x^{(n-1)}(t) \;>\; 0 \;\text{ and }\; x'(t) \;>\; 0 \quad \text{eventually.}$$

Thus, there exist a $t_2 \geq t_1$ and a constant $c_1 > 0$ such that $x[g(t)] \geq c$ for $t \geq t_2$. Hence,

$$x^{(n)}(t) + H(t, c) \;\leq\; 0 \quad \text{for} \quad t \geq t_2$$

and therefore

$$x^{(n-1)}(t) - x^{(n-1)}(t_2) + \int_{t_2}^{t} H(s, c)ds \;\leq\; 0.$$

Now from condition (2.8.7), it follows that

$$0 \;<\; x^{(n-1)}(t) \;\to\; -\infty \quad \text{as} \quad t \to \infty,$$

which is a contradiction.

(II) Suppose $x^{(n-1)}(t) < 0$ eventually. Then, $x^{(n-2)}(t) > 0$ eventually and we have two cases to consider:

(i) $x'(t) > 0$ eventually, and (ii) $x'(t) < 0$ eventually.

(i) Assume $x'(t) > 0$ eventually. Then, there exist a $T \geq t_1$ and a constant $k > 0$ such that $x[g(t)] \geq k_1$ for $t \geq T$. Now from the equation (E_{45}), it follows that

$$x^{(n-1)}(t) - x^{(n-1)}(T) + p(t)x^{(n-2)}(t) - p(T)x^{(n-2)}(T) - \int_{T}^{t} p'(s)x^{(n-2)}(s)ds$$

$$+\int_T^t H(s,k)ds \leq 0,$$

which implies that

$$x^{(n-1)}(t) - p(T)x^{(n-2)}(T) + \int_T^t H(s,k)ds \leq 0$$

and by condition (2.8.7) we arrive at a contradiction.

(ii) Assume $x'(t) < 0$ eventually. Then $x(t) \to a > 0$ as $t \to \infty$. We claim that $a = 0$. There exists a $T_1 \geq t_1$ such that $x[g(t)] \geq a$ for $t \geq T_1$. Now proceeding as in the proof of case (II)–(i) we obtain the desired contradiction. ∎

As an illustration, we give

Example 2.8.1. The equation

$$x^{(4)}(t) + 2x^{(3)}(t) + e^{-\tau}x[t-\tau] = 0, \qquad \tau \geq 0$$

has a nonoscillatory solution $x(t) = e^{-t} \to 0$ as $t \to \infty$. Clearly, for this equation all conditions of Theorem 2.8.6 are satisfied.

A complete oscillation criterion for (E_{45}) has been offered by Grace and Hamedani [88]. They do not employ condition (2.8.1), rather assume

$$\left\{ \begin{array}{l} \text{there exists a continuous function } q : [t_0, \infty) \to \mathbb{R}_+ \text{ such} \\ \text{that } H(t,x) \operatorname{sgn} x \geq q(t)|x| \text{ for } x \neq 0, \text{ and } t \geq t_0. \end{array} \right. \tag{2.8.8}$$

Theorem 2.8.7. Let n be even, condition (2.8.8) holds, $g'(t) \geq 0$, $g(t) \leq t$, and $p'(t) \leq 0$ for $t \geq t_0$. Moreover, suppose that the equation

$$y^{(n)}(t) + q(t)y[g(t)] = 0 \tag{E_{46}}$$

is oscillatory, and for every θ, $0 < \theta < 1$

$$\liminf_{t\to\infty} \int_{g(t)}^t \left[\left(\int_{g(u)}^u \frac{\theta}{(n-2)!} g^{n-2}(s)q(s)ds \right) - p[g(u)] \right] du > \frac{1}{e} \tag{2.8.9}$$

and

$$\liminf_{t\to\infty} \int_{g(t)}^t \left[\left(\int_{g(u)}^u \frac{(g(u)-g(s))^{n-2}}{(n-2)!} q(s)ds \right) - p[g(u)] \right] du > \frac{1}{e}. \tag{2.8.10}$$

Then, the equation (E_{45}) is oscillatory.

In the case, when

$$\int^{\infty} q(s)ds = \infty \tag{2.8.11}$$

the following corollary of Theorem 2.8.7 is immediate.

Corollary 2.8.8. Let n be even, $g'(t) \geq 0$, $g(t) \leq t$, and $p'(t) \leq 0$ for $t \geq t_0$. If conditions (2.8.8), (2.8.10) and (2.8.11) hold, then the equation (E_{45}) is oscillatory.

The oscillation criterion for the equation (E_{45}) when n is odd is also given by Grace and Hamedani [88]. Their result is stated in the following:

Theorem 2.8.9. Let n be odd, condition (2.8.8) holds, $g'(t) \geq 0$, $g(t) < t$, and $p'(t) \leq 0$ for $t \geq t_0$. If condition (2.8.11) holds and the equation (E_{46}) is oscillatory, then the equation (E_{45}) is oscillatory.

Example 2.8.2. The equation

$$x^{(3)} + x^{(2)} + \sqrt{2}\, x\left[t - \frac{7\pi}{4}\right] = 0$$

has an oscillatory solution $x(t) = \sin t$. The equation

$$x^{(3)} + \sqrt{2}\, x\left[t - \frac{7\pi}{4}\right] = 0$$

is oscillatory by a result proved in [198]. Further, since

$$(\sqrt{2})^{1/3}\, \frac{7\pi}{12} > \frac{1}{e}$$

all the conditions of Theorem 2.8.9 are satisfied.

Next, we shall consider the case when $p(t) \leq 0$ eventually, i.e. we shall consider the equation

$$x^{(n)}(t) - p(t)x^{(n-1)}(t) + H(t, x[g(t)]) = 0. \tag{E_{47}}$$

Theorem 2.8.10. Let n be even, $p'(t) \geq 0$, $p''(t) \leq 0$, $0 \leq p(t) \leq p_0$ for $t \geq t_0$, where p_0 is a real constant. If condition (2.8.7) holds for every $c > 0$ and $t \geq t_0$, then every solution $x(t)$ of the equation (E_{47}) is oscillatory, or $x^{(i)}(t) \to \infty$ monotonically as $t \to \infty$ and $i = 0, 1, \cdots, n-2$.

Proof. Let $x(t)$ be an eventually positive solution of the equation (E_{47}). As in the proof of Theorem 2.8.6, we find that $x^{(n-1)}(t)$ is eventually of one sign, and the case $x^{(n-1)}(t) < 0$ eventually is impossible. Thus, we consider the case $x^{(n-1)}(t) > 0$ eventually. Now we consider the following two cases:

(i) $x^{(n-2)}(t) > 0$ eventually, and (ii) $x^{(n-2)}(t) < 0$ eventually.

(i) Suppose $x^{(n-2)}(t) > 0$ eventually. Then, $x^{(i)}(t) > 0$ eventually, $i = 0, 1, 2, \cdots, n-1$. Thus, there exist $k > 0$ and $T \geq t_0$ such that

$$x[g(t)] \geq k \quad \text{for } t \geq T.$$

Integrating (E_{47}) from T to t, we get

$$x^{(n-1)}(t) - x^{(n-1)}(T) - p(t)x^{(n-2)}(t) + p(T)x^{(n-2)}(T) + \int_T^t p'(s)x^{(n-2)}(s)ds$$
$$+ \int_T^t H(s,k)ds \leq 0,$$

which implies that $x^{(n-2)}(t) \to \infty$, and hence $x^{(i)}(t) \to \infty$ as $t \to \infty$, $i = 0, 1, \cdots, n-2$.

(ii) Suppose $x^{(n-2)}(t) < 0$ eventually. Since n is even, we have $x'(t) > 0$ and $x^{(n-3)}(t) > 0$ eventually. Thus, there exist a constant $k_1 > 0$ and $T_1 \geq t_0$ such that

$$x[g(t)] \geq k_1 \quad \text{for } t \geq T_1,$$

and hence

$$x^{(n-1)}(t) - x^{(n-1)}(T) - p(t)x^{(n-2)}(t) + p(T)x^{(n-2)}(T) + \int_{T_1}^t p'(s)x^{(n-2)}(s)ds$$
$$+ \int_{T_1}^t H(s,k_1)ds \leq 0$$

or

$$-x^{(n-1)}(T) + p(T)x^{(n-2)}(T) + p'(t)x^{(n-3)}(t) - p'(T_1)x^{(n-3)}(T_1)$$
$$- \int_{T_1}^t p''(s)x^{(n-3)}(s)ds + \int_{T_1}^t H(s,k_1)ds \leq 0.$$

But, now with condition (2.8.7), we arrive at a contradiction. ∎

Example 2.8.3. The equation

$$x^{(4)}(t) - x^{(3)}(t) + \frac{6}{t}\, x^{1/3}(t) = 0$$

has a nonoscillatory solution $x(t) = t^3$. Clearly, for this equation all conditions of Theorem 2.8.10 are satisfied.

Next, we present the following complete oscillation criterion for the equation (E_{47}).

Theorem 2.8.11. Let n be even, $p'(t) \geq 0$, $p''(t) \leq 0$, $0 \leq p(t) \leq p_0$ for $t \geq t_0$ where p_0 is a real constant, and let $g(t)$ be of mixed type and $g'(t) \geq 0$ for $t \geq t_0$. If conditions (2.8.8) and (2.8.11) hold, and

$$\liminf_{t\to\infty} \int_t^{\rho(t)} \left[p[\rho(s)] + \int_s^{\rho(s)} \frac{(g(u)-\rho(s))^{n-2}}{(n-2)!} q(u)du \right] ds \; > \; \frac{1}{e}, \quad (2.8.12)$$

where $\rho(t) = \min\{\max\{s, g(s)\} : s \geq t\}$ and $g(s) \geq \rho(t)$ for $t < s < \rho(t)$, then the equation (E_{47}) is oscillatory.

Proof. Let $x(t)$ be an eventually positive solution of the equation (E_{47}). Proceeding as in the proof of Theorem 2.8.10, we only need to consider case (i). In this case we have $x^{(i)}(t) > 0$ eventually, $i = 0, 1, \cdots, n-1$. From Taylor's formula, it follows that

$$x(w) \; \geq \; \frac{(w-v)^{n-2}}{(n-2)!} x^{(n-2)}(v) \quad \text{for} \quad w \geq v \geq T \geq t_0.$$

Letting $w = g(u)$ and $v = \rho(t)$ in the above inequality, where $t \leq u \leq \rho(t) \leq g(u)$, we get

$$x[g(u)] \; \geq \; \frac{(g(u)-\rho(t))^{n-2}}{(n-2)!} x^{(n-2)}[\rho(t)], \quad T \leq t \leq u \leq \rho(t).$$

Integrating (E_{47}) from t to $\rho(t)$, we find

$$0 \; \geq \; x^{(n-1)}[\rho(t)] - x^{(n-1)}(t) - p[\rho(t)]x^{(n-2)}[\rho(t)] - p(t)x^{(n-2)}(t)$$
$$+ \int_t^{\rho(t)} p'(u)x^{(n-2)}(u)du + \left(\int_t^{\rho(t)} \frac{(g(u)-\rho(t))^{n-2}}{(n-2)!} q(u)du \right) x^{(n-2)}[\rho(t)]$$

or

$$x^{(n-1)}(t) \geq \left(p[\rho(t)] + \int_t^{\rho(t)} \frac{(g(u)-\rho(t))^{n-2}}{(n-2)!} q(u)du \right) x^{(n-2)}[\rho(t)] \text{ for } t \geq T.$$

Setting $x^{(n-1)}(t) = y(t)$, $t \geq T$ we obtain

$$y'(t) \; \geq \; \left(p[\rho(t)] + \int_t^{\rho(t)} \frac{(g(u)-\rho(t))^{n-2}}{(n-2)!} q(u)du \right) y[\rho(t)]. \quad (2.8.13)$$

But, by Lemma 2.2.10(i), condition (2.8.12) implies that inequality (2.8.13) has no eventually positive solution, which is a contradiction. ∎

We note that Theorem 2.8.11 holds if $g(t) \geq t$, and $g'(t) \geq 0$ for $t \geq t_0$.

For n odd, we present the following result:

Theorem 2.8.12. Let n be odd, $p'(t) \geq 0$, $p''(t) \leq 0$, $0 \leq p(t) \leq p_0$ for $t \geq t_0$ where p_0 is a real constant, and condition (2.8.7) holds. Then, every solution $x(t)$ of the equation (E_{47}), is oscillatory or $x^{(i)}(t) \to \infty$ monotonically as $t \to \infty$, $i = 0, 1, \cdots, n-2$, or $x^{(i)}(t) \to 0$ monotonically as $t \to \infty$, $i = 0, 1, \cdots, n-2$.

The following example illustrates how Theorem 2.8.12 can be used in practice.

Example 2.8.4. Consider the equation

$$x^{(3)}(t) - x^{(2)}(t) + q(t)x[g(t)] = 0, \quad t_0 \geq 0 \tag{E_{48}}$$

where $q, g : [t_0, \infty) \to \mathbb{R}$ are continuous, $q(t) \geq 0$ eventually, and $\lim_{t\to\infty} g(t) = \infty$. We have the following:

(i) when $q(t) = 2/t$ and $g(t) = \sqrt{t}$, the equation (E_{48}) has a solution $x(t) = t^2$,

(ii) when $q(t) = 2$ and $g(t) = t$, the equation (E_{48}) has a solution $x(t) = e^{-t}$.

Clearly, for the equation (E_{48}) all the conditions of Theorem 2.8.12 are satisfied.

Theorem 2.8.13. Let n be odd, $p'(t) \geq 0$, $p''(t) \leq 0$, and $0 \leq p(t) \leq p_0$ for $t \geq t_0$ where p_0 is a real constant. If conditions (2.8.8), (2.8.11) and (2.8.12) are satisfied, then every solution $x(t)$ of the equation (E_{47}) is oscillatory, or $x^{(i)}(t) \to 0$ monotonically as $t \to \infty$, $i = 0, 1, \cdots, n-2$.

It should be remarked that if we multiply (E_{47}) by $\gamma(t) = e^{-\lambda(t)}$, where $\lambda(t)$ is an antiderivative of $p(t)$, then we have

$$\left(\gamma(t)x^{(n-1)}(t)\right)' + \gamma(t)H(t, x[g(t)]) = 0. \tag{E_{49}}$$

In this case, clearly

$$\int^{\infty} \frac{1}{\gamma(s)} ds = \infty, \tag{2.8.14}$$

and hence some special cases of the results of Kitamura [165] as well as of Grace [70,72.73] give complete oscillation criteria for (E_{49}), and so for the equation (E_{47}).

Next, we shall consider the equation

$$x^{(n)}(t) = p(t)x^{(n-1)}(t) + H(t, x[g(t)]), \tag{E_{50}}$$

where the functions g, p and H are as in (E_{47}).

Theorem 2.8.14. Let n be even, $p'(t) \leq 0$ for $t \geq t_0$, and let condition (2.8.7) hold for every constant $c > 0$ and $t_0 \geq 0$. Then, every solution $x(t)$ of the equation (E_{50}) is oscillatory, $x^{(i)}(t) \to \infty$ monotonically as $t \to \infty$, $i = 0, 1, \cdots, n-2$, or $x^{(i)}(t) \to 0$ monotonically as $t \to \infty$, $i = 0, 1, \cdots, n-2$.

Proof. Let $x(t)$ be an eventually positive solution of (E_{50}). As in the proof of Theorem 2.8.6 we find that $x^{(n-1)}(t)$ is eventually of one sign, and the two cases (I) and (II) exist.

(I) Suppose $x^{(n-1)}(t) > 0$ eventually. Then, $x^{(i)}(t) > 0$ eventually for $i = 0, 1, \cdots, n$ and

$$x^{(n)}(t) \;\geq\; H(t, x[g(t)]) \;\geq\; 0 \quad \text{eventually.}$$

Clearly, condition (2.8.7) implies that $x^{(i)}(t) \to \infty$ monotonically as $t \to \infty$, $i = 0, 1, \cdots, n-1$.

(II) Suppose $x^{(n-1)}(t) < 0$ eventually. Then, $x^{(n-2)}(t) > 0$ eventually and the following two cases are considered.

(i) $x'(t) > 0$ eventually, and (ii) $x'(t) < 0$ eventually.

The rest of the proof is easy and hence omitted. ∎

Example 2.8.5. The equation

$$x^{(4)}(t) \;=\; \frac{1}{2}x^{(3)}(t) + qx(t), \qquad t \geq 0$$

has a solution $x(t) = e^t$ if $q = 1/2$, and has a solution $x(t) = e^{-t}$ if $q = 3/2$. Clearly for this equation all the conditions of Theorem 2.8.14 are satisfied.

Theorem 2.8.15. Let n be odd, $p'(t) \leq 0$ for $t \geq t_0$, and let condition (2.8.7) hold for every constant $c > 0$ and $t_0 \geq 0$. Then, every solution $x(t)$ of the equation (E_{50}) is oscillatory, or $x^{(i)}(t) \to \infty$ monotonically as $t \to \infty$, $i = 0, 1, \cdots, n-2$.

Once again, if we multiply (E_{50}) by $\gamma(t) = e^{-\lambda(t)}$, where $\lambda(t)$ is an antiderivative of $p(t)$, then

$$\left(\gamma(t)x^{(n-1)}(t)\right)' \;=\; \gamma(t)H(t, x[g(t)]). \tag{E_{51}}$$

Since condition (2.8.14) is automatically satisfied, we can make use of some special cases of the results of Kitamura [165] to obtain complete oscillation criteria for the equation (E_{51}), and hence for (E_{50}).

Finally, in this section we shall consider the equation

$$x^{(n)}(t) + p(t)x^{(n-1)}(t) \;=\; H(t, x[g(t)]), \tag{E_{52}}$$

where the functions g, p and H are as in (E_{47}).

Theorem 2.8.16. Let $p'(t) \leq 0$ for $t \geq t_0$, and condition (2.8.7) holds for every constant $c > 0$ and $t_0 \geq 0$. Then, every solution $x(t)$ of the equation (E_{52}) is oscillatory, $x^{(i)}(t) \to \infty$ monotonically as $t \to \infty$, $i = 0, 1, \cdots, n-2$, or $x^{(i)}(t) \to 0$ monotonically as $t \to \infty$, $i = 0, 1, \cdots, n-1$.

As an illustration, we have

Example 2.8.6. The equation

$$x^{(n)}(t) + x^{(n-1)}(t) - 2x(t) = 0$$

has a solution $x(t) = e^t$, and each of the following equations

$$x^{(4)}(t) + \frac{1}{2}x^{(3)}(t) - \frac{1}{2}x(t) = 0$$

and

$$x^{(3)}(t) + 2x^{(2)}(t) - x(t) = 0$$

has a solution $x(t) = e^{-t}$. For these equations all the conditions of Theorem 2.8.16 are satisfied.

The last two oscillation criteria for (E_{52}) are due to Grace and Hamedani [88]. In their theorems we shall make use of the following notation:

For $T \geq t_0$, we denote

$$A[u, v] = \exp\left(\int_v^u p(s)ds\right) \quad \text{for} \quad u \geq v \geq T$$

and

$$A^*[z, w] = \int_w^z (A[z, s])^{-1} ds \quad \text{for} \quad z \geq s \geq w \geq T.$$

Theorem 2.8.17. Let n be odd, $g(t)$ is of mixed type, $g'(t) \geq 0$, $p'(t) \geq 0$ for $t \geq t_0$, $\sup_{t \geq t_0} p(t) = \beta > 0$, and condition (2.8.8) holds. If

$$\limsup_{t \to \infty} \int_t^{\rho(t)} A^*[\rho(t), u](g(u) - \rho(t))^{n-1} q(u)du > (n-2)!, \tag{2.8.15}$$

$$\limsup_{t \to \infty} \int_{\tau(t)}^t (\tau(t) - g(s))(g(s))^{n-2} q(s)ds > (n-2)!, \tag{2.8.16}$$

for every θ, $0 < \theta < 1$

$$\liminf_{t \to \infty} \int_{\tau(t)}^t \left[\int_{\tau(u)}^u \frac{\theta}{(n-3)!}(\tau(t) - g(s))(g(s))^{n-3} q(s)ds - \beta\right] du > \frac{1}{e} \tag{2.8.17}$$

and

$$\liminf_{t\to\infty} \int_{\tau(t)}^{t} \left[\int_{\tau(u)}^{u} \frac{(\tau(t)-g(s))^{n-2}}{(n-2)!} q(s)ds - \beta \right] du \; > \; \frac{1}{e},$$

then the equation (E_{52}) is oscillatory.

Theorem 2.8.18. Let n be even, $g(t)$ is of mixed sign, $g'(t) \geq 0$, $p'(t) \geq 0$ for $t \geq t_0$, and $\sup_{t\geq t_0} p(t) = \beta$. If conditions (2.8.8), (2.8.15) – (2.8.17) hold, and

$$\limsup_{t\to\infty} \int_{\tau(t)}^{t} (\tau(t)-g(s))^{n-2} q(s)ds \; > \; (n-1)!,$$

then the equation (E_{52}) is oscillatory.

2.9. Forced Differential Equations

Consider the differential equations

$$x^{(n)}(t) + q(t)f(x[g(t)]) \; = \; e(t), \tag{E_{53}}$$

$$x^{(n)}(t) + H(t, x[g(t)]) \; = \; e(t) \tag{E_{54}}$$

and

$$x^{(n)}(t) + H^*(t, x[g(t)]) \; = \; e(t), \tag{E_{55}}$$

where $g,\ e,\ q : [t_0,\infty) \to \mathbb{R}$, $f : \mathbb{R} \to \mathbb{R}$, $H,\ H^* : [t_0,\infty) \times \mathbb{R} \to \mathbb{R}$ are continuous, $xf(x) > 0$, $xH(t,x) > 0$, $xH^*(t,x) > 0$ for $x \neq 0$, $q(t) \geq 0$ eventually, and $\lim_{t\to\infty} g(t) = \infty$.

Kartsatos' technique which was introduced in [145,146] assumes the existence of a continuous function $\eta : [t_0,\infty) \to \mathbb{R}$ such that $\eta^{(n)}(t) = e(t)$, and reduces the equation (E_{53}) to a homogeneous equation of the form

$$x^{(n)}(t) + q(t)f(x[g(t)]) \; = \; 0. \tag{E_{56}}$$

The basic assumption in Kartsatos results in [145,146] is that $\eta(t)$ is either a small function for large values of t, or it is periodic in t.

In what follows we shall assume the following condition on the forcing term:

$$\text{there exists an } \eta \in C^n[t_0,\infty) \text{ such that } \eta^{(n)}(t) \; = \; e(t). \tag{2.9.1}$$

Kartsatos' results in [145,146] applied to the equation (E_{53}) with n even, $g(t) = t$ take the following form:

Theorem 2.9.1. Let $f'(x) \geq 0$ for $x \neq 0$, n is even, $g(t) = t$, condition (2.9.1) holds, $\eta(t)$ is oscillatory, and $\lim_{t\to\infty} \eta(t) = 0$. If

$$\int_{\pm\epsilon}^{\pm\infty} \frac{du}{f(u)} < \infty \quad \text{and} \quad \int^{\infty} s^{n-1} q(s) ds = \infty, \quad \text{for any } \epsilon > 0,$$

then the equation (E_{53}) is oscillatory.

Theorem 2.9.2. Let $f'(x) \geq 0$ for $x \neq 0$, n is even, $g(t) = t$, condition (2.9.1) holds, and $\eta(t)$ is oscillatory and periodic on $[t_0, \infty)$. Then, (E_{53}) is oscillatory if the equation (E_{56}) is oscillatory.

Teufel's result in [289] when applied to equations of type (E_{53}) with $n = 2$ and $g(t) = t$ reduces to:

Theorem 2.9.3. Let $f'(x) > 0$ for $x \neq 0$, $n = 2$ and $g(t) = t$. Let there exist two constants $\delta_1, \delta_2 > 0$ and a sequence of intervals of the form $[a+mb, a+mb+3\delta_1]$, $m = 1, 2, \cdots$, on which the function $\eta : [t_0, \infty) \to \mathbb{R}$ satisfies $\eta(t) < -\delta_1$, and a sequence of intervals $[c+md, c+md+3\delta_2]$, $m = 1, 2, \cdots$, on which $\eta(t) > \delta_2$. Then, the conditions

$$\int_{t_0}^{\infty} q(s) ds = \infty \quad \text{and} \quad \left| \int_{t_0}^{\infty} \eta(s) ds \right| = o\left(\int_{t_0}^{t} q(s) ds \right)$$

are sufficient for the equation (E_{53}) to be oscillatory.

Theorems 2.9.1 and 2.9.2 have been extended to functional differential equations of type (E_{53}) by True [297], Kusano and Onose [183], Onose [233] and several others. Here, we shall present a result of Kusano and Onose [183].

Theorem 2.9.4. Let $g(t) \leq t$, $g'(t) \geq 0$ for $t \geq t_0$, $f'(x) \geq 0$ for $x \neq 0$, and assume that there exists a function $\phi(y)$ satisfying the following conditions:

$$\phi \in C^1[\mathbb{R}_0, \mathbb{R}], \quad \phi(y) > 0, \quad \phi'(y) \geq 0$$

$$\int_{\pm\epsilon}^{\pm\infty} \frac{dy}{f(y)\phi(y^{1/(n-1)})} < \infty \quad \text{for any } \epsilon > 0$$

and

$$\int^{\infty} \frac{g^{n-1}(s) q(s)}{\phi(g(s))} ds = \infty.$$

In addition assume that condition (2.9.1) holds, and either

(i) $\lim_{t\to\infty} \eta(t) = 0$, or

(ii) there exist constants α and β, and sequences $\{t'_m\}$, $\{t''_m\}$ with the following properties:

$$\lim_{m\to\infty} t'_m = \infty, \quad \lim_{m\to\infty} t''_m = \infty, \quad \eta(t'_m) = \alpha, \quad \eta(t''_m) = \beta$$
$$\alpha \le \eta(t) \le \beta \quad \text{for} \quad t \ge t_0 \ge 0.$$

Let (i) hold, then every solution $x(t)$ of the equation (E_{53}) is oscillatory, or $\lim_{t\to\infty} x(t) = 0$.

Let (ii) hold, then if n is even, every solution $x(t)$ of the equation (E_{53}) is oscillatory while if n is odd, every solution is oscillatory, or

$$\lim_{t\to\infty} [x(t) - \eta(t)] = -\alpha \quad \text{or} \quad -\beta.$$

The following comparison result obtained for the equations (E_{54}) and (E_{55}) with n even and $g(t) = t$ is due to Kartsatos [148].

Theorem 2.9.5. Let n be even, $g(t) = t$, H and H^* be increasing in the second variable, and condition (2.9.1) holds such that $\eta(t)$ is oscillatory, and $\lim_{t\to\infty} \eta(t) = 0$. If

$$H(t,x) \operatorname{sgn} x \le H^*(t,x) \operatorname{sgn} x$$

and the equation (E_{54}) is oscillatory, then the equation (E_{55}) is oscillatory.

Next, we state the following results of Foster [58].

Theorem 2.9.6. Let n be even, H be increasing in the second variable, $g(t) = t$, and condition (2.9.1) holds such that $\eta(t)$ assumes positive and negative values on any interval $[T, \infty)$ for some $T \ge t_0$. Moreover, assume that for any $a > 0$ and $v_0 \ne 0$ all solutions $v(t, a, v_0)$ of the problem

$$v'(t) = \frac{(t-a)^{n-1}}{(n-1)!} H(t, v(t)), \quad v(a) = v_0$$

have finite escape time. Then, if $\lim_{t\to\infty} \eta(t) = 0$, (E_{54}) is oscillatory.

Theorem 2.9.7. Let n be even, $g(t) = t$, H is increasing in the second variable, and condition (2.9.1) holds. Moreover, assume that there exist two sequences $\{s_m\}$, $\{t_m\}$ with $\lim_{m\to\infty} s_m = \infty$ and $\lim_{t\to\infty} t_m = \infty$, and such that

$$\eta(s_m) = \alpha > 0, \quad \eta(t_m) = -\beta < 0$$

and

$$\inf_{s \ge a} \eta(s) = -\beta \quad \text{and} \quad \sup_{s \ge a} \eta(s) = \alpha \quad \text{for every} \quad a > 0.$$

Then, the equation (E_{54}) is oscillatory.

The above results of Foster [58] and those in [57] which cover the sublinear cases are closely related to Theorem 2.9.1, and the following criterion of Kartsatos [145].

Theorem 2.9.8. Let n be even, $g(t) = t$, H is increasing in the second variable, and condition (2.9.1) holds. Then, under any one of the following conditions (E_{54}) is oscillatory:

(i) $\eta(t)$ is oscillatory, and $\lim_{t\to\infty} \eta(t) = 0$,

(ii) there exists $\eta_1 : [t_0, \infty) \to \mathbb{R}$ such that $\eta_1^{(n)}(t) = e(t)$, and such that $\eta(t)$ and $\eta_1(t)$ are oscillatory with

$$\liminf_{t\to\infty} \eta(t) = 0 \quad \text{and} \quad \limsup_{t\to\infty} \eta_1(t) = 0.$$

Next, Kartsatos and Manogian [154] proved the following interesting result:

Theorem 2.9.9. Let n be even, $g(t) = t$, condition (2.9.1) holds, and

$$\limsup_{t\to\infty} \eta(t) > 0, \qquad \liminf_{t\to\infty} \eta(t) < 0.$$

Moreover, assume that for some constant $k > 0$

$$\int_{t_0}^{\infty} s^{n-1} |H(s, u(s) + \eta(s))| ds < \infty$$

for all $u : [t_0, \infty) \to \mathbb{R}$ with $|u(t)| \le k$, $t \ge t_0$. Then, the equation (E_{54}) has at least one oscillatory solution.

The following result applied to (E_{54}) is due to Graef and Spikes [118]. It improves Theorem 2.1 established in [155].

Theorem 2.9.10. Let n be even, $g(t) = t$, and assume that for every $k > 0$ and $t \ge T$

$$\limsup_{t\to\infty} \left\{ \int_T^t (t-s)^{n-1} e(s) ds - kt^{n-1} \right\} > 0$$

and

$$\liminf_{t\to\infty} \left\{ \int_T^t (t-s)^{n-1} e(s) ds - kt^{n-1} \right\} < 0.$$

Then, the equation (E_{54}) is oscillatory.

In Theorem 2.9.10 the oscillation can be due to the forcing term $e(t)$, and no growth condition has been imposed on the function H. The following result of a similar nature was proved in [154].

Theorem 2.9.11. Let condition (2.9.1) hold. Moreover, assume that for every $\lambda > 0$, $T \geq 0$

$$\limsup_{t\to\infty} \int_T^t H(s, \lambda + \eta[g(s)])ds = \infty$$

and

$$\liminf_{t\to\infty} \int_T^t H(s, -\lambda + \eta[g(s)])ds = -\infty.$$

Then, for n even the equation (E_{54}) is oscillatory. For n odd, every solution $x(t)$ of the equation (E_{54}) is oscillatory, or such that $(x(t) - \eta(t)) \to 0$ monotonically as $t \to \infty$. If the above integral conditions hold for $\lambda = 0$, then the equation (E_{54}) is oscillatory for n is odd.

An interesting corollary of Theorem 2.9.11 reads as follows:

Corollary 2.9.12. Let $f(x) = x^{2\mu+1}$, where μ is a nonnegative integer, and condition (2.9.1) is satisfied with $\eta(t)$ oscillatory,

$$\int_T^\infty q(s)ds = \infty \quad \text{and} \quad \int_T^\infty q(s)|\eta[g(s)]|^m ds < \infty$$

for every $m = 1, 2, \cdots, 2\mu + 1$. Then, the conclusion of Theorem 2.9.11 for the case $\lambda \neq 0$ holds.

In the following theorems we shall extend to an arbitrary n the known results of Wong [308] for second order equations.

Theorem 2.9.13. Let n be even, condition (2.9.1) holds, $\eta(t)$ is oscillatory, and

$$\liminf_{t\to\infty} \frac{\eta(t)}{t^{n-1}} = -\infty \quad \text{and} \quad \limsup_{t\to\infty} \frac{\eta(t)}{t^{n-1}} = \infty.$$

Then, the equation (E_{54}) is oscillatory.

Proof. Let $x(t)$ be an eventually positive solution of (E_{54}). Let $x(t) = y(t) + \eta(t)$. Then, the equation (E_{54}) takes the form

$$y^{(n)}(t) + H(t, y[g(t)] + \eta[g(t)]) = 0.$$

We claim that $y(t) > 0$ eventually. To prove it assume that $y(t) < 0$ eventually. Now

$$-\eta(t) < x(t) - \eta(t) = y(t) < 0 \quad \text{eventually},$$

which is a contradiction to the oscillatory character of the function $\eta(t)$. By Lemma 2.2.1 there exist $t_1 \geq t_0 \geq 0$ and a posiitve constant c such that

$$y(t) > 0, \quad y'(t) > 0, \quad y^{(n-1)}(t) > 0 \quad \text{and} \quad y^{(n)}(t) \leq 0 \quad \text{for } t \geq t_1 \tag{2.9.2}$$

and

$$y^{(n-1)}(t) \le c \quad \text{for} \quad t \ge t_1. \tag{2.9.3}$$

Thus, there exists a $t_2 \ge t_1$ and a constant $C > 0$ such that

$$0 \; < \; y(t) \; \le \; Ct^{n-1} \quad \text{for} \quad t \ge t_2$$

or

$$\limsup_{t\to\infty} \frac{y(t)}{t^{n-1}} \; \le \; C.$$

On the other hand, we have $y(t) + \eta(t) = x(t) > 0$ eventually, or $y(t) > -\eta(t)$. Now, dividing this inequality by t^{n-1} and taking lim sup on both sides, we get

$$\infty \; > \; C \; \ge \; \limsup_{t\to\infty} \frac{y(t)}{t^{n-1}} \; > \; \limsup_{t\to\infty} -\frac{\eta(t)}{t^{n-1}} \; = \; -\liminf_{t\to\infty} \frac{\eta(t)}{t^{n-1}} \; = \; \infty,$$

which is a contradiction. ∎

Theorem 2.9.14. Let n be even, H is nondecreasing in the second variable, and condition (2.9.1) holds with $\eta(t)$ oscillatory. If

$$\int_T^{\infty} H(s, \eta_+[g(s)])ds \; = \; \infty \quad \text{and} \quad \int_T^{\infty} H(s, \eta_-[g(s)])ds \; = \; \infty, \quad T \ge t_0$$

where $\eta_+(t) = \max\{\eta(t), 0\}$ and $\eta_-(t) = \min\{\eta(t), 0\}$. Then, the equation (E_{54}) is oscilatory.

Proof. Let $x(t)$ be an eventually positive solution of the equation (E_{54}). As in the proof of Theorem 2.9.13 we obtain (2.9.2). We note that for all $t \ge t_1$, $y(t) + \eta(t) > \eta_+(t)$. To see this, we write $y(t) + \eta(t) = y(t) + \eta_+(t) - \eta_-(t)$, and observe that

(i) for $\eta_+(t) = 0$, $y(t) + \eta(t) = y(t) - \eta_-(t) = x(t) > 0 = \eta_+(t)$, and

(ii) for $\eta_-(t) = 0$, $y(t) + \eta(t) = y(t) + \eta_+(t) > \eta_+(t)$.

Since H is nondecreasing in the second variable, we have

$$H(t, y[g(t)] + \eta[g(t)]) \; \ge \; H(t, \eta_+[g(t)]).$$

Integrating the equation (E_{54}) from t_1 to t, we find

$$y^{(n-1)}(t) - y^{(n-1)}(t_1) + \int_{t_1}^{t} H(s, y[g(s)] + \eta[g(s)])ds \; = \; 0$$

or

$$y^{(n-1)}(t) \; \le \; y^{(n-1)}(t_1) - \int_{t_1}^{t} H(s, \eta_+[g(s)])ds \to -\infty \quad \text{as} \quad t \to \infty,$$

which is a contradiction. ∎

Theorem 2.9.15. Let n be even, $g(t) \leq t$, $g'(t) \geq 0$ for $t \geq t_0 \geq 0$, G and H is a pair of continuous components of f with H being nondecreasing, and assume that condition (2.9.1) holds with $\eta(t)$ oscillatory, and $\eta[g(t)]/g^{n-1}(t)$ bounded. If, for every $\mu > 0$

$$\int_T^\infty q(s)G\left(\mu g^{n-1}(s)\right) H(\eta_+[g(s)])ds = \infty$$

and

$$\int_T^\infty q(s)G\left(\mu g^{n-1}(s)\right) H(\eta_-[g(s)])ds = \infty, \quad T \geq t_0$$

then the equation (E_{53}) is oscillatory.

Proof. Let $x(t)$ be an eventually positive solution of the equation (E_{53}). As in the proof of Theorem 2.9.14 we obtain (2.9.2) and (2.9.3). Now, it is easy to see that there exist a $t_2 \geq t_1$ and a constant $k > 0$ such that

$$x[g(t)] \leq kg^{n-1}(t) \quad \text{for } t \geq t_2.$$

Since

$$f(x[g(t)]) = G(x[g(t)])H(x[g(t)]),$$

we have

$$f(x[g(t)]) \leq G\left(kg^{n-1}(t)\right) H(y[g(t)] + \eta[g(t)]), \quad t \geq t_2.$$

Thus,

$$y^{(n)}(t) + q(t)G\left(kg^{n-1}(t)\right) H(y[g(t)] + \eta[g(t)]) \leq 0. \quad t \geq t_2.$$

The rest of the proof is similar to that of Theorem 2.9.14. ∎

The following comparison result is deduced from the work of Grace and Lalli [105].

Theorem 2.9.16. Let $t_0 \geq 0$, $g(t) \leq t$, $g'(t) \geq 0$ for $t \geq t_0$, condition (2.9.1) holds with $\eta(t)$ oscillatory, and $\lim_{t\to\infty} \eta(t) = 0$. In addition, let G and H be a pair of continuous components of f with H being the nondecreasing. Moreover, assume that

$$Q(t) \leq q(t)G\left(\pm kg^{n-1}(t)\right) \quad \text{for all} \quad k > 0$$

and

$$h(x) \operatorname{sgn} x \leq H(x) \operatorname{sgn} x \quad \text{for all} \quad x,$$

where $Q : [t_0, \infty) \to \mathbb{R}_0$ and $h : \mathbb{R} \to \mathbb{R}$ are continuous, $xh(x) > 0$ and $h'(x) \geq 0$ for $x \neq 0$. If the equation

$$y^{(n)}(t) + Q(t)h(y[g(t)]) = 0$$

is almost oscillatory, then so does the equation (E_{53}).

2.10. Forced Equations with Middle Term of Order $(n-1)$

We shall consider the equation

$$x^{(n)}(t) + p(t)x^{(n-1)}(t) + H(t, x[g(t)]) = e(t), \quad n \text{ even} \qquad (E_{57})$$

where $e,\ g,\ p : [t_0, \infty) \to \mathbb{R}$, $H : [t_0, \infty) \times \mathbb{R} \to \mathbb{R}$ are continuous, $t_0 \geq 0$, $\lim_{t\to\infty} g(t) = \infty$, $xH(t,x) > 0$ for $x \neq 0$ and H is increasing in the second variable. Moreover, by $\eta(t)$ we denote a fixed solution of the equation

$$\eta^{(n)}(t) + p(t)\eta^{(n-1)}(t) = e(t) \quad \text{for} \quad t \geq t_0. \qquad (E_{58})$$

Let $x(t) = y(t) + \eta(t)$, where $x(t)$ is a solution of the equation (E_{57}), to obtain

$$y^{(n)}(t) + p(t)y^{(n-1)}(t) + H(t, y[g(t)] + \eta[g(t)]) = 0. \qquad (E_{59})$$

In [150], Kartsatos proved the following two oscillation criteria for the equation (E_{57}) with $g(t) = t$.

Theorem 2.10.1. Let condition (2.8.1) hold with $p(t) \geq 0$ for $t \geq t_0$. If

$$\limsup_{t\to\infty} \eta(t) = \infty \quad \text{and} \quad \liminf_{t\to\infty} \eta(t) = -\infty,$$

then every bounded solution of the equation (E_{57}) is oscillatory.

Theorem 2.10.2. Let condition (2.8.1) hold with $p(t) \geq 0$ for $t \geq t_0$, and assume that $\eta(t)$ is oscillatory and $\eta(t) \to 0$ as $t \to \infty$. If (E_{45}) is oscillatory, then the equation (E_{57}) is oscillatory.

Next, we state an extension of Kartsatos result [153, Theorem 4] which is due to Kosmala [172].

Theorem 2.10.3. Let $g(t) = t$, condition (2.8.1) holds with $p(t) \leq 0$, and the function $\eta(t)$ is nonnegative and satisfies the equation (E_{58}) with $\liminf_{t\to\infty} \eta(t) = 0$. Suppose that the equation (E_{45}) has no eventually positive solution. If the equation (E_{57}) has two eventually positive solutions $u(t)$ and $v(t)$, then

$$\lim_{t\to\infty} [u(t) - v(t)] = 0$$

and $u(t) - v(t)$ oscillates.

In [172], Kosmala also studied the oscillation of (E_{57}) without finding a function $\eta(t)$ which satisfies the equation (E_{58}). His main result is as follows:

Theorem 2.10.4. Let condition (2.8.1) hold, H is increasing in the second variable, and the equation

$$x^{(n)}(t) + p(t)x^{(n-1)}(t) + \mu H(t, x(t)) = 0 \qquad (E_{60})$$

is oscillatory for all $\mu > 0$. Suppose that

$$x^{(n)}(t) + p(t)x^{(n-1)}(t) + H(t, x(t)) \leq e(t)$$

has an eventually positive solution $x_1(t)$, and

$$x^{(n)}(t) + p(t)x^{(n-1)}(t) + H(t, x(t)) \geq e(t)$$

has an eventually negative solution $x_2(t)$ such that

$$\lim_{t\to\infty} x_1(t) = 0 = \lim_{t\to\infty} x_2(t).$$

Then, the equation (E_{57}) is oscillatory.

As a corollary of Theorem 2.10.4, we have

Corollary 2.10.5. Let condition (2.8.1) hold, H is increasing in the second variable, and the equation (E_{60}) has all bounded solutions oscillatory for all $\mu > 0$. Let $x_1(t)$ and $x_2(t)$ be two eventually positive solutions of the equation (E_{57}) with $\lim_{t\to\infty} x_1(t) = 0 = \lim_{t\to\infty} x_2(t)$. Then, $x_1(t) - x_2(t)$ is oscillatory.

If we multiply the equation (E_{57}) by $a(t) = e^{\lambda(t)}$ where $\lambda(t)$ is an antiderivative of $p(t)$, we obtain

$$\left(a(t)x^{(n-1)}(t)\right)' + H_1(t, x[g(t)]) = E(t),$$

where $H_1(t,x) = a(t)H(t,x)$ and $E(t) = a(t)e(t)$. In this case if we let $\eta(t)$ satisfy

$$\left(a(t)\eta^{n-1}(t)\right)' = E(t),$$

we obtain, instead of (E_{59}), the equation

$$\left(a(t)y^{(n-1)}(t)\right)' + H_1\left(t, y[g(t)] + \eta[g(t)]\right) = 0. \qquad (E_{61})$$

The oscillatory behavior of more general equations which include (E_{61}) as a special case will be discussed in later sections.

2.11. Superlinear Forced Equations

Here we shall present some oscillation criteria for the nth order superlinear forced differential equation

$$x^{(n)}(t) - q(t)|x(t)|^{\lambda}\text{sgn } x(t) = e(t), \quad \lambda > 1 \qquad (E_{62})$$

where $n \geq 1$, e, $q : [t_0, \infty) \to \mathbb{R}$ are continuous, and $q(t) > 0$ for $t \geq t_0$.

Theorem 2.11.1. Suppose there exists an $(n-1)$ times–differentiable function $H : D = \{(t,s) : t \geq s \geq t_0\} \to \mathbb{R}$ such that

$$H(t,t) = 0, \quad t \geq t_0, \quad H(t,s) > 0 \quad \text{for} \quad (t,s) \in D \qquad (2.11.1)$$

$$\begin{cases} h_i(t,s) = -\dfrac{\partial^i H(t,s)}{\partial s^i}, \quad h_i(t,t) = 0 \text{ for } t \geq t_0, \ i = 1,2,\cdots,n-1, \\ h_{n-1}(t,s) \quad \text{is nonnegative continuous function on } D, \text{ and} \\ 0 \leq \liminf\limits_{t\to\infty} \dfrac{h_i(t,t_0)}{H(t,t_0)} < \infty \quad \text{for } i = 1,2,\cdots,n-2. \end{cases} \qquad (2.11.2)$$

If

$$\limsup_{t\to\infty} \frac{1}{H(t,t_0)} \int_{t_0}^{t} [H(t,s)e(s) - Q(t,s)]ds = \infty \qquad (2.11.3)$$

and

$$\liminf_{t\to\infty} \frac{1}{H(t,t_0)} \int_{t_0}^{t} [H(t,s)e(s) - Q(t,s)]ds = -\infty, \qquad (2.11.4)$$

where

$$Q(t,s) = (\lambda-1)\lambda^{\lambda/(1-\lambda)} \left(\frac{h_{n-1}^{\lambda}(t,s)}{H(t,s)}\right)^{1/(\lambda-1)} q^{1/(1-\lambda)}(s) \quad \text{for} \quad t \geq s \geq t_0,$$

then the equation (E_{62}) is oscillatory.

Proof. Let $x(t)$ be a nonoscillatory solution of the equation (E_{62}), say $x(t) > 0$ for $t \geq t_0$. Multiplying the equation (E_{62}) by $H(t,s)$ for $t \geq s \geq t_0$ and integrating from t_0 to t, we get

$$\int_{t_0}^{t} H(t,s)e(s)ds = \int_{t_0}^{t} H(t,s)x^{(n)}(s)ds - \int_{t_0}^{t} H(t,s)q(s)x^{\lambda}(s)ds.$$

Now, since

$$\begin{aligned}\int_{t_0}^{t} H(t,s)x^{(n)}(s)ds &= -H(t,t_0)x^{(n-1)}(t_0) + \int_{t_0}^{t} h_1(t,s)x^{(n-1)}(s)ds \\ &= -H(t,t_0)x^{(n-1)}(t_0) - \sum_{i=1}^{n-2} h_i(t,t_0)x^{(n-i-1)}(t_0) \\ &\quad + \int_{t_0}^{t} h_{n-1}(t,s)x(s)ds\end{aligned}$$

and in view of the hypotheses there exists a finite number c such that for $t \geq t_0$

$$-H(t,t_0)x^{(n-1)}(t_0) - \sum_{i=1}^{n-2} h_i(t,t_0)x^{(n-i-1)}(t_0) \ \leq\ cH(t,t_0),$$

we find

$$\int_{t_0}^{t} H(t,s)e(s)ds \ \leq\ cH(t,t_0)+\int_{t_0}^{t} h_{n-1}(t,s)x(s)ds-\int_{t_0}^{t} H(t,s)q(s)x^{\lambda}(s)ds. \tag{2.11.5}$$

Set

$$A \ =\ [H(t,s)q(s)]^{1/\lambda}x(s)$$

and

$$B \ =\ \left(\frac{1}{\lambda}h_{n-1}(t,s)[H(t,s)q(s)]^{-1/\lambda}\right)^{1/(\lambda-1)},$$

and apply Lemma 2.2.13 in (2.11.5), to obtain

$$\frac{1}{H(t,t_0)}\int_{t_0}^{t}\left[H(t,s)e(s) - (\lambda-1)\lambda^{\lambda/(1-\lambda)}\left(\frac{h_{n-1}^{\lambda}(t,s)}{H(t,s)}\right)^{1/(\lambda-1)} \times\ q^{1/(1-\lambda)}(s)\right]ds \ \leq\ c.$$

Taking lim sup as $t \to \infty$ in the above inequality, we obtain a contradiction to condition (2.11.3). ∎

Corollary 2.11.2. Let the function H be as in Theorem 2.11.1 such that (2.11.1) and (2.11.2) hold. If

$$\limsup_{t\to\infty}\frac{1}{H(t,t_0)}\int_{t_0}^{t} H(t,s)e(s)ds \ =\ \infty, \tag{2.11.6}$$

$$\liminf_{t\to\infty}\frac{1}{H(t,t_0)}\int_{t_0}^{t} H(t,s)e(s)ds \ =\ -\infty \tag{2.11.7}$$

and

$$\lim_{t\to\infty} \frac{1}{H(t,t_0)} \int_{t_0}^{t} \left(\frac{h_{n-1}^{\lambda}(t,s)}{H(t,s)} \right)^{1/(\lambda-1)} q^{1/(1-\lambda)}(s)ds \;<\; \infty, \tag{2.11.8}$$

then the equation (E_{62}) is oscillatory.

Corollary 2.11.3. Let $H(t,s) = (t-s)^m$, $m \geq n+1$ and $t \geq s \geq t_0$. If

$$\limsup_{t\to\infty} \frac{1}{t^m} \int_{t_0}^{t} [(t-s)^m e(s) - Q^*(t,s)]ds \;=\; \infty \tag{2.11.9}$$

and

$$\liminf_{t\to\infty} \frac{1}{t^m} \int_{t_0}^{t} [(t-s)^m e(s) - Q^*(t,s)]ds \;=\; -\infty, \tag{2.11.10}$$

where

$$\begin{aligned} Q^*(t,s) \;=\;& (\lambda-1)\lambda^{\lambda/(1-\lambda)}(m(m-1)\cdots(m-n+1))^{\lambda/(\lambda-1)} \\ & \times (t-s)^{(m(\lambda-1)-n)/(\lambda-1)} q^{1/(1-\lambda)}(s) \quad \text{for} \quad t \geq s \geq t_0, \end{aligned}$$

then the equation (E_{62}) is oscillatory.

Remark 2.11.1. When $\lambda = 1$, the inequality (2.11.5) reduces to

$$\int_{t_0}^{t} H(t,s)e(s)ds \;\leq\; cH(t,t_0) - \int_{t_0}^{t} [H(t,s)q(s) - h_{n-1}(t,s)]x(s)ds,$$

and thus, we have the following result:

Theorem 2.11.4. Let the function H be as in Theorem 2.11.1 such that (2.11.1) and (2.11.2) hold, and let

$$H^*(t,s) \;=\; H(t,s)q(s) - h_{n-1}(t,s) \;\geq\; 0 \quad \text{for} \quad t \geq s \geq t_0. \tag{2.11.11}$$

If

$$\limsup_{t\to\infty} \frac{1}{H^*(t,t_0)} \int_{t_0}^{t} H^*(t,s)e(s)ds \;=\; \infty \tag{2.11.12}$$

and

$$\liminf_{t\to\infty} \frac{1}{H^*(t,t_0)} \int_{t_0}^{t} H^*(t,s)e(s)ds \;=\; -\infty, \tag{2.11.13}$$

then the equation (E_{62}) with $\lambda = 1$ is oscillatory.

The following examples illustrate how Theorem 2.11.4 can be used in practice.

Example 2.11.1. Each of the following equations

$$x'(t) - x(t) = e^t \cos t, \quad t \geq 0$$

$$x'(t) - e^{-t}x^3(t) = -e^{2t}\sin^3 t + e^t(\sin t + \cos t), \quad t \geq 0$$

has an oscillatory solution $x(t) = e^t \sin t$. Notice conditions of Theorem 2.11.4 and Corollary 2.11.3 are respectively satisfied with $H(t,s) = (t-s)^2$.

Example 2.11.2. Consider the equations

$$x''(t) - x(t) = e^t(2\cos t - \sin t), \quad t \geq 0$$

$$x''(t) - e^{-t}x^3(t) = -e^{2t}\sin^3 t + 2e^t\cos t, \quad t \geq 0.$$

It is easy to see that all the conditions of Theorem 2.11.4 and Corollary 2.11.3 are respectively satisfied with $H(t,s) = (t-s)^3$, and hence all solutions of these equations are oscillatory. One such solution for both equations is $x(t) = e^t \sin t$.

Remark 2.11.2. From the above examples it is clear that the forcing term e is not necessarily small. Also, the forcing terms in the above equations in fact generate oscillations.

2.12. Sublinear Forced Equations

Here we shall offer oscillation criteria for the perturbed–forced sublinear differential equation

$$\delta x^{(n)}(t) + q(t)|x(t)|^\lambda \operatorname{sgn} x(t) = p(t)x(t) + e(t), \qquad (E_{63};\delta)$$

where $n \geq 1$, $\delta = \pm 1$, $0 < \lambda < 1$, p, $q : [t_0, \infty) \to \mathbb{R}_0$ and $e : [t_0, \infty) \to \mathbb{R}$ are continuous, $t_0 \geq 0$.

Theorem 2.12.1. Suppose that conditions of Theorem 2.11.1 are satisfied with

$$Q(t,s) = (1-\lambda)\lambda^{\lambda/(1-\lambda)} P^{\lambda/(\lambda-1)}(t,s)\left(H(t,s)q(s)\right)^{1/(1-\lambda)},$$

where

$$P(t,s) = [p(s)H(t,s) - h_{n-1}(t,s)]$$

is nonnegative for $(t,s) \in D$. Then, the equation $(E_{63};1)$ is oscillatory.

Proof. The proof is similar to that of Theorem 2.11.1. ∎

Corollary 2.12.2. If in Corollary 2.11.2 condition (2.11.8) is replaced by

$$\lim_{t\to\infty} \frac{1}{H(t,t_0)} \int_{t_0}^{t} Q(t,s)ds < \infty,$$

where Q is as in Theorem 2.12.1, then the equation $(E_{63};1)$ is oscillatory.

Theorem 2.12.3. Let the function H be as in Theorem 2.11.1 such that (2.11.1) and (2.11.2) hold. If

$$\limsup_{t\to\infty} \frac{1}{H(t,t_0)} \int_{t_0}^{t} [H(t,s)e(s) - Q^*(t,s)]\, ds \;=\; \infty$$

and

$$\liminf_{t\to\infty} \frac{1}{H(t,t_0)} \int_{t_0}^{t} [H(t,s)e(s) - Q^*(t,s)]\, ds \;=\; -\infty,$$

where

$$Q^*(t,s) \;=\; (1-\lambda)\lambda^{\lambda/(1-\lambda)} P^{*\lambda/(\lambda-1)}(t,s)(H(t,s)q(s))^{1/(1-\lambda)}$$

and

$$P^*(t,s) \;=\; [H(t,s)p(s) + h_{n-1}(t,s)] \;>\; 0 \quad \text{for} \quad (t,s) \in D,$$

then the equation $(E_{63};-1)$ is oscillatory.

We note that the above results are applicable to $(E_{63};\delta)$ for any $n \geq 1$ and fail if either $\lambda \geq 1$, $p = 0$, or $e = 0$. For $\lambda = 1$ and $p = 0$, i.e. for the equation

$$x^{(n)}(t) + q(t)x(t) \;=\; e(t), \qquad (E_{64})$$

we have the following oscillation result.

Theorem 2.12.4. Let the function H be as in Theorem 2.11.1 such that conditions (2.11.1) and (2.11.2) hold. If for every $c > 0$

$$\limsup_{t\to\infty} \frac{1}{H(t,t_0)} \int_{t_0}^{t} [H(t,s)e(s) - c(H(t,s)q(s) - h_{n-1}(t,s))]\, ds \;=\; \infty$$

and

$$\liminf_{t\to\infty} \frac{1}{H(t,t_0)} \int_{t_0}^{t} [H(t,s)e(s) - c(H(t,s)q(s) - h_{n-1}(t,s))]\, ds \;=\; -\infty,$$

then every bounded solution of (E_{64}) is oscillatory.

The following examples illustrate how the above theorems can be used in practice.

Example 2.12.1. Consider the equations

$$x'(t) + x^{1/3}(t) \;=\; x(t) + e^t \cos t + e^{t/3} \sin^{1/3} t, \quad t \geq 0$$

and

$$-x'(t) + x^{1/3}(t) \;=\; x(t) + (-e^t \cos t - 2e^t \sin t + e^{t/3} \sin^{1/3} t), \quad t \geq 0.$$

Let $H(t,s) = (t-s)^2$. Clearly, all the conditions of Theorems 2.12.1 and 2.12.3 respectively are satisfied and hence the above equations are oscillatory. One such solution is $x(t) = e^t \sin t$.

Example 2.12.2. Consider the equation

$$x^{(n)}(t) + x(t) = e^t \sin t, \quad t \geq 0.$$

Let $H(t,s) = (t-s)^{n+1}$. One can easily see that all the conditions of Theorem 2.12.4 are satisfied and hence every bounded solution of the above is oscillatory.

2.13. Perturbed Functional Equations

Here we shall study the oscillatory behavior of perturbed functional differential equations of the form

$$\delta x^{(n)}(t) + H(t, x[g(t)]) = P(t, x[g(t)]), \qquad (E_{65};\delta)$$

where $n \geq 1$, $\delta = \pm 1$, $g : [t_0, \infty) \to \mathbb{R}$, H, $P : [t_0,\infty) \times \mathbb{R} \to \mathbb{R}$ are continuous, $t_0 \geq 0$ and $\lim_{t\to\infty} g(t) = \infty$.

We shall assume that there exist continuous functions a, p, $q : [t_0, \infty) \to \mathbb{R}_0$, and positive constants λ and μ, $\gamma = \mu - \lambda > 1$ such that

$$H(t,x) \operatorname{sgn} x \leq a(t)|x|^{\lambda+1} \quad \text{for} \quad x \neq 0, \ t \geq t_0 \qquad (2.13.1)$$

$$P(t,x) \operatorname{sgn} x \geq p(t)|x|^{\mu} + q(t)|x|^{\lambda} \quad \text{for} \quad x \neq 0, \ t \geq t_0 \qquad (2.13.2)$$

and

$$Q(t) = q(t) - \beta a^{\gamma/(\gamma-1)}(t) p^{1/(1-\gamma)}(t) \geq 0 \quad \text{for} \quad t \geq t_0 \qquad (2.13.3)$$

and $Q(t) \not\equiv 0$ on any ray of the form $[t^*, \infty)$ for some $t^* \geq t_0$, where $\beta = (\gamma - 1)\gamma^{\gamma/(1-\gamma)}$.

The oscillatory behavior of $(E_{65};1)$ with n even and $g(t) = t$ has been discussed by Kartsatos [150–152]. In [150], Kartsatos raised some problems regarding the oscillation of the equation $(E_{65};\delta)$ without assuming, either

$$\lim_{t\to\infty} \frac{H(t, u[g(t)])}{P(t, u[g(t)])} = 0 \qquad (*)$$

or

$$\lim_{t\to\infty} \sup_{|u|\leq K} \frac{H(t,u)}{P(t,u)} = 0. \qquad (**)$$

The main purpose of this section is to provide sufficient conditions for the oscillation of the equation $(E_{65};\delta)$ without necessarily requiring the assumption (*) or (**).

Theorem 2.13.1. Let n be even, and conditions (2.13.1) – (2.13.3) hold. If the equation

$$x^{(n)}(t) + Q(t)|x[\sigma(t)]|^{\lambda}\text{sgn}\; x[\sigma(t)] \;=\; 0 \qquad (E_{66})$$

is oscillatory, where $0 < \sigma(t) = \min\{t, g(t)\} \to \infty$, as $t \to \infty$, then the equation $(E_{65};-1)$ is oscillatory.

Proof. Let $x(t)$ be a nonoscillatory solution of the equation $(E_{65};-1)$, say $x(t) > 0$ and $x[g(t)] > 0$ for $t \geq t_0$. Using conditions (2.13.1) and (2.13.2) in the equation $(E_{65};-1)$, we get

$$\begin{aligned} 0 &\geq x^{(n)}(t) - a(t)x^{\lambda+1}[g(t)] + p(t)x^{\mu}[g(t)] + q(t)x^{\lambda}[g(t)] \\ &= x^{(n)}(t) + q(t)x^{\lambda}[g(t)] - x^{\lambda}[g(t)]\,[a(t)x[g(t)] - p(t)x^{\gamma}[g(t)]] \;\text{ for } t \geq t_0. \end{aligned} \tag{2.13.4}$$

Now we set

$$A \;=\; p^{1/\gamma}(t)x[g(t)] \quad \text{and} \quad B \;=\; \left(\frac{a(t)}{\gamma}p^{-1/\gamma}(t)\right)^{1/(\gamma-1)}$$

and apply Lemma 2.2.13, to get

$$0 \;\geq\; x^{(n)}(t) + q(t)x^{\lambda}[g(t)] - \beta a^{\gamma/(\gamma-1)}(t)p^{1/(1-\gamma)}(t)x^{\lambda}[g(t)],$$

or

$$x^{(n)}(t) + Q(t)x^{\lambda}[g(t)] \;\leq\; 0, \qquad t \geq T_1 \geq t_0. \tag{2.13.5}$$

Since n is even, we see that $x(t)$ is an increasing function for $t \geq T_1$ and (2.13.5) reduces to

$$x^{(n)}(t) + Q(t)x^{\lambda}[\sigma(t)] \;\leq\; 0 \quad \text{for} \quad t \geq T \geq T_1.$$

This inequality in view of Lemma 2.5.1 leads to a contradiction. ∎

Remark 2.13.1. For $a(t) = p(t) = q(t)$ the condition (2.13.3) holds for all $\lambda > 0$ and $\mu > \lambda + 1$. However, for this case conditions of the type (*) and (**) are not valid.

Remark 2.13.2. From the proof of Theorem 2.13.1, we see that the equation $(E_{65};-1)$ under the assumptions (2.13.1) – (2.13.3) is reduced to an inequality of the type (2.13.5). Now for any $n \geq 1$ and any $\lambda > 0$, one can apply the results of Kitamura [165] to this inequality and obtain complete oscillation criteria for the equation $(E_{65};-1)$, or make use of comparison results of Kusano and Naito [180] and compare the oscillatory and asymptotic behavior of equations of type (E_{66}) to that of $(E_{65}; -1)$.

Next, we present the following oscillation criterion for the equation $(E_{65}; 1)$, $\lambda = 1$ and n is odd. The other cases for any $n \geq 1$ and $\lambda > 0$ can be obtained similarly.

Theorem 2.13.2. Let $\lambda = 1$, n is odd and conditions (2.13.1) - (2.13.3) hold. If

$$\int^{\infty} s\sigma^{n-2}(s) g^{-\epsilon}(s) Q(s) ds \; = \; \infty \quad \text{for some} \quad \epsilon > 0, \tag{2.13.6}$$

$$\int^{\infty} g^{n-1}(s) Q(s) ds \; = \; \infty \tag{2.13.7}$$

and

$$\limsup_{t \to \infty} \int_t^{\rho(t)} \frac{(g(s) - \rho(t))^j}{j!} \frac{(\rho(t) - s)^{n-j-1}}{(n-j-1)!} Q(s) ds \; > \; 1 \tag{2.13.8}$$

for some $j = 0, 1, \cdots, n-1$ where $0 < \sigma(t) = \min\{t, g(t)\} \to \infty$ as $t \to \infty$ and $\rho(t) = \min\{\max\{s, g(s)\} : s \geq t\}$, then the equation $(E_{65}; 1)$ is oscillatory.

Proof. Let $x(t)$ be a nonoscillatory solution of the equation $(E_{65}; 1)$, say $x(t) > 0$ and $x[g(t)] > 0$ for $t \geq t_0$. Using conditions (2.13.1) and (2.13.2) in the equation $(E_{65}; 1)$, we obtain

$$\begin{aligned} 0 \; &\leq \; x^{(n)}(t) + a(t) x^{\lambda+1}[g(t)] - p(t) x^{\mu}[g(t)] - q(t) x^{\lambda}[g(t)] \\ &= \; x^{(n)}(t) - q(t) x^{\lambda}[g(t)] + x^{\lambda}[g(t)]\,[a(t) x[g(t)] - q(t) x^{\gamma}[g(t)]] \; \text{ for } t \geq t_0. \end{aligned}$$

Now we let A and B be as in the proof of Theorem 2.13.1 and apply Lemma 2.2.13, to get

$$x^{(n)}(t) - Q(t) x^{\lambda}[g(t)] \; \geq \; 0 \quad \text{for} \quad t \geq T \geq t_0. \tag{2.13.9}$$

The rest of the proof follows by applying Theorem 2.4.19(v). ∎

As an application we consider the following equation which arises in the study of nonlinear neural networks

$$\begin{aligned} \delta \frac{dx(t)}{dt} \; &= \; - q(t) |x[g(t)]|^{\lambda} \text{sgn } x[g(t)] + a(t) |\tanh x[g(t)]|^{\lambda} \tanh x[g(t)] \\ &\quad - p(t) |x[g(t)]|^{\mu} \text{sgn } x[g(t)], \end{aligned} \tag{$E_{67}; \delta$}$$

where $\delta = \pm 1$, λ and μ are real constants, $\lambda > 0$ and $\mu > \lambda + 1$, the functions a, g, p and q are as in the equation $(E_{65}; \delta)$ and conditions

(2.13.1) and (2.13.2) hold. As in the proof of Theorems 2.13.1 and 2.13.2 one can easily see that equations $(E_{67}; 1)$ and $(E_{67}; -1)$ are reduced respectively to the following inequalities

$$\left\{\frac{dx(t)}{dt} + Q(t)|x[g(t)]|^{\lambda}\right\} \operatorname{sgn} x[g(t)] \ \le\ 0 \tag{2.13.10}$$

and

$$\left\{\frac{dx(t)}{dt} - Q(t)|x[g(t)]|^{\lambda}\right\} \operatorname{sgn} x[g(t)] \ \ge\ 0, \tag{2.13.11}$$

where $Q(t)$ is defined in (2.13.3).

Now, (2.13.10) is oscillatory if one of the following conditions holds:

(I) $\lambda = 1$, $g(t) \le t$ and $g'(t) \ge 0$ for $t \ge t_0$, and

$$\liminf_{t\to\infty} \int_{g(t)}^{t} Q(s)ds \ >\ \frac{1}{e}$$

(II) $0 < \lambda < 1$, and

$$\int_{R_g} Q(s)ds = \infty,$$

where $R_g = \{t \in [t_0, \infty) : t_0 \le g(t) \le t\}$.

Also, (2.13.11) is oscillatory if one of the following conditions holds:

(III) $\lambda = 1$, $g(t) \ge t$, $g'(t) \ge 0$ for $t \ge t_0$, and

$$\liminf_{t\to\infty} \int_{t}^{g(t)} Q(s)ds \ >\ \frac{1}{e}$$

(IV) $\lambda > 1$, and

$$\int_{A_g} Q(s)ds \ =\ \infty,$$

where $A_g = \{t \in [t_0, \infty) : g(t) \ge t\}$.

Thus, the oscillation of the equation $(E_{67}; \delta)$, $\delta = \pm 1$ follows from those for inequalities (2.13.10) and (2.13.11).

2.14. Comparison of Neutral Equations with Nonneutral Equations

We consider the neutral equations

$$(x(t) + p(t)x[\tau(t)])^{(n)} + \delta q(t) f\left(x[g(t)]\right) \ =\ 0, \tag{$E_{68}; \delta$}$$

$$(x(t) - p(t)x[\tau(t)])^{(n)} + \delta q(t) f\left(x[g(t)]\right) = 0 \qquad (E_{69};\delta)$$

and

$$(x(t) + p(t)x[\tau(t)])^{(n)} + \delta q(t) f\left(x[g(t)]\right) = e(t), \qquad (E_{70};\delta)$$

where $p,\ q,\ g,\ \tau : [t_0,\infty) \to \mathbb{R}$, $f : \mathbb{R} \to \mathbb{R}$ are continuous, $p(t) \geq 0$ and $q(t) \geq 0$ eventually, $xf(x) > 0$ for $x \neq 0$ and $\lim_{t\to\infty} g(t) = \infty$, and $\lim_{t\to\infty} \tau(t) = \infty$.

The following results of Gopalsamy, Grace and Lalli [62] are concerned with the oscillation of the equation $(E_{68};1)$ when n is even.

Theorem 2.14.1. Assume that n is even, $0 \leq p(t) \leq 1$, $\tau(t) < t$, $\tau'(t) > 0$, $0 < \sigma(t) = \min\{t, g(t)\}$, $\sigma'(t) \geq 0$ for $t \geq t_0$, $\lim_{t\to\infty}\sigma(t) = \infty$, and

$$f(x) \operatorname{sgn} x \ \geq\ |x|^\gamma \quad \text{for } x \neq 0 \quad \text{and} \quad \gamma > 0. \qquad (2.14.1)$$

If the equation

$$y^{(n)}(t) + q(t)\left(1 - p[\sigma(t)]\right)^\gamma |y[\sigma(t)]|^\gamma \operatorname{sgn} y[\sigma(t)] = 0$$

is oscillatory, then the equation $(E_{68};1)$ is oscillatory.

Theorem 2.14.2. Assume that n is even, $1 \leq p_1 \leq p(t) \leq p_2$ where p_1 and p_2 are real numbers, $\tau(t) > t$, $\tau'(t) > 0$, $0 < \sigma(t) = \min\{t, \tau^{-1}\circ g(t)\}$, $\sigma'(t) > 0$ for $t \geq t_0$, $\lim_{t\to\infty}\sigma(t) = \infty$, and condition (2.14.1) holds. If the equation

$$z^{(n)}(t) + q(t)\left(\frac{p_1 - 1}{p_1 p_2}\right)^\gamma |z[\sigma(t)]|^\gamma \operatorname{sgn} z[\sigma(t)] = 0$$

is oscillatory, then the equation $(E_{68};1)$ is oscillatory.

In the next result we shall assume that the function $p(t)$ satisfies

$$-p_1 < p(t) < 0 \quad \text{for some} \quad p_1, \quad 0 < p_1 < 1. \qquad (2.14.2)$$

Theorem 2.14.3. Assume that n is even, conditions (2.14.1) and (2.14.2) hold, $\tau(t) \leq t$, $\tau'(t) > 0$, $0 < \sigma(t) = \min\{t, \tau^{-1} \circ g(t)\}$, $\sigma'(t) > 0$ for $t \geq t_0$, and $\lim_{t\to\infty}\sigma(t) = \infty$. If the equation

$$w^{(n)}(t) + q(t)|w[\sigma(t)]|^\gamma \operatorname{sgn} w[\sigma(t)] = 0$$

is oscillatory and all bounded solutions of the equation

$$z^{(n)}(t) - q(t)\left(-p[\sigma(t)]\right)^{-\gamma} |z[\sigma(t)]|^\gamma \operatorname{sgn} z[\sigma(t)] = 0$$

are oscillatory, then the equation $(E_{68};1)$ is oscillatory.

It is clear that by applying the results of earlier sections to the equations appearing in the above theorems we can deduce many new oscillation criteria for the equation $(E_{68};1)$, n is even. Other recent comparison results of Grace [83] are stated in the following:

Theorem 2.14.4. Let n be even, $0 \le p(t) \le p_1 \le 1$, $\tau(t) < t$, $\tau'(t) > 0$, $0 < \sigma(t) = \min\{t, g(t)\}$, $\sigma'(t) > 0$ for $t \ge t_0$, and $\lim_{t\to\infty}\sigma(t) = \infty$. Moreover, assume that $f \in C(R_{t_0})$, $t_0 > 0$ and let the functions G and H be a pair of continuous components of f with H nondecreasing, and

$$H(x)\ \mathrm{sgn}\ x \ \ge\ |x|^{\gamma} \quad \text{for } x \ne 0,\ \gamma > 0. \tag{2.14.3}$$

If for every constant $c \ge 1$, the equation

$$(\gamma(t)y'(t))' + \frac{(1-p_1)^{\gamma}}{2\,(n-2)!} G\left(cg^{n-1}(t)\right) q(t)Q(t)y(t) \ = \ 0$$

is oscillatory, where $\gamma(t) = 1/(\sigma^{n-2}(t)\sigma'(t))$, and

$$Q(t) = \begin{cases} a_1, & \text{any positive constant if } \gamma > 1 \\ a_2, & \text{any constant, } 0 < a_2 < 1 \text{ if } \gamma = 1 \\ a_3\sigma^{(\gamma-1)(n-1)}(t), & a_3 \text{ is any constant, } 0 < a_3 < 1 \text{ if } 0 < \gamma < 1, \end{cases}$$

then the equation $(E_{68};1)$ is oscillatory.

Theorem 2.14.5. Let n be even, $1 \le p_1 \le p(t) \le p_2$ for $t \ge t_0$ and p_1 and p_2 are real numbers, $\tau(t) > t$, $\tau'(t) > 0$, $0 < \sigma(t) = \min\{t, \tau^{-1} \circ g(t)\}$, $\sigma'(t) > 0$ for $t \ge t_0$, and $\lim_{t\to\infty}\sigma(t) = \infty$. Moreover, assume that $f : C(\mathbb{R}_{t_0})$, $t_0 > 0$ and let the functions G and H be as in Theorem 2.14.4 and condition (2.14.3) holds. If for every $c \ge 1$, the equation

$$(\gamma(t)y'(t))' + \frac{b^{\gamma}}{2\,(n-2)!} G\left(cg^{n-1}(t)\right) q(t)Q(t)y(t) \ = \ 0$$

is oscillatory, where $b = (p_1 - 1)/p_1p_2$, and the functions $\gamma(t)$ and $Q(t)$ are as in Theorem 2.14.4, then the equation $(E_{68};1)$ is oscillatory.

Theorem 2.14.6. Let n be even, $\tau(t) < t$, $\tau'(t) > 0$, $0 < \sigma(t) = \min\{t, \tau^{-1} \circ g(t)\}$, $\sigma'(t) > 0$ for $t \ge t_0$, $\lim_{t\to\infty}\sigma(t) = \infty$, and let conditions (2.14.1) and (2.14.2) hold. If the equation

$$(\gamma(t)y'(t))' + \frac{1}{2\,(n-2)!} q(t)Q(t)y(t) \ = \ 0$$

is oscillatory, where the functions $\gamma(t)$ and $Q(t)$ are as in Theorem 2.14.4, and all bounded solutions of the equation

$$w^{(n)}(t) - q(t)\left(|w[\sigma(t)]|^{\gamma}\right)\ \mathrm{sgn}\ w[\sigma(t)] \ = \ 0$$

are oscillatory, then the equation $(E_{68};1)$ is oscillatory.

Once again, we can apply earlier results to the equations appearing in Theorems 2.14.4 – 2.14.6 to obtain new oscillation criteria for the even order equations of type $(E_{68}; 1)$.

The following theorem due to Erbe, Kong and Zhang [54] provides comparison criterion for a special case of the equation $(E_{69}; 1)$, namely

$$(x(t) - p(t)x[t-\tau])^{(n)} + q(t)x[t-\beta] = 0, \qquad (E_{71})$$

where τ and β are positive real numbers.

Theorem 2.14.7. Let n be odd, $0 \leq p(t) \leq 1$ and suppose that there exist two nonnegative integers m and N with $m \leq N$, and there exists an $i_0 \in \{m, m+1, \cdots, N\}$ such that for all large t

$$q(s)\prod_{j=1}^{i_0} p(s-\beta-(j-1)\tau) \neq 0 \quad \text{for} \quad s \in [t, t+\beta+i_0\tau].$$

If every solution of the equation

$$y^{(n)}(t) + \sum_{i=m}^{N} q(t)\left(\prod_{j=1}^{i} p(t-\beta-(j-1)\tau)\right) y[t-\beta-i\tau] = 0$$

is oscillatory, then every solution of the equation (E_{71}) is oscillatory.

The following interesting result is also proved in [54].

Theorem 2.14.8. Let n be odd, $g(t) = t - \beta(t)$, $\beta \in C\,[[t_0, \infty), \mathbb{R}_0]$, $\tau(t) = t - \tau$, $\tau > 0$, $f'(x) \geq 0$, $\liminf_{x\to\infty} f(x)/x \geq k > 0$, there exist p_1 and p_2 such that $0 \leq p_1 \leq p(t) \leq p_2 \leq 1$, and either

$$\int_{t_0}^{\infty} q(s)(s-\beta(s))^{n-2} ds = \infty \quad \text{for} \quad n \geq 3$$

or

$$\int_{t_0}^{\infty} q(s) ds = \infty \quad \text{for} \quad n = 1.$$

If there exists a positive integer N such that every bounded solution of

$$y^{(n)}(t) + kq(t)\sum_{j=0}^{N} p_1^j y[t - j\alpha - \beta(t)] = 0$$

is oscillatory, then the equation $(E_{69}; 1)$ is also oscillatory.

The oscillation of solutions of $(E_{68}; \delta)$ and $(E_{69}; \delta)$, $\delta = \pm 1$ has been studied by several researchers. Here we refer the reader to the papers of Grammatikopoulos, Grove and Ladas [121], Grammatikopoulos, Ladas and Meimaridou [122], Ladas and Sficas [195,196], Györi and Ladas [129], and the references therein.

Now we shall consider a special case of the equation $(E_{68};1)$, namely

$$(y(t)+p(t)y[t-\tau])^{(n)}+q(t)y[t-\sigma] \;=\; 0, \qquad (E_{72})$$

where p and q are as in $(E_{68};\delta)$, and $\tau,\ \sigma\in\mathbb{R}_0$.

The following three theorems for (E_{72}) are proved by Grammatikopoulos et. al. [122].

Theorem 2.14.9. Asume that n is even, $-1\le p(t)<0$, and suppose that there exists a positive constant γ such that

$$\frac{q(t)}{p(t+\tau-\sigma)} \;\le\; -\gamma \quad\text{and}\quad \gamma^{1/n}\left(\frac{\sigma-\tau}{n}\right) \;>\; \frac{1}{e}.$$

Then, the equation (E_{72}) is oscillatory.

Theorem 2.14.10. Assume that n is odd, $p(t)\le -1$, and

$$\int^{\infty} q(s)ds \;=\; \infty.$$

Suppose also that there exists a positive constant γ such that

$$\frac{q(t)}{p(t+\tau-\sigma)} \;\le\; -\gamma.$$

Then, the equation (E_{72}) is oscillatory.

Theorem 2.14.11. Assume that n is odd, $p(t)\ge -1$, and

$$\int^{\infty} q(s)ds \;=\; \infty.$$

Then, the equation (E_{72}) is oscillatory.

Now let

$$\tau^0(t) \;=\; t,\ \ \tau^i(t) \;=\; \tau\left(\tau^{i-1}(t)\right),\ \ \tau^{-i}(t) \;=\; \tau^{-1}\left(\tau^{-(i-1)}(t)\right),\ \ i=1,2,\cdots$$

where τ^{-1} denotes the inverse function of $\tau(t)$, and define

$$P_0(t) \;=\; 1, \qquad P_i(t) \;=\; \prod_{j=1}^{i-1} p\left(\tau^j(t)\right), \qquad i=1,2,\cdots.$$

Jeros and Kusano [137] considered $(E_{69};\delta)$, $\delta=\pm1$, $f(x)=x$ and observed the following:

If $x(t)$ is a nonoscillatory solution of the equation $(E_{69};\delta)$ satisfying $x(t)y(t) > 0$ for all large t, where $y(t) = x(t) - p(t)x[\tau(t)]$, then for any integer $k \geq 0$ there is a $t_k \geq t_0$ such that

$$|x(t)| \;\geq\; \sum_{i=0}^{k} P_i(t)\left|y\left[\tau^i(t)\right]\right| \quad \text{for} \quad t \geq T_k \tag{2.14.4}$$

and, when $x(t)$ is a nonoscillatory solution of $(E_{69};\delta)$ satisfying $x(t)y(t) < 0$ for all large t, then for any integer $m \geq 1$ there is a $T_m \geq t_0$ such that

$$|x(t)| \;\geq\; \sum_{j=1}^{m} \frac{y\left[\tau^{-j}(t)\right]}{P_j\left[\tau^{-j}(t)\right]} \quad \text{for} \quad t \geq T_m. \tag{2.14.5}$$

Let $x(t)$ be a nonoscillatory solution of $(E_{69};\delta)$ and assume (2.14.4) holds, then for any integer $k \geq 0$ the function $y(t)$ satisfies

$$\left\{\delta y^{(n)}(t) + q(t)\sum_{i=0}^{k} P_i[g(t)]y\left[\tau^i[g(t)]\right]\right\} \operatorname{sgn} y(t) \;\leq\; 0 \tag{2.14.6}$$

for $t \geq T_k$ and if (2.14.5) holds, then for any integer $m \geq 1$ the function $y(t)$ is a nonoscillatory solution of

$$\left\{\delta y^{(n)}(t) - q(t)\sum_{j=1}^{m} \left[P_j\left(\tau^{-1}[g(t)]\right)\right]^{-1} y\left[\tau^{-j}[g(t)]\right]\right\} \operatorname{sgn} y(t) \;\geq\; 0 \tag{2.14.7}$$

for $t \geq T_m$.

The problem is thus reduced to finding conditions guaranteeing the nonexistence of nonoscillatory solutions of (2.14.6) and (2.14.7) for some $k \geq 0$ and $m \geq 1$ respectively. For this, we note that a suitable combination of the results presented earlier can be employed.

Next we shall state the results of Jeros and Kusano [137]. Let

$$g_i^*(t) = \max\{t,\, g_i(t)\}, \qquad \alpha[g_i](t) = \min_{s\geq t} g_i^*(s)$$
$$g_{i*}(t) = \min\{t,\, g_i(t)\}, \qquad \rho[g_i](t) = \max_{t_0\leq s\leq t} g_{i*}(s).$$

Theorem 2.14.12. Let n be even. Then, the equation $(E_{69};1)$ is oscillatory if there exist a nonnegative integer i and positive integers j, k and m such that

$$\int_{t_0}^{\infty} \left[(\tau^i \circ g)_*(s)\right]^{n-1} \left[(\tau^i \circ g)(s)\right]^{-\epsilon} q(s)P_i[g(s)]ds \;=\; \infty \quad \text{for some } \epsilon > 0, \tag{2.14.8}$$

$$\int_{t_0}^{\infty} s\left[(\tau^{-j}\circ g)_*(s)\right]^{n-3}\left[(\tau^{-j}\circ g)(s)\right]^{1-\epsilon_1}\frac{q(s)}{P_j(\tau^{-j}\circ g(s))}ds = \infty$$

for some $\epsilon_1 > 0$, (2.14.9)

$$\limsup_{t\to\infty}\int_{\rho[\tau^{-k}\circ g(t)]}^{t}\frac{\{s-\rho\left[\tau^{-k}\circ g\right](t)\}^{n-\nu-1}}{(n-\nu-1)!}\frac{\{\rho\left[\tau^{-k}\circ g\right](t)-\left(\tau^{-k}\circ g\right)(s)\}^{\nu}}{\nu!}$$

$$\times\frac{q(s)}{P_k\left(\tau^{-k}\circ g(s)\right)}ds > 1 \quad (2.14.10)$$

for some $\nu\in\{0,1,2,\cdots\}$, and

$$\limsup_{t\to\infty}\int_{t}^{\alpha[\tau^{-m}\circ g](t)}\frac{\{(\tau^{-m}\circ g)(s)-\alpha\left[\tau^{-m}\circ g\right](t)\}^{n-\mu-1}}{(n-\mu-1)!}\frac{\{\alpha\left[\tau^{-m}\circ g\right](t)-s\}^{\mu}}{\mu!}$$

$$\times\frac{q(s)}{P_m\left(\tau^{-m}\circ g(s)\right)}ds > 1 \quad (2.14.11)$$

for some $\mu\in\{0,1,2,\cdots\}$.

Theorem 2.14.13. Let n be odd. Then, the equation $(E_{69};1)$ is oscillatory if there exist nonnegative integers i and k and positive integers j and m such that

$$\int_{t_0}^{\infty}\left[\left(\tau^{i}\circ g\right)_*(s)\right]^{n-2}\left[\left(\tau^{i}\circ g\right)(s)\right]^{1-\epsilon}q(s)P_i[g(s)]ds = \infty \text{ for some } \epsilon > 0, \quad (2.14.12)$$

$$\int_{t_0}^{\infty} s\left[\left(\tau^{-j}\circ g\right)_*(s)\right]^{n-2}\left[\tau^{-j}\circ g(s)\right]^{-\epsilon_1}\frac{q(s)}{P_j\left(\tau^{-1}\circ g(s)\right)}ds = \infty$$

for some $\epsilon_1 > 0$, (2.14.13)

$$\limsup_{t\to\infty}\int_{\rho[\tau^{k}\circ g](t)}^{t}\frac{\{s-\rho\left[\tau^{k}\circ g\right](t)\}^{n-\nu-1}}{(n-\nu-1)!}\frac{\{\rho\left[\tau^{k}\circ g\right](t)-\left(\tau^{k}\circ g\right)(s)\}^{\nu}}{\nu!}$$

$$\times\, q(s)P_k(g(s))ds > 1 \quad (2.14.14)$$

for some $\nu\in\{0,1,2,\cdots\}$, and condition (2.14.11) holds for some $\mu\in\{0,1,2,\cdots\}$.

Theorem 2.14.14. Let n be even. Then, the equation $(E_{69};-1)$ is oscillatory if there exist nonnegative integers i, k and m and a positive integer j such that

$$\int_{t_0}^{\infty} s\left[\left(\tau^{i}\circ g\right)_*(s)\right]^{n-3}\left[\left(\tau^{i}\circ g\right)(s)\right]^{1-\epsilon}q(s)P_i(g(s))ds = \infty$$

for some $\epsilon > 0$, (2.14.15)

$$\int_{t_0}^{\infty} \left[\left(\tau^{-j} \circ g\right)_* (s)\right]^{n-1} \left[\left(\tau^{-j} \circ g\right)(s)\right]^{-\epsilon_1} \frac{q(s)}{P_j\left(\tau^{-1} \circ g(s)\right)} ds = \infty$$

$$\text{for some } \epsilon_1 > 0, \qquad (2.14.16)$$

condition (2.14.14) holds for some $\nu \in \{0,1,2,\cdots\}$, and

$$\limsup_{t\to\infty} \int_t^{\alpha[\tau^m \circ g](t)} \frac{\{(\tau^m \circ g)(s) - \alpha\left[\tau^m \circ g\right](t)\}^{n-\mu-1}}{(n-\mu-1)!} \frac{\{\alpha\left[\tau^m \circ g\right](t) - s\}^{\mu}}{\mu!}$$

$$\times\, q(s)P_m(g(s))ds > 1 \qquad (2.14.17)$$

for some $\mu \in \{0,1,2,\cdots\}$.

Theorem 2.14.15. Let n be odd. Then, the equation $(E_{69};-1)$ is oscillatory if there exist nonnegative integers i and m and positive integers j and k such that

$$\int_{t_0}^{\infty} s\left[\left(\tau^{i} \circ g\right)_* (s)\right]^{n-2} \left[\left(\tau^{i} \circ g\right)(s)\right]^{-\epsilon} q(s)P_i(g(s))ds = \infty \text{ for some } \epsilon > 0, \qquad (2.14.18)$$

$$\int_{t_0}^{\infty} \left[\left(\tau^{-j} \circ g\right)_* (s)\right]^{n-2} \left[\left(\tau^{-j} \circ g\right)(s)\right]^{1-\epsilon_1} \frac{q(s)}{P_j\left(\tau^{-j} \circ g(s)\right)} ds = \infty \qquad (2.14.19)$$

for some $\epsilon_1 > 0$, and conditions (2.14.10) and (2.14.17) are satisfied for some $\nu \in \{0,1,2,\cdots\}$ and $\mu \in \{0,1,2,\cdots\}$ respectively.

The above oscillation criteria become simpler if $p(t)$ and $\tau(t)$ in $(E_{69};\delta)$ satisfy the additional conditions

$$0 < p(t) \le p_1 < 1 \quad \text{and} \quad \tau(t) < t \quad \text{for} \quad t \ge t_0 \qquad (2.14.20)$$

or

$$p_2 \ge p(t) \ge p_3 > 1 \quad \text{and} \quad \tau(t) > t \quad \text{for} \quad t \ge t_0, \qquad (2.14.21)$$

where p_1, p_2 and p_3 are real constants.

If (2.14.20) holds, conditions (2.14.9), (2.14.11), (2.14.13), (2.14.16) and (2.14.19) are excluded from Theorems 2.14.12 – 2.14.15.

If (2.14.21) holds, conditions (2.14.8), (2.14.12), (2.14.17) and (2.14.18) can be deleted from Theorems 2.14.12 – 2.14.15.

Next, we shall consider a special case of $(E_{69};1)$, namely

$$(x(t) - x[t-\tau])^{(n)} + q(t)f(x[g(t)]) = 0, \qquad (E_{73})$$

where τ is a positive real number, functions g, q and f are as in $(E_{69};1)$, $g(t) < t$, $g'(t) \geq 0$ for $t \geq t_0$, $f'(x) \geq 0$ for $x \neq 0$, and

$$-f(-xy) \geq f(xy) \geq f(x)f(y) \quad \text{for} \quad xy > 0. \tag{2.14.22}$$

The following two results for the oscillation of (E_{73}) are due to Grace [85].

Theorem 2.14.16. Let n be odd, $g(t) = t - \sigma$, σ is a positive constant. If

$$\int^{\infty} q(s)ds < \infty \quad \text{and} \quad \int^{\infty} q(s)f\left(s\int_s^{\infty} q(u)du\right)ds = \infty,$$

then the equation (E_{73}) is oscillatory.

Theorem 2.14.17. Let θ, $0 < \theta < 1$ be such that the first order differential equation

$$w'(t) + f\left(\frac{\theta}{(n-1)!}\right)q(t)f\left(g^{n-1}(t)\right)f(w[g(t)]) = 0$$

is oscillatory. Moreover, assume that $g'(t) < t$ and all bounded solutions of the differential equation

$$y^{(n)}(t) + (-1)^{n+1}q(t)f(y[g^*(t)]) = 0$$

are oscillatory, where

$$g^*(t) = \begin{cases} g(t) & \text{if } n \text{ is odd} \\ g(t) + \tau & \text{if } n \text{ is even.} \end{cases}$$

Then, the equation (E_{73}) is oscillatory.

For a special case of the equation (E_{73}), i.e. for the equation

$$(x(t) - x[t-\tau])^{(n)} + q(t)x[t-\sigma] = 0, \quad n \text{ odd} \tag{E_{74}}$$

where τ and σ are positive constants, Yu, Wang and Zhang [326] extended Theorem 2.14.16. Their result is as follows:

Theorem 2.14.18. If

$$\int^{\infty} s^{n-1}q(s)ds = \infty,$$

then the equation (E_{74}) is oscillatory.

The next result which extends even Theorem 2.14.18 is due to Yu [324].

Theorem 2.14.19. If

$$\int_{t_0}^{\infty} s^n q(s) \int_s^{\infty} (u-s)^{n-1} q(u)duds = \infty,$$

then the equation (E_{74}) is oscillatory.

Recently, Tang and Shen [286] have obtained a oscillation result for the equation (E_{74}) by comparing it with an even order ordinary differential equation. Their result is stated in the following:

Theorem 2.14.20. If the equation

$$y^{(n+1)}(t) + \frac{1}{\tau} q(t+\sigma_n) y(t) = 0 \qquad (E_{75})$$

with $\sigma_n = \sigma \operatorname{sgn}(n-1)$ is oscillatory, then the equation (E_{74}) is also oscillatory.

We can apply previous results to equation (E_{75}) to obtain many interesting criteria for the oscillation of odd order equations of type (E_{74}).

For the forced neutral equation $(E_{70};\delta)$, as in previous sections we assume that there exists a function $\eta \in C^n[[t_0,\infty),\mathbb{R}]$ such that $\eta(t)$ satisfies the equation $\eta^{(n)}(t) = e(t)$.

Now, let $x(t)$ be a solution of the equation $(E_{70};\delta)$, and set

$$y(t) + \eta(t) = x(t) + p(t)x[\tau(t)].$$

Then, the equation $(E_{70};\delta)$ reduces to the homogeneous equation

$$y^{(n)}(t) + \delta q(t) f(x[\tau(t)]) = 0. \qquad (E_{76};\delta)$$

Finally, we note that on imposing suitable conditions on the function η, as in earlier sections (see also Grace [83]), $(E_{75};\delta)$ can be reduced to a equation of the type $(E_{68};\delta)$.

2.15. Comparison of Neutral Equations with Equations of the Same Form

We consider the neutral differential equations

$$(x(t) - p_1(t)x[t-\tau_1])^{(n)} + \delta q(t)x[t-\sigma_1] = 0 \qquad (E_{77};\delta)$$

and

$$(y(t) - p_2(t)y[t-\tau_2])^{(n)} + \delta q(t)y[t-\sigma_2] = 0 \qquad (E_{78};\delta)$$

where $\delta = \pm 1$, $n \geq 1$, p_1, p_2, q_1, $q_2 \in C[[t_0,\infty),\mathbb{R}_0]$, τ_1, $\tau_2 \in \mathbb{R}_+$ and σ_1, $\sigma_2 \in \mathbb{R}_0$.

In [192] Ladas and Qian established a comparison result which basically says that if every solution of $(E_{77};1)$ oscillates, then every solution of the equation $(E_{78};1)$ also oscillates. We present their result in the following:

Theorem 2.15.1. Assume $n \geq 1$ is an odd integer. Suppose that for all large t

(i) $p_1 \in C^1[[t_0,\infty),\mathbb{R}_0]$, $p_1(t)$ is bounded, $p_1'(t) \geq 0$,

(ii) $p_1(t) \leq p_2(t-\sigma_2)\dfrac{q_2(t)}{q_2(t-\tau_2)}$, $q_1(t) \leq q_2(t)$ and $0 \leq p_2(t) \leq 1$,

(iii) $q_3(t) > 0$ and $\displaystyle\int_{t_0}^{\infty} q_2(s)ds = \infty$,

(iv) $\tau_1 \leq \tau_2$, $\sigma_1 \leq \sigma_2$, and either

(v) $p_1(t) > 0$ or $\sigma_1 > 0$ and $q_1(t) \not\equiv 0$ on any interval of length σ_1.

If every solution of the equation $(E_{77};1)$ is oscillatory, then every solution of $(E_{78};1)$ is oscillatory.

Next, they considered the following nonlinear neutral delay differential equation of odd order

$$(x(t) - p(t)g(x[t-\tau]))^{(n)} + \delta q(t)h(x[t-\sigma]) = 0, \qquad (E_{79};\delta)$$

where p, $q \in C[[t_0,\infty),\mathbb{R}]$, g, $h \in C[\mathbb{R},\mathbb{R}]$, $\tau \in \mathbb{R}_+$ and $\sigma \in \mathbb{R}_0$.

Theorem 2.15.2. Assume that

(i) $\liminf_{t\to\infty} p(t) = p_0 \in (0,1)$, $\limsup_{t\to\infty} p(t) = P_0 \in (0,1)$, and $\lim_{t\to\infty} q(t) = q_0 \in \mathbb{R}_+$,

(ii) $0 \leq g(u)/u \leq 1$ for $u \neq 0$, $\lim_{u\to\infty} g(u)/u = 1$,

(iii) $uh(u) > 0$ for $u \neq 0$, $|h(u)| \geq h_0 > 0$ for $|u|$ sufficiently large, and $\lim_{u\to\infty} h(u)/u = 1$.

If every solution of the equation

$$(y(t) - p_0 y[t-\tau])^{(n)} + q_0 y[t-\sigma] = 0 \qquad (E_{80})$$

oscillates, then every solution of the equation $(E_{79};1)$ also oscillates.

The following result of Ladas and Qian [192] provides a partial converse to Theorem 2.15.2.

Theorem 2.15.3. Consider the equation $(E_{79}; 1)$ with $g(x) = x$, and assume that there exist positive constants p_0, q_0 and δ_1 such that

(i) $0 < p(t) \le p_0 < 1$, $0 \le q(t) \le q_0$, and
(ii) either $0 \le h(u) \le u$ for $0 \le u \le \delta_1$, or $0 \ge h(u) \ge u$ for $-\delta_1 \le u \le 0$.

Suppose also that $h(u)$ is nondecreasing in u and that the characteristic equation of $(E_{80}; 1)$, i.e.

$$\lambda^n - p_0 \lambda^n e^{-\lambda\tau} + q_0 e^{-\lambda\sigma} = 0 \tag{2.15.1}$$

has a real root. Then, the equation $(E_{79}; 1)$ has a nonoscillatory solution.

Combining Theorems 2.15.2 and Theorem 2.15.3 we get the following:

Corollary 2.15.4. Consider the equation $(E_{79}; 1)$ with $g(x) = x$ and assume that there exist positive constants h_0, p_0, q_0 and δ such that

(i) $0 < p(t) \le p_0 = \lim_{t\to\infty} p(t) < 1$, $0 \le q(t) \le q_0 = \lim_{t\to\infty} q(t)$,
(ii) $uh(u) > 0$ for $u \ne 0$, $|h(u)| \ge h_0 > 0$ for $|u|$ sufficiently large, and $\lim_{u\to\infty} h(u)/u = 1$, and either
(iii) $0 \le h(u) \le u$ for $0 \le u \le \delta$, or $0 \ge h(u) \ge u$ for $-\delta \le u \le 0$,

and $h(u)$ is nondecreasing in u. Then, every solution of $(E_{79}; 1)$ oscillates if and only if every solution of the linear equation (E_{80}) oscillates, equivalently if and only if the equation (2.15.1) has no negative real roots.

Next, Gopalsamy, Lalli and Zhang [63] considered a special case of $(E_{69}; 1)$, namely the equation

$$(x(t) - p(t)x[t-\tau])^{(n)} + q(t)x[t-\sigma(t)] = 0, \quad n \text{ odd} \tag{E_{81}}$$

where p, q, $\sigma : [t_0, \infty) \to \mathbb{R}$, $\tau \in \mathbb{R}_+$, $\lim_{t\to\infty}(t - \sigma(t)) = \infty$, $q(t) \not\equiv 0$, and proved the following:

Theorem 2.15.5. Assume that $0 \le p(t) \le 1$, and

$$p(t) + q(t)\sigma(t) > 0 \quad \text{for} \quad t \ge t_0 \ge 0.$$

Then, every solution of the equation (E_{81}) is oscillatory if and only if the differential inequality

$$(x(t) - p(t)x[t-\tau])^{(n)} + q(t)x[t-\sigma(t)] \le 0$$

has no eventually positive solution.

Next, they compared the equation (E_{81}) with the equation

$$(y(t) - p^*(t)y[t-\tau])^{(n)} + q^*(t)y[t-\sigma(t)] = 0, \quad n \text{ odd} \tag{E_{82}}$$

where p^*, $q^* \in C[[t_0, \infty), \mathbb{R}]$.

Theorem 2.15.6. Assume that the assumptions of Theorem 2.15.5 hold, $p(t) \le p^*(t) \le 1$, and $q(t) \le q^*(t)$ for $t \ge t_0$. Then, every solution of the equation (E_{81}) is oscillatory implies the same for (E_{82}).

For even order equations of type $(E_{77}; -1)$ and $(E_{78}; -1)$ and τ_1, τ_2, σ_1, $\sigma_2 \in \mathbb{R}_0$, Ladas and Qian [193] obtained the following results.

Theorem 2.15.7. Assume that

(i) $p_1(t)$, $p_2(t)$ are bounded, $p_1'(t) \ge 0$ and $q_2(t) > 0$ for $t \ge t_0$,

(ii) $\tau_1 \le \tau_2$ and $\sigma_1 \le \sigma_2$,

(iii) $p_1(t) \le p_2[t - \sigma_2]\dfrac{q_2(t)}{q_2[t - \tau_2]}$ and $q_1(t) \le q_2(t)$ for $t \ge t_0$,

(iv) $\displaystyle\int_{t_0}^{\infty} q_2(s)ds = \infty$,

and either $p_1(t) > 0$ for $t \ge t_0$, or $\sigma_1 > 0$ and $q_1(t) \not\equiv 0$ on any interval of length σ_1.

Suppose that every bounded solution of $(E_{77}; -1)$ oscillates. Then, every bounded solution of $(E_{78}; -1)$ also oscillates.

Theorem 2.15.8 (Linearized Oscillations). Consider the equation $(E_{79};$ $-1)$ and assume that

(i) $\limsup_{t\to\infty} p(t) = P_0 \in (0, 1)$, $\liminf_{t\to\infty} p(t) = p_0 \in (0, 1)$, and $\lim_{t\to\infty} q(t) = q_0 \in (0, 1)$,

(ii) $0 \le g(u)/u \le 1$ for $u \ne 0$, $\lim_{u\to 0} g(u)/u = 1$,

(iii) $uh(u) > 0$ for $u \ne 0$, $\lim_{u\to 0} h(u)/u = 1$.

Suppose that every bounded solution of the linearized equation

$$(y(t) - p_0 y[t - \tau])^{(n)} - q_0 y[t - \sigma] = 0 \qquad (E_{83})$$

oscillates. Then, every bounded solution of $(E_{79}; -1)$ oscillates.

The following theorem deals with a partial converse of Theorem 2.15.8.

Theorem 2.15.9. Consider the equation $(E_{79}; -1)$ with $g(x) = x$, and assume that there exist positive constants p_0, q_0 and δ_1 such that

(i) $0 \le p(t) \le p_0$ and $0 \le q(t) \le q_0$ for $t \ge t_0$, and that

(ii) either $0 \le h(u) \le u$ for $0 \le u \le \delta_1$, or $0 \ge h(u) \ge u$ for $-\delta_1 \le u \le 0$.

Suppose also that $h(u)$ is nondecreasing in a neighborhood of the origin, and the characteristic equation of (E_{83}), i.e.

$$\lambda^n - p_0\lambda^n e^{-\lambda\tau} - q_0 e^{-\lambda\sigma} \; = \; 0$$

has a root in the interval $(-\infty, 0]$. Then, the equation $(E_{79}; -1)$ has a bounded nonoscillatory solution.

Combining Theorems 2.15.8 and 2.15.9 we have the following:

Corollary 2.15.10. Consider the equation $(E_{79}; -1)$ with $g(x) = x$, and assume that there exist constants p_0, $q_0 \in \mathbb{R}_+$ such that

(i) $0 < p(t) = p_0 = \lim_{t\to\infty} p(t) < 1$ for $t \geq t_0$, and

(ii) $0 \leq q(t) \leq q_0 = \lim_{t\to\infty} q(t)$ for $t \geq t_0$.

Suppose that $\lim_{u\to 0} h(u)/u = 1$, $uh(u) > 0$ for $u \neq 0$, and that $h(u)$ is nondecreasing in a neighborhood of the origin. Then, every bounded solution of the equation $(E_{79}; -1)$ oscillates if and only if every bounded solution of (E_{83}) oscillates.

2.16. Neutral Differential Equations of Mixed Type

We consider the neutral differential equations of mixed type of the form

$$(x(t) + ax[t-\tau] - bx[t+\sigma])^{(n)} + \delta(q(t)x[t-g] + p(t)x[t+h]) \; = \; 0, \quad (E_{84}; \delta)$$

$$(x(t) - ax[t-\tau] + bx[t+\sigma])^{(n)} + \delta(q(t)x[t-g] + p(t)x[t+h]) \; = \; 0, \quad (E_{85}; \delta)$$

$$(x(t) + ax[t-\tau] + bx[t+\sigma])^{(n)} + \delta(q(t)x[t-g] + p(t)x[t+h]) \; = \; 0 \quad (E_{86}; \delta)$$

and

$$(x(t) - ax[t-\tau] - bx[t+\sigma])^{(n)} + \delta(q(t)x[t-g] + p(t)x[t+h]) \; = \; 0, \quad (E_{87}; \delta)$$

where $\delta = \pm 1$, a, b, g, h, τ and σ are nonnegative real numbers, and p, $q : [t_0, \infty) \to \mathbb{R}_0$ are continuous functions.

Recently, Grace [78,80,82], and Agarwal and Grace [11] have established several easily verifiable sufficient conditions for the oscillation of equations $(E_{84}; \delta) - (E_{87}; \delta)$ involving coefficients and arguments only. We list these results here first for the case when $q(t) = q$ and $p(t) = p$, where p and q are nonnegative real numbers, and present a sample proof of one of these results.

Theorem 2.16.1. Suppose that $g > \tau$, and

$$\left(\frac{p}{1+a}\right)^{1/n} \left(\frac{\sigma}{n}\right) e \; > \; 1. \tag{2.16.1}$$

If

$$\left(\frac{q}{b}\right)^{1/n}\left(\frac{g+\sigma}{n}\right)e \;>\; 1 \quad \text{for } n \text{ odd} \tag{2.16.2}$$

and

$$\left(\frac{q}{1+a}\right)^{1/n}\left(\frac{g-\tau}{n}\right)e \;>\; 1 \quad \text{for } n \text{ even,} \tag{2.16.3}$$

then the equation $(E_{84};-1)$ is oscillatory.

Proof. Let $x(t)$ be an eventually positive solution of $(E_{84};-1)$, say $x(t)>0$ for $t \ge t_0 \ge 0$. Set

$$z(t) \;=\; x(t)+ax[t-\tau]-bx[t+\sigma].$$

Then,

$$z^{(n)}(t) \;=\; qx[t-g]+px[t+h] \;\ge\; 0 \quad \text{eventually} \tag{2.16.4}$$

and hence we see that $z^{(i)}(t)$, $i=0,1,\cdots,n$ are eventually of one sign. There are two possibilities to consider:

(i) $z(t)<0$ eventually, and (ii) $z(t)>0$ eventually.

(i) Assume that $z(t)<0$ for $t \ge t_1 \ge t_0$. In this case, we get

$$0 \;<\; u(t) \;=\; -z(t) \;=\; bx[t+\sigma]-ax[t-\tau]-x(t) \;\le\; bx[t+\sigma] \quad \text{for } t \ge t_1.$$

Thus, $x[t+\sigma] \ge (1/b)u(t)$, or

$$x(t) \;\ge\; \frac{1}{b}u[t-\sigma] \quad \text{for } t \ge t_2 \ge t_1. \tag{2.16.5}$$

From (2.16.4), we have

$$u^{(n)}(t)+qx[t-g] \;\le\; 0 \quad \text{for } t \ge t_2. \tag{2.16.6}$$

Using (2.16.5) in (2.16.6), we get

$$u^{(n)}(t)+\frac{q}{b}u[t-\sigma-g] \;\le\; 0 \quad \text{for } t \ge t_2. \tag{2.16.7}$$

It is easy to check that $u^{(n-1)}(t)>0$ for $t \ge t_3 \ge t_2$, and either

(I) $u'(t)>0$ for $t \ge t_3$, or (II) $u'(t)<0$ for $t \ge t_3$.

(I) Suppose $u'(t)>0$ for $t \ge t_3$. There exist a $t_4 \ge t_3$ and a positive constant α such that

$$u[t-\sigma-g] \;\ge\; \alpha \quad \text{for } t \ge t_4. \tag{2.16.8}$$

Using (2.16.8) in (2.16.7) and integrating from t_4 to t, we have

$$0 < u^{(n-1)}(t) \le u^{(n-1)}(t_4) - \frac{q}{b}\alpha(t-t_4) \to -\infty \quad \text{as} \quad t \to \infty,$$

which is a contradiction.

(II) $u'(t) < 0$ for $t \ge t_3$. This is the case when n is odd, and hence we see that

$$(-1)^i u^{(i)}(t) > 0 \quad \text{for} \quad i = 0, 1, \cdots, n \quad \text{and} \quad t \ge T \ge t_4. \tag{2.16.9}$$

But, in view of Lemma 2.2.11(ii) and condition (2.16.2), inequality (2.16.7) has no solution such that (2.16.9) holds, which is a contradiction.

(ii) Assume $z(t) > 0$ for $t \ge t_1 \ge t_0$. Set

$$w(t) = z(t) + az[t-\tau] - bz[t+\sigma]. \tag{2.16.10}$$

Then,

$$w^{(n)}(t) = qz[t-g] + pz[t+h] \tag{2.16.11}$$

and since the function $w(t)$ satisfies the equation $(E_{84}; -1)$, we obtain

$$(w(t) + aw[t-\tau] - bw[t+\sigma])^{(n)} = qw[t-g] + pw[t+h]. \tag{2.16.12}$$

Using the procedure of case (i), we observe that $w(t) > 0$ for $t \ge t_2 \ge t_1$. Next, we have two cases to consider:

$(\bar{\text{I}})$ $z'(t) > 0$ for $t \ge t_3 \ge t_2$, and $(\bar{\text{II}})$ $z'(t) < 0$ for $t \ge t_3$.

$(\bar{\text{I}})$ Let $z'(t) > 0$ for $t \ge t_3$. From the equation (2.16.10) we have $w^{(i)}(t) > 0$ for $i = n, n+1$ and $t \ge t_3$, and hence we see that

$$w^{(i)}(t) > 0, \quad i = 0, 1, \cdots, n+1 \quad \text{and} \quad t \ge t_3. \tag{2.16.13}$$

Now, using the fact that $w^{(n)}(t)$ is eventually increasing, we find

$$\begin{aligned}(1+a)w^{(n)}(t) &\ge w^{(n)}(t) + aw^{(n)}[t-\tau] - bw^{(n)}[t+\sigma] \\ &= qw[t-g] + pw[t+h] \\ &\ge pw[t+h] \quad \text{for} \quad t \ge t_3\end{aligned}$$

and hence

$$w^{(n)}(t) \ge \frac{p}{1+a} w[t+h] \quad \text{for} \quad t \ge t_3. \tag{2.16.14}$$

But, in view of Lemma 2.2.11(i) and condition (2.16.1), inequality (2.16.14) has no solution such that (2.16.13) holds, which is a contradiction.

$(\bar{\text{II}})$ Let $z'(t) < 0$ for $t \ge t_3$. From (2.16.4), n must be even. We claim that $w'(t) < 0$ for $t \ge t_4 \ge t_3$. To prove it, assume that $w'(t) > 0$ for

$t \geq t_4$. Then from (2.16.11), we see that $w^{(n)}(t) > 0$ and $w^{(n+1)}(t) < 0$ for $t \geq t_3$. Using this fact in (2.16.12), one can easily see that

$$(1+a)w^{(n)}[t-\tau] \;\geq\; pw[t+h],$$

or

$$w^{(n)}(t) \;\geq\; \frac{p}{1+a}w[t+h+\tau] \quad \text{for} \quad t \geq t_4.$$

Since $w(t)$ is an increasing function, we have

$$w^{(i)}(t) \;>\; 0, \quad i = 0, 1, \cdots, n \quad \text{and} \quad t \geq t_4$$

and

$$w^{(n)}(t) \;\geq\; \frac{p}{1+a}w[t+h] \quad \text{for} \quad t \geq t_4$$

and again, we are led to a contradiction. Thus, $w'(t) < 0$ for $t \geq t_4$ and from (2.16.11), we have

$$(-1)^i w^{(i)}(t) \;>\; 0, \quad i = 0, 1, \cdots, n+1 \quad \text{for} \quad t \geq t_4.$$

Now using the fact that the function $w^{(n)}(t)$ is decreasing for $t \geq t_4$ in (2.16.12), we obtain

$$\begin{aligned}(1+a)w^{(n)}[t-\tau] \;&\geq\; w^{(n)}(t) + aw^{(n)}[t-\tau] - bw^{(n)}[t+\sigma] \\ &=\; qw[t-g] + pw[t+h] \\ &\geq\; qw[t-g] \quad \text{for} \quad t \geq t_4\end{aligned}$$

and hence

$$w^{(n)}(t) \;\geq\; \frac{q}{1+a}w[t-(g-\tau)] \quad \text{for} \quad t \geq t_4.$$

The rest of the proof is similar to that of case (II). ∎

Theorem 2.16.2. Suppose that $b > 0$, $h > \sigma$, $g > \tau$, and

$$\left(\frac{p}{b}\right)^{1/n}\left(\frac{h-\sigma}{n}\right)e \;>\; 1.$$

If

$$\left(\frac{q}{1+a}\right)^{1/n}\left(\frac{g-\tau}{n}\right)e \;>\; 1 \quad \text{for} \quad n \quad \text{odd}$$

and

$$\left(\frac{q}{b}\right)^{1/n}\left(\frac{g+\sigma}{n}\right)e \;>\; 1 \quad \text{for} \quad n \quad \text{even},$$

then the equation $(E_{84}; 1)$ is oscillatory.

Theorem 2.16.3. Let $a > 0$, $h > \sigma$, $g > \tau$, and

$$\left(\frac{p}{1+b}\right)^{1/n}\left(\frac{h-\sigma}{n}\right)e > 1.$$

If

$$\left(\frac{q}{a}\right)^{1/n}\left(\frac{g-\tau}{n}\right)e > 1 \quad \text{for } n \text{ odd}$$

and

$$\left(\frac{q}{1+b}\right)^{1/n}\left(\frac{g}{n}\right)e > 1 \quad \text{for } n \text{ even},$$

then the equation $(E_{85}; -1)$ is oscillatory.

Theorem 2.16.4. Let $a > 0$, $g > \tau$, and

$$\left(\frac{p}{a}\right)^{1/n}\left(\frac{h+\sigma}{n}\right)e > 1.$$

If

$$\left(\frac{q}{1+b}\right)^{1/n}\left(\frac{g}{n}\right)e > 1 \quad \text{for } n \text{ odd}$$

and

$$\left(\frac{q}{a}\right)^{1/n}\left(\frac{g-\tau}{n}\right)e > 1 \quad \text{for } n \text{ even},$$

then the equation $(E_{85}; 1)$ is oscillatory.

Theorem 2.16.5. Suppose that $g > \tau$, $h > \sigma$, and

$$\left(\frac{p}{1+a+b}\right)^{1/n}\left(\frac{h-\sigma}{n}\right)e > 1.$$

If

$$\left(\frac{q}{1+a+b}\right)^{1/n}\left(\frac{g-\tau}{n}\right)e > 1 \quad \text{for } n \text{ even},$$

then the equation $(E_{86}; -1)$ is oscillatory.

Theorem 2.16.6. Suppose that $g > \sigma$. If

$$\left(\frac{q}{1+a+b}\right)^{1/n}\left(\frac{g-\tau}{n}\right)e > 1 \quad \text{for } n \text{ odd},$$

then the equation $(E_{86}; 1)$ is oscillatory.

Theorem 2.16.7. Let $a + b > 0$, $g > \tau$, and

$$p^{1/n}\left(\frac{\sigma}{n}\right)e > 1.$$

If

$$\left(\frac{q}{a+b}\right)^{1/n}\left(\frac{g-\tau}{n}\right)e > 1 \quad \text{for} \quad n \quad \text{odd}$$

and

$$q^{1/n}\left(\frac{g}{n}\right)e > 1 \quad \text{for} \quad n \quad \text{even},$$

then the equation $(E_{87};-1)$ is oscillatory.

Theorem 2.16.8. Let $a+b>0$, $g>\tau$, $h>\sigma$, and

$$\left(\frac{p}{a+b}\right)^{1/n}\left(\frac{h-\sigma}{n}\right)e > 1.$$

If

$$q^{1/n}\left(\frac{g}{n}\right)e > 1 \quad \text{for} \quad n \quad \text{odd}$$

and

$$\left(\frac{q}{a+b}\right)^{1/n}\left(\frac{g-\tau}{n}\right)e > 1 \quad \text{for} \quad n \quad \text{even},$$

then the equation $(E_{87};1)$ is oscillatory.

Next, we shall consider equations $(E_{84};\delta)$ – $(E_{87};\delta)$, $\delta = \pm 1$ when $\tau=\sigma$ and the functions p and q are periodic of period τ. Before, we state our results we denote the following:

$$I(u,v,i,p) = \int_v^u \frac{(s-v)^i(u-s)^{n-i-1}}{i!\,(n-i-1)!}p(s)ds, \qquad u \geq v$$

for some $i=0,1,\cdots,n-1$, and

$$J(u,v,j,q) = \int_v^u \frac{(u-s)^j(s-v)^{n-j-1}}{j!\,(n-j-1)!}q(s)ds, \qquad u \geq v$$

for some $j=0,1,\cdots,n-1$.

Theorem 2.16.9. Let $b>0$, $g>\tau$, and

$$\limsup_{t\to\infty} I(t+h,t,i,p) > 1+a$$

holds for some $i=0,1,\cdots,n-1$. If

$$\limsup_{t\to\infty} J(t,t-(\tau+g),j,q) > b \quad \text{for} \quad n \quad \text{odd}$$

holds for some $j=0,1,\cdots,n-1$, and

$$\limsup_{t\to\infty} J(t,t-(g-\tau),k,q) > 1+a \quad \text{for} \quad n \quad \text{even}$$

holds for some $k = 0, 1, \cdots, n-1$, then the equation $(E_{84}; -1)$ is oscillatory.

Theorem 2.16.10. Let $b > 0$, $h > \tau$, $g > \tau$, and

$$\limsup_{t \to \infty} I(t + (h - \tau), t, i, p) > b$$

is satisfied for some $i = 0, 1, \cdots, n-1$. If

$$\limsup_{t \to \infty} J(t, t - g, j, q) > 1 + a \quad \text{for} \quad n \quad \text{odd}$$

is satisfied for some $j = 0, 1, \cdots, n-1$, and

$$\limsup_{t \to \infty} J(t, t - (g + \tau), k, q) > b \quad \text{for} \quad n \quad \text{even}$$

is satisfied for some $k = 0, 1, \cdots, n-1$, then the equation $(E_{84}; 1)$ is oscillatory.

Theorem 2.16.11. Let $a > 0$, $g > \tau$, $h > \tau$, and

$$\limsup_{t \to \infty} I(t, t + (h - \tau), i, p) > 1 + b$$

holds for some $i = 0, 1, \cdots, n-1$. If

$$\limsup_{t \to \infty} J(t, t - (g - \tau), j, q) > a \quad \text{for} \quad n \quad \text{odd}$$

holds for some $j = 0, 1, \cdots, n-1$, and

$$\limsup_{t \to \infty} J(t, t - g, k, q) > 1 + b \quad \text{for} \quad n \quad \text{even}$$

holds for some $k = 0, 1, \cdots, n-1$, then the equation $(E_{85}; -1)$ is oscillatory.

Theorem 2.16.12. Let $a > 0$, $g > \tau$, and

$$\limsup_{t \to \infty} I(t + h + \tau, t, i, p) > a$$

holds for some $i = 0, 1, \cdots, n-1$. If

$$\limsup_{t \to \infty} J(t, t - (g + \tau), j, q) > 1 + b \quad \text{for} \quad n \quad \text{odd}$$

holds for some $j = 0, 1, \cdots, n-1$, and

$$\limsup_{t \to \infty} J(t, t - (g - \tau), k, q) > a \quad \text{for} \quad n \quad \text{even}$$

holds for some $k = 0, 1, \cdots, n-1$, then the equation $(E_{85}; 1)$ is oscillatory.

Theorem 2.16.13. Let $g > \tau$, $h > \tau$, and

$$\limsup_{t\to\infty} I(t+h-\tau, t, i, p) \;>\; 1+a+b$$

holds for some $i = 0, 1, \cdots, n-1$. If

$$\limsup_{t\to\infty} J(t, t-(g-\tau), j, q) \;>\; 1+a+b \quad \text{for} \quad n \quad \text{even}$$

holds for some $j = 0, 1, \cdots, n-1$, then the equation $(E_{86}; -1)$ is oscillatory.

Theorem 2.16.14. If

$$\limsup_{t\to\infty} J(t, t-(g+\tau), i, q) \;>\; 1+a+b \quad \text{for} \quad n \quad \text{odd}$$

holds for some $i = 0, 1, \cdots, n-1$, then the equation $(E_{86}; 1)$ is oscillatory.

Theorem 2.16.15. Let $a + b > 0$, and

$$\limsup_{t\to\infty} I(t+h, t, i, p) \;>\; 1$$

holds for some $i = 0, 1, \cdots, n-1$. If

$$\limsup_{t\to\infty} J(t, t-(g+\tau), j, q) \;>\; a+b \quad \text{for} \quad n \quad \text{odd}$$

holds for some $j = 0, 1, \cdots, n-1$, and

$$\limsup_{t\to\infty} J(t, t-g, k, q) \;>\; 1 \quad \text{for} \quad n \quad \text{even}$$

holds for some $k = 0, 1, \cdots, n-1$, then the equation $(E_{87}; -1)$ is oscillatory.

Theorem 2.16.16. Let $a + b > 0$, $g > \tau$, $h > \tau$, and

$$\limsup_{t\to\infty} I(t+(h-\tau), t, i, p) \;>\; a+b \tag{2.16.15}$$

holds for some $i = 0, 1, \cdots, n-1$. If

$$\limsup_{t\to\infty} J(t, t-g, j, q) \;>\; 1 \quad \text{for} \quad n \quad \text{odd} \tag{2.16.16}$$

holds for some $j = 0, 1, \cdots, n-1$, and

$$\limsup_{t\to\infty} J(t, t-(g-\tau), k, q) \;>\; a+b \quad \text{for} \quad n \quad \text{even} \tag{2.16.17}$$

holds for some $k = 0, 1, \cdots, n-1$, then the equation $(E_{87}; 1)$ is oscillatory.

Proof. Let $x(t)$ be an eventually positive solution of the equation $(E_{87};1)$, say $x(t) > 0$ for $t \geq t_0 \geq 0$. Set

$$y(t) = x(t) - ax[t-\tau] - bx[t+\tau].$$

Then,

$$y^{(n)}(t) = -q(t)x[t-g] - p(t)x[t+h] \leq 0 \quad \text{for} \quad t \geq t_1 \geq t_0,$$

which implies that $y^{(i)}(t)$, $i = 0, 1, \cdots, n-1$ are eventually of one sign. Next, we consider the following two cases:

(I) $y(t) < 0$ for $t \geq t_1$, and (II) $y(t) > 0$ for $t \geq t_1$.

(I) Suppose that $y(t) < 0$ for $t \geq t_1$. Set

$$0 < u(t) = -y(t) = ax[t-\tau] + bx[t+\tau] - x(t)$$

and hence, we obtain

$$u^{(n)}(t) = q(t)x[t-g] + p(t)x[t+h] \quad \text{for} \quad t \geq t_1.$$

Now, the following two cases are considered:

(i) $u'(t) > 0$ for $t \geq t_1$, and (ii) $u'(t) < 0$ for $t \geq t_1$.

(i) Assume that $u'(t) > 0$ for $t \geq t_1$. Set

$$V(t) = au[t-\tau] + bu[t+\tau] - u(t), \quad t \geq t_1. \tag{2.16.18}$$

Then,

$$V^{(n)}(t) = q(t)u[t-g] + p(t)u[t+h]$$

and

$$(aV[t-\tau] + bV[t+\tau] - V(t))^{(n)} = q(t)V[t-g] + p(t)V[t+h]. \tag{2.16.19}$$

We shall show that $V^{(i)}(t) > 0$ eventually, $i = 0, 1, \cdots, n$. There exist positive constants c_1 and c_2 and a $T \geq t_1$ such that

$$u[t-g] \geq c_1 \quad \text{and} \quad u[t+h] \geq c_2 \quad \text{for} \quad t \geq T.$$

Thus,

$$V^{(n)}(t) \geq c_1 q(t) + c_2 p(t) \quad \text{for} \quad t \geq T$$

and hence

$$V^{(n-1)}(t) \to \infty \quad \text{and} \quad V^{(i)}(t) \to \infty \quad \text{as} \quad t \to \infty, \quad i = 0, 1, \cdots, n-1.$$

Therefore, we conclude that

$$V^{(i)}(t) > 0 \quad \text{eventually}, \quad i = 0, 1, \cdots, n. \tag{2.16.20}$$

Using (2.16.20) in the equation (2.16.19), we see that

$$(a+b)V^{(n)}[t+\tau] \;\ge\; p(t)V[t+h]$$

or

$$V^{(n)}(t) \;\ge\; \frac{p(t)}{a+b}V[t+h-\tau] \quad \text{for} \quad t \ge T^* \ge T$$

and by Lemma 2.2.12(i) and condition (2.16.15), we arrive at a desired contradiction.

(ii) Assume that $u'(t) < 0$ for $t \ge t_1$. First, we claim that $u(t) \to 0$ monotonically as $t \to \infty$. Otherwise, $u(t) \to c > 0$ as $t \to \infty$. There exists a $T_1 \ge t_1$ such that

$$u[t-g] \;\ge\; \frac{c}{2} \quad \text{and} \quad u[t+h] \;\ge\; \frac{c}{2} \quad \text{for} \quad t \ge T_1.$$

Thus,

$$V^{(n)}(t) \;\ge\; \frac{c}{2}[q(t)+p(t)] \quad \text{for} \quad t \ge T_1$$

and hence

$$V^{(i)}(t) \to \infty \quad \text{as} \quad t \to \infty, \;\; i = 0, 1, \cdots, n-1.$$

On the other hand from (2.16.18) we see that $V(t) < \infty$ for $t \ge T_1$, which is a contradiction. Therefore, we conclude that $u(t) \to 0$ and $V^{(i)}(t) \to 0$ monotonically as $t \to \infty$, $i = 0, 1, \cdots, n-1$. It is easy to check that n must be even, and

$$(-1)^i V^{(i)}(t) \;>\; 0, \quad i = 0, 1, \cdots, n-1 \quad \text{and} \quad t \ge T_2 \ge T_1.$$

Using the fact that the function $u(t)$ is decreasing on $[T_2, \infty)$ and the functions $p(t)$ and $q(t)$ are periodic of period τ on $[T_2, \infty)$, we have

$$\begin{aligned} V^{(n)}[t-\tau] &= q[t-\tau]u[t-\tau-g] + p[t-\tau]u[t-\tau+h] \\ &= q(t)u[t-\tau-g] + p(t)u[t-\tau+h] \\ &\ge q(t)u[t-g] + p(t)u[t+h] \;=\; V^{(n)}(t) \quad \text{for} \quad t \ge T_2. \end{aligned}$$

Using this fact in (2.16.19), we obtain

$$(a+b)V^{(n)}[t-\tau] \;\ge\; q(t)V[t-g]$$

or

$$V^{(n)}(t) \;\ge\; \frac{q(t)}{a+b}V[t-(g-\tau)], \quad t \ge T_2.$$

By Lemma 2.2.12(ii) and condition (2.16.17), we obtain the desired contradiction.

(II) Suppose that $y(t) > 0$ for $t \geq t_1$. Then,

$$y(t) \leq x(t) \quad \text{for} \quad t \geq t_1,$$

and hence

$$y^{(n)}(t) + q(t)y[t-g] \leq 0 \quad \text{for} \quad t \geq t_1. \tag{2.16.21}$$

Next, we consider the two cases:

(A) $y'(t) > 0$ for $t \geq t_2 \geq t_1$, and (B) $y'(t) < 0$ for $t_1 \geq t_2$.

(A) Assume that $y'(t) > 0$ for $t \geq t_2$. Then, there exist a positive constant c and a $t_3 \geq t_2$ such that

$$y[t-g] \geq c \quad \text{for} \quad t \geq t_3. \tag{2.16.22}$$

Using (2.16.22) in (2.16.21) and integrating from t_3 to t, we get

$$0 < y^{(n-1)}(t) \leq -y^{(n-1)}(t_3) - \int_{t_3}^{t} c\, q(s)ds \to -\infty \quad \text{as} \quad t \to \infty,$$

which is a contradiction.

(B) Assume that $y'(t) < 0$ for $t \geq t_2$. This is the case when n is odd. It is easy to check that y satisfies

$$(-1)^i y^{(i)}(t) > 0, \quad i = 0, 1, \cdots, n-1, \quad t \geq t_3 \geq t_2.$$

By Lemma 2.2.12(ii) and condition (2.16.16) we arrive at the desired contradiction. ∎

2.17. Functional Differential Equations Involving Quasi–derivatives

Consider the functional differential equation

$$L_n x(t) + F(t, x[g(t)]) = 0, \quad n \quad \text{is even} \tag{E_{88}}$$

where

$$L_0 x(t) = x(t)$$
$$L_k x(t) = \frac{1}{a_k(t)}\left(L_{k-1}x(t)\right)', \quad k = 1, 2, \cdots, n$$

$a_i : [t_0, \infty) \to \mathbb{R}_+$, $i = 1, 2, \cdots, n-1$, $a_n \equiv 1$, $g : [t_0, \infty) \to \mathbb{R}$ and $F : [t_0, \infty) \times \mathbb{R} \to \mathbb{R}$, $t_0 \geq 0$ are continuous, $g(t) \to \infty$ as $t \to \infty$, and

$$\int^{\infty} a_i(s)ds = \infty, \quad i = 1, 2, \cdots, n-1.$$

In what follows we shall assume that there exist a continuous function $q : [t_0, \infty) \to \mathbb{R}_0$, $q(t)$ not identically zero for all large t, a differentiable function $\sigma : [t_0, \infty) \to \mathbb{R}_+$, and a positive constant γ such that

$$F(t,x) \operatorname{sgn} x \geq q(t)|x|^\gamma \quad \text{for} \quad x \neq 0 \quad \text{and} \quad t \geq t_0 \tag{2.17.1}$$

and

$$\sigma(t) \leq \min\{t, g(t)\}, \quad \sigma'(t) > 0 \quad \text{and} \quad \sigma(t) \to \infty \quad \text{as} \quad t \to \infty. \tag{2.17.2}$$

The domain of L_n, $D(L_n)$ is defined to be the set of all functions $x : [t_0, \infty) \to \mathbb{R}$ such that $L_j x(t)$, $j = 0, 1, \cdots, n$ exist and continuous on $[t_0, \infty)$.

Here we shall provide sufficient conditions for the oscillation of (E_{88}) when $\gamma > 0$. We shall also present two criteria for the oscillation of (E_{88}) by comparing it with a second order linear ordinary differential equation, and a first order delay differential equation whose oscillatory character is known. Finally, we shall discuss the oscillatory behavior of (E_{88}) when n is odd.

To formulate our results we shall need the following notation: For continuous functions $p_i : [t_0, \infty) \to \mathbb{R}$, $i = 1, 2, \cdots$ we define

$$I_0 = 1$$

$$I_i(t, s; p_i, \cdots, p_1) = \int_s^t p_i(u) I_{i-1}(u, s; p_{i-1}, \cdots, p_1) du, \quad i = 1, 2, \cdots.$$

It is easy to verify that for $i = 1, 2, \cdots, n-1$

$$I_i(t, s; p_1, \cdots, p_i) = (-1)^i I_i(s, t; p_i, \cdots, p_1)$$

and

$$I_i(t, s; p_1, \cdots, p_i) = \int_s^t p_i(u) I_{i-1}(t, u; p_1, \cdots, p_{i-1}) du.$$

For any $T \geq t_0$ and all $t \geq T$, we let

$$I[t, T] = \max_{1 \leq i \leq n-1} I_i(t, T; a_i, \cdots, a_1),$$

$$I^*[v, u] = I_{n-1}(v, u; a_{n-1}, \cdots, a_1)$$

$$= \int_u^v a_1(s_1) \int_{s_1}^v a_2(s_2) \cdots \int_{s_{n-2}}^v a_{n-1}(s_{n-1}) ds_{n-1} \cdots ds_1, \quad v \geq u \geq T,$$

$$A(t) = \int_t^\infty q(s) ds,$$

$$\begin{aligned} w[t,T] &= \min\left\{\int_T^t I_{n-\ell-1}(t,s;a_{n-1},\cdots,a_{\ell+1})a_\ell(s)\right. \\ &\qquad \left. \times\, I_{\ell-1}(t,s;a_1,\cdots,a_{\ell-1})ds\ :\ \ell=1,3,\cdots,n-1\right\}, \\ w_1[t,T] &= \min\left\{a_1(t)\int_T^t I_{n-\ell-1}(t,s;a_{n-1},\cdots,a_{\ell+1})a_\ell(s)\right. \\ &\qquad \left. \times\, I_{\ell-2}(t,s;a_2,\cdots,a_{\ell-1})ds\ :\ \ell=3,5,\cdots,n-1\right\} \text{ if } n>2 \\ &= a_1(t) \text{ if } n=2, \\ w_2[t,T] &= a_1(s)I_{n-2}(t,s;a_{n-1},\cdots,a_2) \text{ for } t\ge s\ge T, \\ w_3[t,T] &= a_1(t)I_{n-2}(t,T;a_2,\cdots,a_{n-1}), \end{aligned}$$

and

$$w^*[t,T] = \min\{w_1[t/2,T];w_2[t,t/2];w_3[t/2,T]\}, \quad t\ge 2T.$$

We shall also need the following three lemmas:

Lemma 2.17.1. If $x\in D(L_n)$ is of constant sign and not identically zero for all large t, then there exists a $t_x\ge t_0$ and an integer ℓ, $0\le\ell\le n$ with $n+\ell$ even for $x(t)L_nx(t)$ nonnegative, or $n+\ell$ odd for $x(t)L_nx(t)$ nonpositive, and such that for every $t\ge t_x$

$$\ell \ge 0 \quad \text{implies} \quad x(t)L_kx(t) > 0, \quad k=0,1,\cdots,\ell$$

and

$$\ell \le n-1 \quad \text{implies} \quad (-1)^{\ell+k}x(t)L_kx(t) > 0, \quad k=\ell,\ell+1,\cdots,n.$$

The above result is a generalization of Kiguradze's Lemma 2.2.1.

Lemma 2.17.2 [64,69]. If the function x is as in Lemma 2.17.1, and $\lim_{t\to\infty}x(t)\ne 0$ also $L_{n-1}x(t)L_nx(t)\le 0$ for every $t\ge t_x$, then for all large $T\ge t_x$ with $t\ge 2T$,

$$x(t) \ge w[t,T]L_{n-1}x(t)$$

and

$$x'\left[\frac{t}{2}\right] \ge w^*[t,T]L_{n-1}x(t).$$

Lemma 2.17.3 [239]. If the function x is as in Lemma 2.17.1, and satisfies $(-1)^iL_ix(t)>0$ for $t\ge T$ (large), $i=0,1,\cdots,n$, then

$$x(u) \ge (-1)^{n-1}I^*[v,u]L_{n-1}x(v) \quad \text{for } v\ge u\ge T.$$

Our first oscillation result for the equation (E_{88}) when $\gamma>0$ is contained in the following theorem.

Theorem 2.17.4. Let conditions (2.17.1) and (2.17.2) hold, and that $A(t) \geq 0$ for $t \geq t_0$. If for every $T \geq t_0$ and $\sigma(t) > 2T$ and all constants $c_i > 0$, $i = 1,2$

$$\int^{\infty} w^*[\sigma(s),T]\frac{\sigma'(s)}{2}\left\{A(s)+\gamma c_1\int_s^{\infty} A^2(\tau)w^*[\sigma(\tau),T]\frac{\sigma'(\tau)}{2}d\tau\right\}ds = \infty,$$

when $\gamma > 1$ (2.17.3)

$$\limsup_{t\to\infty} w[\sigma(t)/2,T]\left\{A(t)+\int_t^{\infty} A^2(s)w^*[\sigma(s),T]\frac{\sigma'(s)}{2}ds\right\} > 1,$$

when $\gamma = 1$ (2.17.4)

$$\limsup_{t\to\infty} w[\sigma(t)/2,T]A^{1/\gamma}(t)\left\{1+\frac{\gamma}{c_2A(t)}\int_t^{\infty} A^{1+1/\gamma}(s)w^*[\sigma(s),T]\frac{\sigma'(s)}{2}ds\right\}$$

$$> c_2, \text{ when } A(t) > 0 \text{ and } 0 < \gamma < 1, \quad (2.17.5)$$

then (E_{88}) is oscillatory.

Proof. Let $x(t)$ be a nonoscillatory solution of (E_{88}), say $x(t) > 0$ for $t \geq t_0 \geq 0$. By Lemma 2.17.1, there exists a $t_1 \geq t_0$ such that

$$L_{n-1}x(t) > 0 \quad \text{and} \quad x'(t) > 0 \quad \text{for} \quad t \geq t_1. \quad (2.17.6)$$

Using condition (2.17.1) in (E_{88}), we get

$$L_nx(t)+q(t)x^{\gamma}[g(t)] \leq 0 \quad \text{for} \quad t \geq t_1.$$

There exist a $t_2 \geq t_1$ and $k_i > 0$, $i = 1,2$ so that $\sigma(t) > 2t_1$ for $t \geq t_2$,

$$x[\sigma(t)/2] \geq k_1 \quad \text{and} \quad L_{n-1}x(t) \leq k_2 \quad \text{for} \quad t \geq t_2, \quad (2.17.7)$$

$$L_nx(t)+q(t)x^{\gamma}[\sigma(t)/2] \leq 0 \quad \text{for} \quad t \geq t_2. \quad (2.17.8)$$

Integrating (2.17.8) from t to $u \geq t \geq t_2$, we obtain

$$x^{-\gamma}[\sigma(u)/2]L_{n-1}x(u) - x^{-\gamma}[\sigma(t)/2]L_{n-1}x(t) + \gamma\int_t^u q(s)ds$$

$$+\gamma\int_t^{\infty} x^{-\gamma-1}[\sigma(s)/2]x'[\sigma(s)/2]\frac{\sigma'(s)}{2}L_{n-1}x(s)ds \leq 0.$$

Letting $u \to \infty$, we find

$$L_{n-1}x(t) \geq A(t)x^{\gamma}[\sigma(t)/2] + \gamma x^{\gamma}[\sigma(t)/2]$$

$$\times \int_t^{\infty} x^{-\gamma-1}[\sigma(s)/2]x'[\sigma(s)/2]\frac{\sigma'(s)}{2}L_{n-1}x(s)ds. \quad (2.17.9)$$

Now we have three possible cases to consider:

(I) Let $\gamma > 1$. Then, by Lemma 2.17.2 there exists a $t_3 \geq t_2$ such that for $t \geq t_3$,

$$x'[\sigma(t)/2] \geq w^*[\sigma(t), t_1]L_{n-1}x(t) \qquad (2.17.10)$$

and

$$x[\sigma(t)/2] \geq w[\sigma(t)/2, t_1]L_{n-1}x(t). \qquad (2.17.11)$$

Using (2.17.7) and (2.17.10) in (2.17.9), we obtain

$$x'[\sigma(t)/2] \geq w^*[\sigma(t), t_1]x^\gamma[\sigma(t)/2]\left\{A(t) + \gamma k_1^{\gamma-1}\int_t^\infty A^2(s) \times w^*[\sigma(s), t_1]\frac{\sigma'(s)}{2}ds\right\} \quad \text{for } t \geq t_3$$

or

$$\frac{x'[\sigma(t)/2]\sigma'(t)/2}{x^\gamma[\sigma(t)/2]} \geq w^*[\sigma(t), t_1]\frac{\sigma'(t)}{2}\left\{A(t) + \gamma k_1^{\gamma-1}\int_t^\infty A^2(s) \times w^*[\sigma(s), t_1]\frac{\sigma'(s)}{2}ds\right\} \quad \text{for } t \geq t_3.$$

Integrating the above inequality from t_3 to $t \geq t_3$, we get

$$\int_{t_3}^t w^*[\sigma(s), t_1]\frac{\sigma'(s)}{2}\left\{A(s) + \gamma k_1^{\gamma-1}\int_s^\infty A^2(\tau)w^*[\sigma(\tau), t_1]\frac{\sigma'(\tau)}{2}d\tau\right\}ds \leq \frac{1}{\gamma - 1}x^{1-\gamma}[\sigma(t_3)/2] < \infty \quad \text{as} \quad t \to \infty$$

which is a contradiction.

(II) Let $\gamma = 1$. From (2.17.9) it follows that

$$L_{n-1}x(t) \geq A(t)x[\sigma(t)/2] \quad \text{for} \quad t \geq t_2. \qquad (2.17.12)$$

Using (2.17.10) – (2.17.12) in (2.17.9), we have

$$1 \geq w[\sigma(t)/2, t_1]\left\{A(t) + \int_t^\infty A^2(s)w^*[\sigma(s), t_1]\frac{\sigma'(s)}{2}ds\right\}.$$

Taking $\limsup$ on both sides of the above inequality as $t \to \infty$, we obtain the desired contradiction.

(III) Let $0 < \gamma < 1$. From (2.17.7) and (2.17.9), we have

$$k_2 \geq L_{n-1}x(t) \geq A(t)x^\gamma[\sigma(t)/2] \quad \text{for} \quad t \geq t_2 \qquad (2.17.13)$$

or

$$x[\sigma(t)/2] \;\le\; k_2^{1/\gamma}(A(t))^{-1/\gamma} \tag{2.17.14}$$

and

$$x^{\gamma-1}[\sigma(t)/2] \;\ge\; k_2^{(\gamma-1)/\gamma}(A(t))^{(1-\gamma)/\gamma} \quad \text{for} \quad t \ge t_2. \tag{2.17.15}$$

Also, from (2.17.9) – (2.17.11), we find

$$\begin{aligned} &L_{n-1}x(t)x^{1-\gamma}[\sigma(t)/2] \\ &\quad\ge\; A(t)x[\sigma(t)/2] + \gamma x[\sigma(t)/2] \\ &\quad\quad\times \int_t^\infty x^{-\gamma-1}[\sigma(s)/2]x'[\sigma(s)/2]\frac{\sigma'(s)}{2}L_{n-1}x(s)ds \\ &\quad\ge\; A(t)w[\sigma(t)/2,t_1]L_{n-1}x(t) + \gamma w[\sigma(t)/2,t_1]L_{n-1}x(t) \\ &\quad\quad\times \int_t^\infty A^2(s)w^*[\sigma(s),t_1]\frac{\sigma'(s)}{2}x^{\gamma-1}[\sigma(s)/2]ds \end{aligned} \tag{2.17.16}$$

or

$$x^{1-\gamma}[\sigma(t)/2] \ge w[\sigma(t)/2,t_1]\left\{A(t) + \frac{\gamma}{k}\int_t^\infty w^*[\sigma(s),t_1]\frac{\sigma'(s)}{2}A^{1+1/\gamma}(s)ds\right\},$$

where $k = k_2^{(1-\gamma)/\gamma}$. Using (2.17.14) in the above inequality, we obtain

$$k(A(t))^{(\gamma-1)/\gamma} \ge w[\sigma(t)/2,t_1]\left\{A(t) + \frac{\gamma}{k}\int_t^\infty w^*[\sigma(s),t_1]\frac{\sigma'(s)}{2}A^{1+1/\gamma}(s)ds\right\}$$

or

$$k \ge w[\sigma(t)/2,t_1](A(t))^{1/\gamma}\left\{1 + \frac{\gamma}{kA(t)}\int_t^\infty w^*[\sigma(s),t_1]\frac{\sigma'(s)}{2}A^{1+1/\gamma}(s)ds\right\}.$$

Taking lim sup on both sides of the above inequality as $t \to \infty$, we obtain a contradiction to (2.17.5). This completes the proof. ∎

Remark 2.17.1. Condition (2.17.5) in Theorem 2.17.4 can be replaced by: for all constants $\alpha_i > 0$, $i = 1,2$ and for all large $T \ge t_0$ with $\sigma(t) > 2T$

$$\begin{aligned} \limsup_{t\to\infty} w[\sigma(t)/2,T]A^{1/\gamma}(t)\left\{1 + \frac{\alpha_1}{A(t)}\int_t^\infty A^2(s)w^*[\sigma(s),T]\right. \\ \left.\times\,\frac{\sigma'(s)}{2}\left(I[\sigma(t)/2,T]\right)^{\gamma-1}ds\right\} \;>\; \alpha_2. \end{aligned} \tag{2.17.5'}$$

In fact, on integrating the second inequality in (2.17.7), $(n-1)$ times from t_1 to t, we find that there exist a constant $k_2 > 0$ and a $T_1 \ge t_2$ such that

$$x[\sigma(t)/2] \;\le\; k_3 I[\sigma(t)/2,t_1] \quad \text{for} \quad t \ge T_1. \tag{2.17.17}$$

Using (2.17.15) and (2.17.17) in (2.17.16) and proceeding as in Case III we arrive at a contradiction.

In our next result we shall obtain the oscillatory behavior of the equation (E_{88}) by comparing it with a second order linear ordinary differential equation whose oscillatory character is known.

Theorem 2.17.5. Let conditions (2.17.1) and (2.17.2) hold. If for all large $T \geq t_0$ with $\sigma(t) > 2T$ for $t \geq T_1$ for some $T_1 \geq T$,

$$(r(t)y'(t))' + q(t)Q(t)y(t) = 0 \tag{E_{89}}$$

is oscillatory, where $r(t) = (2/\sigma'(t)w^*[\sigma(t), T])$, and

$$Q(t) = \begin{cases} c_1 \text{ is any positive constant} & \text{when } \gamma > 1 \\ 1 & \text{when } \gamma = 1 \\ m(t) & \text{when } 0 < \gamma < 1 \end{cases} \tag{2.17.18}$$

where

$$m(t) = \begin{cases} c_2 A^{(1-\gamma)/\gamma}(t), \quad A(t) > 0 \quad \text{for} \quad t \geq t_0, \quad \text{or} \\ c_3 I^{\gamma-1}[\sigma(t)/2, T], \end{cases}$$

and c_2 and c_3 are any positive constants, then (E_{88}) is oscillatory.

Proof. Let $x(t)$ be a nonoscillatory solution of (E_{88}), say $x(t) > 0$ for $t \geq t_0 \geq 0$. As in Theorem 2.17.4, we find that (2.17.6) holds for $t \geq t_1$, (2.17.7) holds for $t \geq t_2$ and (2.17.10) holds for $t \geq t_3$. We define a function W by

$$W(t) = -\frac{L_{n-1}x(t)}{x[\sigma(t)/2]} \quad \text{for} \quad t \geq t_3.$$

Then, for $t \geq t_3$, we have

$$W'(t) = \frac{F(t, x[g(t)])}{x[\sigma(t)/2]} + \frac{L_{n-1}x(t)x'[\sigma(t)/2]\sigma'(t)/2}{x^2[\sigma(t)/2]} = f(t) + \frac{1}{p(t)}W^2(t), \tag{2.17.19}$$

where

$$f(t) = \frac{F(t, x[g(t)])}{x[\sigma(t)/2]} \quad \text{and} \quad p(t) = \frac{L_{n-1}x(t)}{x'[\sigma(t)/2]\sigma'(t)/2}. \tag{2.17.20}$$

The Riccati equation (2.17.19) has a solution on $[t_3, \infty)$. It is well known [285] that this is equivalent to the nonoscillation of the linear second order equation

$$(p(t)u'(t))' + f(t)u(t) = 0. \tag{E_{90}}$$

Using (2.17.1) and (2.17.10) in (2.17.20), we obtain

$$p(t) = \frac{2L_{n-1}x(t)}{\sigma'(t)x'[\sigma(t)/2]} \leq \frac{2}{\sigma'(t)w^*[\sigma(t),T]} = r(t), \quad t \geq t_3$$

and

$$f(t) = \frac{F(t,x[g(t)])}{x[\sigma(t)/2]} \geq q(t)x^{\gamma-1}[\sigma(t)/2], \quad t \geq t_3.$$

We consider the following three cases:

(I) Let $\gamma > 1$. From (2.17.7) it follows that

$$x^{\gamma-1}[\sigma(t)/2] \geq k_1^{\gamma-1} \quad \text{for} \quad t \geq t_3.$$

(II) Let $\gamma = 1$. Then, $x^{\gamma-1}[\sigma(t)/2] = 1$ for all $t \geq t_3$.

(III) Let $0 < \gamma < 1$. From (2.17.17) it follows that

$$x^{\gamma-1}[\sigma(t)/2] \geq k_3^{\gamma-1}I^{\gamma-1}[\sigma(t)/2, t_1] \quad \text{for} \quad t \geq t_3,$$

or else (2.17.15) holds.

Thus, an application of Sturm–Picone comparison theorem (see [173, 186]) to equation (E_{90}) yields the nonoscillation of the second order linear differential equation

$$(r(t)u'(t))' + q(t)Q(t)u(t) = 0,$$

where

$$Q(t) = \begin{cases} k_1^{\gamma-1} & \text{when } \gamma > 1 \\ 1 & \text{when } \gamma = 1 \\ k_2^{(\gamma-1)/\gamma}(A(t))^{(1-\gamma)/\gamma}, \quad \text{or} & \\ k_3^{\gamma-1}I^{\gamma-1}[\sigma(t)/2, t_1] & \text{when } 0 < \gamma < 1. \end{cases}$$

This contradicts the hypothesis that the equation (E_{89}) is oscillatory. Hence, our proof is complete. ∎

The following results for the oscillation of the equation (E_{88}) are immediate consequences of Theorem 2.17.5 and Theorem 5 in [98] and Theorem 1 in [247].

Corollary 2.17.6. Let conditions (2.17.1) and (2.17.2) hold, and assume that there exists a differentiable function $\rho : [t_0, \infty) \to \mathbb{R}_+$ such that

$$\limsup_{t\to\infty} \int_{t_0}^{t} \left\{ \rho(s)Q(s)q(s) - \frac{\rho'^2(s)}{2\sigma'(s)\rho(s)w^*[\sigma(s),T]} \right\} ds = \infty$$

for all large $T \geq t_0$ with $\sigma(t) > 2T$, $t \geq T_1 \geq T$, where the function $Q(t)$ is defined in (2.17.18). Then, the equation (E_{88}) is oscillatory.

Corollary 2.17.7. Let conditions (2.17.1) and (2.17.2) hold, and for all large $T \geq t_0$ with $\sigma(t) > 2T$ for $t \geq T_1 \geq T$,

$$\liminf_{t\to\infty} \left(\int_t^\infty q(s)Q(s)ds\right)\left(\int_{t_0}^t \frac{\sigma'(s)}{2} w^*[\sigma(s), T]ds\right) > \frac{1}{4},$$

where the function $Q(t)$ is defined in (2.17.18). Then, the equation (E_{88}) is oscillatory.

In the following result, we shall reduce the problem of oscillation of (E_{88}) to the problem of oscillation of a first order delay differential equation.

Theorem 2.17.8. Let conditions (2.17.1) and (2.17.2) hold. If for every large $T \geq t_0$ with $\sigma(t) > 2T$ for $t \geq T_1$ for some $T_1 \geq T$ the equation

$$y'(t) + q(t)w^\gamma[\sigma(t), T]|y[\sigma(t)]|^\gamma \operatorname{sgn} y[\sigma(t)] = 0 \qquad (E_{91})$$

is oscillatory. Then, the equation (E_{88}) is oscillatory.

Proof. Let $x(t)$ be a nonoscillatory solution of (E_{88}), say $x(t) > 0$ for $t \geq t_0 \geq 0$. As in Theorem 2.17.4, we obtain (2.17.6) which is satisfied for $t \geq t_1$. By Lemma 2.17.2, there exists a $t_2 \geq t_1$ with $\sigma(t) > 2t_1$ for $t \geq t_2$, and

$$x[\sigma(t)] \geq w[\sigma(t), t_1]L_{n-1}x[\sigma(t)] \quad \text{for} \quad t \geq t_2. \qquad (2.17.21)$$

Using (2.17.1) and (2.17.21) in (E_{88}), we get

$$u'(t) + q(t)w^\gamma[\sigma(t), T]u^\gamma[\sigma(t)] \leq 0 \quad \text{for} \quad t \geq t_2, \qquad (2.17.22)$$

where $u(t) = L_{n-1}x(t)$, $t \geq t_2$.

Integrating (2.17.22) from $t_2 \leq t \leq \eta$ and letting $\eta \to \infty$, we obtain

$$u(t) \geq \int_t^\infty q(s)w^\gamma[\sigma(s), t_1]u^\gamma[\sigma(s)]ds \quad \text{for} \quad t \geq t_2.$$

The function $u = L_{n-1}x$ is obviously strictly decreasing on $[t_2, \infty)$. Hence, by Theorem 2.5.7 we conclude that there exists a positive solution y of (E_{91}) with $\lim_{t\to\infty} y(t) = 0$. This contradiction completes the proof. ∎

Now shall we apply the results established in [93] and [169] to obtain the following oscillation criterion for the equation (E_{88}).

Corollary 2.17.9. Let conditions (2.17.1) and (2.17.2) hold. If for every large $T \geq t_0$ with $\sigma(t) > T$ for $t \geq T_1$ for some $T_1 \geq T$,

$$\liminf_{t\to\infty} \int_{\sigma(t)}^t q(s)w[\sigma(s), T]ds > \frac{1}{e} \quad \text{when} \quad \gamma = 1$$

and

$$\int^{\infty} q(s)w^{\gamma}[\sigma(s),T]ds = \infty \quad \text{when} \quad 0<\gamma<1,$$

then the equation (E_{88}) is oscillatory.

Next, we shall present results for the oscillation of the equation (E_{88}) when n is odd.

Theorem 2.17.10. Let condition (2.17.1) hold, and

$$g(t) < t \quad \text{and} \quad g'(t) > 0 \quad \text{for} \quad t \geq t_0 \tag{2.17.23}$$

and let the conditions of either Theorems 2.17.4, 2.17.5 or 2.17.8 hold with $\sigma(t)/2$ in the function w and $\sigma(t)$ in the function w^* or the function w be replaced by $g(t)$. Moreover, assume that every solution of the equation

$$z'(t)+q(t)\left(I^*[(t+g(t))/2,g(t)]\right)^{\gamma}\left|z\left[\frac{t+g(t)}{2}\right]\right|^{\gamma} \operatorname{sgn} z\left[\frac{t+g(t)}{2}\right] = 0 \tag{E_{92}}$$

is oscillatory. Then, the equation (E_{88}) with n odd is oscillatory.

Proof. Let $x(t)$ be a nonoscillatory solution of (E_{88}) and assume that $x(t)>0$ for $t \geq t_0 \geq 0$. By Lemma 2.17.1 there exists a $t_1 \geq t_0$ such that the following two cases hold for $t \geq t_1$

(i) $L_{n-1}x(t) > 0, \cdots, L_2x(t)>0, x'(t)>0$, and
(ii) $(-1)^i L_i x(t) > 0$, $i=0,1,\cdots,n$.

Let (i) hold. By Lemma 2.17.2 there exists a $t_2 \geq t_1$ such that $g(t)>t_1$ for $t \geq t_2$, and

$$x[g(t)] \geq w[g(t),t_1]L_{n-1}x[g(t)] \geq w[g(t),t_1]L_{n-1}x(t) \quad \text{for} \quad t \geq t_2$$

also

$$x'[g(t)] \geq x'[g(t)/2] \geq w^*[g(t),t_1]L_{n-1}x[g(t)] \geq w^*[g(t),t_1]L_{n-1}x(t) \quad \text{for} \quad t \geq t_2.$$

The rest of the proof is similar to that of Theorems 2.17.4, 2.17.5 and 2.17.8.

Assume (ii) holds. By Lemma 2.17.3 there exists a $T_1 \geq t_1$ such that

$$x[g(t)] \geq I^*[(t+g(t))/2,g(t)]L_{n-1}x\left[\frac{t+g(t)}{2}\right] \quad \text{for} \quad t > (t+g(t))/2 > g(t) \geq T_1. \tag{2.17.24}$$

Using (2.17.24) in (E_{88}), we get

$$\begin{aligned} 0 &\geq L_n x(t) + q(t) x^\gamma[g(t)] \\ &\geq L_n x(t) + q(t)\left(I^*[(t+g(t))/2, g(t)]\right)^\gamma \left(L_{n-1} x\left[\frac{t+g(t)}{2}\right]\right)^\gamma. \end{aligned}$$

Setting $V(t) = L_{n-1}x(t)$, $t \geq T_1$ in this inequality, we obtain

$$V'(t) + q(t)\left(I^*[(t+g(t))/2, g(t)]\right)^\gamma V^\gamma\left[\frac{t+g(t)}{2}\right] \leq 0 \quad \text{for} \quad t \geq T_1.$$

The rest of the proof is similar to that of Theorem 2.17.8. ∎

For the oscillatory behavior of all bounded solutions of (E_{88}) with n odd we can easily extract the following result from Theorem 2.17.10.

Corollary 2.17.11. Let conditions (2.17.1) and (2.17.23) hold, and assume that the equation (E_{92}) is oscillatory. Then, every bounded solution of the equation (E_{88}) with n odd is oscillatory.

In addition, from Theorem 2.17.10 and Corollary 2.17.9, we have the following result.

Corollary 2.17.12. Let n be odd, conditions (2.17.1) and (2.17.23) hold, and assume that for all large T with $g(t) > T$ for $t \geq T_1$ for some $T_1 \geq T$ either

$$\liminf_{t\to\infty} v(t) > \frac{1}{e} \quad \text{when} \quad \gamma = 1,$$

where

$$v(t) = \min\left\{\int_{g(t)}^t q(s) w[g(s), T] ds, \int_{(t+g(t))/2}^t q(s) I^*[(s+g(s))/2, g(s)] ds\right\},$$

or

$$\int^\infty r(s) ds = \infty \quad \text{when} \quad 0 < \gamma < 1,$$

where

$$r(t) = \min\left\{q(t) w^\gamma[g(t), T],\ q(t)\left(I^*[(t+g(t))/2, g(t)]\right)^\gamma\right\}.$$

Then, the equation (E_{88}) is oscillatory.

The following example illustrates how Corollary 2.17.12 can be used in practice.

Example 2.17.1. Consider the odd order functional differential equation

$$L_n x(t) + t^{(2-2n)\gamma-1}\left|x\left[\frac{t}{2}\right]\right|^\gamma \operatorname{sgn} x\left[\frac{t}{2}\right] = 0, \quad t > 0 \quad \text{and} \quad 0 < \gamma \leq 1 \qquad (E_{93})$$

where

$$L_0x(t) = x(t), \quad L_kx(t) = \frac{1}{t}\left(L_{k-1}x(t)\right)', \quad k = 1,2,\cdots,n, \quad a_n(t) = 1$$

$$\text{and } a_i(t) = t, \; i = 1,2,\cdots,n-1.$$

Now, for all large T, $t \geq T$

$$w[t,T] = \frac{\beta}{2^{n-1}(n-1)} t^{2n-2}\left(1 - \frac{T^2}{t^2}\right)^{n-1}$$

$$\beta = \min_{\ell}\left\{\frac{1}{(\ell-1)!\,(n-\ell-1)!} : \ell = 1,3,\cdots,n-1\right\}$$

and

$$I^*[v,u] = \frac{1}{2^{n-1}(n-1)!}(v^2-u^2)^{n-1}, \quad v \geq u \geq T.$$

It is easy to check that all the conditions of Corollary 2.17.12 are satisfied and hence the equation (E_{93}) is oscillatory for $0 < \gamma \leq 1$.

2.18. Neutral and Damped Functional Differential Equations Involving Quasi–derivatives

Following the notations of the previous section here we shall present criteria for the oscillation of the neutral functional differential equation

$$L_n\left(x(t) + c(t)x[\tau(t)]\right) + F(t, x[g(t)]) = 0, \quad n \text{ is even} \qquad (E_{94})$$

where $c, \tau : [t_0, \infty) \to \mathbb{R}$ are continuous, $\tau'(t) > 0$ for $t \to \infty$ and $\lim_{t\to\infty} \tau(t) = \infty$; and the damped functional differential equation

$$L_nx(t) + H\left(t, x[g(t)], \frac{d}{dt}x[h(t)]\right) = 0, \quad n \text{ is even} \qquad (E_{95})$$

where g is as in equation (E_{88}), $h : [t_0, \infty) \to \mathbb{R}$ and $H : [t_0, \infty) \times \mathbb{R}^2 \to \mathbb{R}$ are continuous, and $\lim_{t\to\infty} h(t) = \infty$, by comparing with the oscillation of equations of type (E_{88}) of order n for (E_{94}), and of order n and $(n-1)$ for (E_{95}). Then, we can apply the results obtained for (E_{88}) to investigate the oscillatory character of equations (E_{94}) and (E_{95}).

Theorem 2.18.1. Let conditions (2.17.1) and (2.17.2) hold, and $0 < c(t) \leq \lambda < 1$ and $\tau(t) < t$ for $t \geq t_0$. If the equation

$$L_nw(t) + (1-\lambda)^\gamma q(t)\,|w[\sigma(t)]|^\gamma \operatorname{sgn} w[g(t)] = 0, \quad n \text{ is even} \qquad (E_{96})$$

is oscillatory, then the equation (E_{94}) is oscillatory.

Theorem 2.18.2. Let condition (2.17.1) hold, $1 < \mu \leq c(t) \leq \nu$ and $\tau(t) > t$ for $t \geq t_0$, and assume that there exists a differentiable function $\sigma_* : [t_0, \infty) \to \mathbb{R}_+$ such that $\sigma_*(t) \leq \min\{\tau^{-1} \circ g(t), t\}$, $\sigma_*(t) \to \infty$ as $t \to \infty$, and $\sigma'_*(t) > 0$ for $t \geq t_0$, where $\tau^{-1}(t)$ denotes the inverse function of $\tau(t)$. If the equation

$$L_n z(t) + \left(\frac{\mu - 1}{\mu\nu}\right)^\gamma q(t)\,|z[\sigma_*(t)]|^\gamma \operatorname{sgn} z[\sigma_*(t)] \;=\; 0, \quad n \text{ is even} \quad (E_{97})$$

is oscillatory, then the equation (E_{94}) is oscillatory.

Proofs of Theorems 2.18.1 and 2.18.2. Let $x(t)$ be a nonoscillatory solution of (E_{94}), say $x(t) > 0$ for $t \geq t_0 \geq 0$. Define $y(t) = x(t) + c(t)x[\tau(t)]$. Then, $y(t) > 0$ for $t \geq t_1$ for some $t_1 \geq t_0$ and equation (E_{94}) takes the form

$$L_n y(t) + q(t)x^\gamma[g(t)] \;\leq\; 0 \quad \text{for} \quad t \geq t_1. \tag{2.18.1}$$

By Lemma 2.17.1, there exists a $t_2 \geq t_1$ such that $y'(t) > 0$ for $t \geq t_2$. Now we have the following two cases to consider:

(I) $0 < c(t) \leq \lambda < 1$ and $\tau(t) < t$, and

(II) $1 < \mu \leq c(t) \leq \nu$ and $\tau(t) > t$.

Assume (I) holds, then

$$\begin{aligned} x(t) &= y(t) - c(t)x[\tau(t)] \\ &= y(t) - c(t)[y[\tau(t)] - c[\tau(t)]x[\tau \circ \tau(t)]] \\ &\geq y(t) - c(t)y[\tau(t)] \;\geq\; (1-\lambda)y(t), \quad t \geq t_2. \end{aligned} \tag{2.18.2}$$

Next, we assume (II) holds, then

$$\begin{aligned} x(t) &= \frac{y[\tau^{-1}(t)] - x[\tau^{-1}(t)]}{c[\tau^{-1}(t)]} \\ &= \frac{y[\tau^{-1}(t)]}{c[\tau^{-1}(t)]} - \frac{1}{c[\tau^{-1}(t)]}\left[\frac{y[\tau^{-1} \circ \tau^{-1}(t)]}{c[\tau^{-1} \circ \tau^{-1}(t)]} - \frac{x[\tau^{-1} \circ \tau^{-1}(t)]}{c[\tau^{-1} \circ \tau^{-1}(t)]}\right] \\ &\geq \frac{y[\tau^{-1}(t)]}{c[\tau^{-1}(t)]} - \frac{y[\tau^{-1} \circ \tau^{-1}(t)]}{c[\tau^{-1}(t)]c[\tau^{-1} \circ \tau^{-1}(t)]} \\ &\geq \frac{\mu - 1}{\mu\nu} y[\tau^{-1}(t)] \quad \text{for} \quad t \geq t_2. \end{aligned} \tag{2.18.3}$$

Using (2.18.2) and (2.18.3) in (2.18.1), we obtain

$$L_n y(t) + (1-\lambda)^\gamma q(t) y^\gamma[\sigma(t)] \;\leq\; 0 \quad \text{for} \quad t \geq t_2$$

and

$$L_n y(t) + \left(\frac{\mu - 1}{\mu\nu}\right)^\gamma q(t) y^\gamma[\sigma_*(t)] \;\leq\; 0 \quad \text{for} \quad t \geq t_2$$

respectively. But in view of [66,100,180], it follows that the equation (E_{96}) (respectively, equation (E_{97})) has an eventually positive solution, which is a contradiction. ∎

Next, we shall consider the equation (E_{95}) and assume that there exist a continuous function $q : [t_0, \infty) \to \mathbb{R}_0$, $q(t)$ is not identically zero for all large t, and positive constants γ and λ such that

$$H(t,x,y) \operatorname{sgn} x \;\geq\; q(t)|x|^{\gamma}|y|^{\lambda} \quad \text{for} \quad x \neq 0,\ t \geq t_0. \tag{2.18.4}$$

Also, we shall assume that

$$h(t) \;<\; t \quad \text{and} \quad h'(t) \;\geq\; 0 \quad \text{for} \quad t \geq t_0. \tag{2.18.5}$$

In the following result, we shall give sufficient conditions for the oscillation of (E_{95}) by comparing it with the equations

$$L_n w(t) + c(a_1[h(t)]h'(t))^{\lambda} q(t)|w[g(t)]|^{\gamma} \operatorname{sgn} w[g(t)] \;=\; 0 \tag{E_{98}}$$

and

$$M_m z(t) + C(a_1[h(t)]h'(t))^{\lambda} q(t)|z[h(t)]|^{\gamma} \operatorname{sgn} z[g(t)] \;=\; 0, \tag{E_{99}}$$

where

$$M_0 z(t) \;=\; z(t),\ M_k z(k) \;=\; \frac{1}{b_k(t)}\left(M_{k-1}z(t)\right)',\ k = 1,2,\cdots,m;\ m = n-1$$

$b_k(t) = a_{k+1}(t)$, $k = 1,2,\cdots,n-1$ and c and C are any positive constants.

Theorem 2.18.3. Let conditions (2.17.2), (2.18.4) and (2.18.5) hold. If for every $c > 0$, (E_{98}) is oscillatory, and for every $C > 0$ every bounded solution of (E_{99}) is oscillatory, then the equation (E_{95}) is oscillatory.

Proof. Let $x(t)$ be a nonoscillatory solution of (E_{95}), say $x(t) > 0$ for $t \geq t_0 \geq 0$. By Lemma 2.17.1, there exists a $t_1 \geq t_0$ such that the following two cases hold for $t \geq t_1$,

(i) $L_{n-1}x(t) > 0, \cdots, L_2 x(t) > 0$ and $x'(t) > 0$,
(ii) $L_{n-1}x(t) > 0, \cdots, L_2 x(t) < 0$ and $x'(t) > 0$.

Assume (i) holds. We note that the function $L_1 x(t)$ is increasing for $t \geq t_1$. Thus, there exist a $t_2 \geq t_1$ and a constant $c_1 > 0$ such that

$$\frac{d}{dt}x[h(t)] \;=\; x'[h(t)]h'(t) \;=\; a_1[h(t)]h'(t)L_1x[h(t)] \;\geq\; c_1 a_1[h(t)]h'(t),$$

$$t \geq t_2. \tag{2.18.6}$$

Using (2.18.4) and (2.18.6) in (E_{95}), we obtain

$$L_n x(t) + c_1^{\lambda}\left(a_1[h(t)]h'(t)\right)^{\lambda} q(t)(x[g(t)])^{\gamma} \leq 0 \quad \text{for} \quad t \geq t_2.$$

But, in view of [66,100,180] it follows that the equation

$$L_n x(t) + c_1^{\lambda}\left(a_1[h(t)]h'(t)\right)^{\lambda} q(t)(x[g(t)])^{\gamma} = 0 \quad \text{for} \quad t \geq t_2$$

has a positive nonoscillatory solution, which is a contradiction.

Next, assume that (ii) holds. Since $x(t)$ is an increasing function for $t \geq t_1$, there exist a $T_2 \geq T_1$ and a constant $c_2 > 0$ such that

$$x[g(t)] \geq c_2 \quad \text{for} \quad t \geq T_2. \tag{2.18.7}$$

Using (2.18.4) and (2.18.7) in (E_{95}), we find

$$L_n x(t) + c_2^{\gamma} q(t)\left(\frac{d}{dt}x[h(t)]\right)^{\gamma} \leq 0 \quad \text{for} \quad t \geq T_2,$$

or

$$L_n x(t) + c_2^{\gamma} q(t)\left(a_1[h(t)]h'(t)\right)^{\lambda}\left(L_1 x[h(t)]\right)^{\lambda} \leq 0 \quad \text{for} \quad t \geq T_2.$$

Setting $v(t) = L_1 x(t)$ for $t \geq T_2$, we obtain

$$M_m v(t) + c_2^{\gamma} q(t)\left(a_1[h(t)]h'(t)\right)^{\lambda}\left(v[h(t)]\right)^{\lambda} \leq 0 \quad \text{for} \quad t \geq T_2.$$

Clearly $v(t)$ is a positive and decreasing function for $t \geq T_2$. Applying Corollary 1 in [241] we see that the equation

$$M_m v(t) + c_2^{\gamma} q(t)\left(a_1[h(t)]h'(t)\right)^{\lambda}\left(v[h(t)]\right)^{\lambda} = 0, \quad t \geq T_2$$

has a bounded, eventually positive and decreasing solution, which is a contradiction. ■

The following two corollaries are immediate consequences of Theorem 2.18.3 together with Theorems 2.17.5, 2.17.8 and Corollary 2.17.11.

Corollary 2.18.4. Let conditions (2.17.2), (2.18.4) and (2.18.5) hold, and assume that for all large $T \geq t_0$ with $\sigma(t) > T$ for $t \geq T_1$ for some $T_1 \geq T$ the equation

$$(r(t)y'(t))' + q(t)\left(a_1[h(t)]h'(t)\right)^{\lambda} Q^*(t)y(t) = 0$$

is oscillatory, where the function $r(t)$ is the same as in (E_{89}), and

$$Q^*(t) = \begin{cases} c_1, \ c_1 \text{ is any positive constant} & \text{if } \gamma > 1 \\ c_2, \ c_2 \text{ is any positive constant} & \text{if } \gamma = 1 \\ c_3 I^{\gamma-1}[\sigma(t)/2, T], \ c_3 \text{ is any positive constant} & \text{if } 0 < \gamma < 1. \end{cases}$$

Moreover, assume that for every positive constant C the equation

$$z'(t) + CI^{**}[(t+h(t))/2, h(t)]\left(a_1[h(t)]h'(t)\right)^{\lambda} q(t)\left|z\left[\frac{t+h(t)}{2}\right]\right|^{\lambda} \times \operatorname{sgn} z\left[\frac{t+h(t)}{2}\right] = 0 \quad (E_{100})$$

is oscillatory, where

$$I^{**}[v,u] = \int_u^v a_2(s_2)\int_{s_2}^v a_3(s_3)\cdots\int_{s_{n-2}}^v a_{n-1}(s_{n-1})ds_{n-1}\cdots ds_2.$$

Then, the equation (E_{95}) is oscillatory.

Corollary 2.18.5. Let conditions (2.17.2), (2.18.4) and (2.18.5) hold and assume that for every positive constants c and C and all large T with $\sigma(t) > T$ for $t \geq T_1$ for some $T_1 \geq T$ equation (E_{100}) is oscillatory, and the equation

$$y'(t) + c\left(a_1[h(t)]h'(t)\right)^{\lambda} w[\sigma(t),T]q(t)|y[\sigma(t)]|^{\gamma}\operatorname{sgn} y[\sigma(t)] = 0$$

is oscillatory. Then, the equation (E_{95}) is oscillatory.

The following example illustrates how the above results can be used in practice.

Example 2.18.1. Consider the damped functional differential equation

$$L_n x(t) + q(t)|x[t+\sin t]|^{\gamma}\left|\frac{d}{dt}x\left[\frac{t}{2}\right]\right|^{\lambda} \operatorname{sgn} x[t+\sin t] = 0, \quad t > 0 \ (E_{101})$$

where the differential operator L_n is the same as in (E_{93}), n is even, γ and λ are positive constants, and $q : [t_0,\infty) \to \mathbb{R}_+$ is continuous. For all large T and $t \geq T$, we define the functions w and I^* as in Example 2.17.1. And,

$$w_1[t,T] = \min_{\ell}\left\{\frac{t(t^2-T^2)^{n-2}}{2^{n-2}(n-2)(n-\ell-1)!(\ell-2)!} : \ell = 3,5,\cdots,n-3\right\},$$

$$w_2[t,s] = \frac{s(t^2-s^2)^{n-2}}{2^{n-2}(n-2)!}, \quad t \geq s \geq T,$$

$$w_3[t,T] = \frac{t(t^2-T^2)^{n-2}}{2^{n-2}(n-2)!}, \qquad I[t,T] = \frac{(t^2-T^2)^{n-1}}{2^{n-2}(n-1)!}$$

also

$$I^{**}[v,u] = \frac{(v^2-u^2)^{n-2}}{2^{n-2}(n-2)!}, \quad v \geq u \geq T.$$

Now, we let $T = 1$ and $T_1 = 3$, $\sigma(t) = t - 1$ and $h(t) = t/2$. Then,

$$w^*[t,1] = \frac{t^{2n-3}}{2^{3n-5}(n-2)!}\left(1 - \frac{4t^2}{T^2}\right)^{n-2}, \quad t \geq 3.$$

Hence, all the conditions of Corollaries 2.17.7 and 2.18.4 are satisfied if

$$q(t) = \begin{cases} \min\left\{t^{(3-2n)\lambda-1},\ t^{(2-2n)\gamma-1+\epsilon-\lambda}\right\} & \text{if } 0 < \gamma < 1 \text{ and } 0 < \lambda \leq 1 \\ \min\left\{t^{(3-2n)\lambda-1},\ t^{1-2n+\epsilon-\lambda}\right\} & \text{if } \gamma \geq 1 \text{ and } 0 < \lambda \leq 1, \end{cases}$$

for $t \geq 1$ and for some $\epsilon > 0$, and hence the equation (E_{101}) is oscillatory.

2.19. Forced Functional Differential Equations Involving Quasi–derivatives

Here we shall provide oscillation criteria for the advanced–forced differential equation

$$L_n x(t) + \delta f(t, x[g_1(t)], \cdots, x[g_m(t)]) = Q(t). \qquad (E_{102};\delta)$$

The results presented here for $(E_{102};\delta)$ are not valid for the corresponding ordinary differential equations, i.e. oscillations are caused by the advanced arguments.

In $(E_{102};\delta)$, $\delta = \pm 1$, the quasi–operator L_n is the same as in (E_{88}) and $g_i : \mathbb{R}_0 \to \mathbb{R}_0$, $i = 1,2,\cdots,m$, $Q : \mathbb{R}_0 \to \mathbb{R}$, $f : R_0 \times \mathbb{R}^m \to \mathbb{R}$ are continuous and $g_i(t) \geq t$ on $\mathbb{R}_0$, $i = 1,2,\cdots,m$.

We shall always assume that there exists a continuous function $\eta : \mathbb{R}_0 \to \mathbb{R}$ such that

$$L_n\eta(t) = Q(t), \quad \eta(t) \text{ is oscillatory on } \mathbb{R}_0 \text{ and } \lim_{t\to\infty} \eta(t) = 0. \qquad (2.19.1)$$

We also introduce the following conditions:

(C_1) There exist continuous functions $q_i : \mathbb{R}_0 \to \mathbb{R}_0$ and $F_i : \mathbb{R}_0 \to \mathbb{R}_0$, $i = 1,2,\cdots,m$ such that for some $i \in \{1,2,\cdots,m\}$, $q_i(t)$ is not identically zero on any ray of the form $[t^*, \infty)$ for some $t^* \geq t_0 \geq 0$, and

$$f(t, x_1, x_2, \cdots, x_m) \operatorname{sgn} x_1 \geq \sum_{i=1}^m q_i(t) F_i(|x_i|)$$

for $t \in \mathbb{R}_0$ and $x_1 x_i > 0$, $i = 1,2,\cdots,m$.

(C_2) There exist continuous functions $q : \mathbb{R}_0 \to \mathbb{R}_0$ and $F_i : \mathbb{R}_0 \to \mathbb{R}_0$, $i = 1, 2, \cdots, m$ such that $q(t)$ is not identically zero on any ray of the form $[t^*, \infty)$ for some $t^* \geq t_0 \geq 0$, and

$$f(t, x_1, x_2, \cdots, x_m) \operatorname{sgn} x_1 \geq q(t) \prod_{i=1}^{m} F_i(|x_i|)$$

for $t \in \mathbb{R}_0$ and $x_1 x_i > 0$, $i = 1, 2, \cdots, m$.

We shall discuss oscillation of only those solutions x of $(E_{102}; \delta)$ which satisfy $\sup\{|x(t)| : t \geq T\} > 0$ for all $T \geq T_x$. As in Secition 2.2 the equation $(E_{102}; \delta)$ is said to be *almost oscillatory* if

(i) for $\delta = 1$ and n even, every solution of the equation $(E_{102}; 1)$ is oscillatory,

(ii) for $\delta = 1$ and n odd, every solution $x(t)$ of the equation $(E_{102}; 1)$ is either oscillatory or $L_j x(t) \to 0$ as $t \to \infty$, $j = 0, 1, \cdots, n-1$,

(iii) for $\delta = -1$ and n even, every solution $x(t)$ of the equation $(E_{102}; -1)$ is oscillatory, $L_j x(t) \to 0$ as $t \to \infty$, $j = 0, 1, \cdots, n-1$, or $|L_j x(t)| \to \infty$ as $t \to \infty$, $j = 0, 1, \cdots, n-1$,

(iv) for $\delta = -1$ and n odd, every solution $x(t)$ of the equation $(E_{102}; -1)$ is oscillatory or $|L_j x(t)| \to \infty$ as $t \to \infty$, $j = 0, 1, \cdots, n-1$.

Now, using the notations of Section 2.17 we have the following generalization of Taylor's formula with remainder.

Lemma 2.19.1. Let $x \in D(L_n)$, then for $t, s \in \mathbb{R}_0$ and $0 \leq i < k \leq n$

(i) $$L_i x(t) = \sum_{j=i}^{k-1} I_{j-i}(t, s; a_{i+1}, \cdots, a_j) L_j x(s) + \int_s^t I_{k-i-1}(t, u; a_{i+1}, \cdots, a_{k-1}) a_k(u) L_k x(u) du.$$

(ii) $$L_i x(t) = \sum_{j=i}^{k-1} (-1)^{j-i} I_{j-1}(s, t; a_j, \cdots, a_{i+1}) L_j x(s) + (-1)^{k-i} \int_t^s I_{k-i-1}(u, t; a_{k-1}, \cdots, a_{i+1}) a_k(u) L_k x(u) du.$$

We shall also need the following result:

Lemma 2.19.2 [300,301]. Consider the integro–differential inequality with advanced arguments

$$y'(t) \geq \int_t^{\infty} Q(t, s) \prod_{i=1}^{m} |y[g_i(s)]|^{\alpha_i} ds, \tag{2.19.2}$$

where $Q : \mathbb{R}_0^2 \to \mathbb{R}_0$ and $g_i : \mathbb{R}_0 \to \mathbb{R}_0$, $i = 1, 2, \cdots, m$ are continuous functions with $g_i(t) \geq t$, $i = 1, 2, \cdots, m$, α_i are nonnegative numbers with $\sum_{i=1}^m \alpha_i = 1$. If

$$\sum_{i=1}^{m} \alpha_i \liminf_{t\to\infty} \int_t^{g_i(t)} \int_s^\infty Q(s,u)duds > \frac{1}{e}, \tag{2.19.3}$$

then the inequality (2.19.2) has no eventually positive solutions.

It will be convenient to make use of the following notations: For any $T \geq t_0$ and all $t \geq T$, we let

$$\begin{aligned}
&\alpha_i[t,s] = I_i(t,s;a_1,\cdots,a_i), \qquad i = 1,2,\cdots,n-1 \\
&\alpha[t,T] = \max_{1\leq i\leq n-1} \alpha_i[t,T] \\
&\beta_i[t,s] = I_{n-i-1}(t,s;a_{n-1},\cdots,a_{i+1}), \qquad i = 0,1,2,\cdots,n-1 \\
&\gamma_1[t,T] = a_1(t), \\
&\gamma_i[t,s] = a_1(s)I_{i-1}(t,s;a_2,\cdots,a_i), \qquad i = 2,3,\cdots,n-1 \\
&R_i[t,T] = \int_T^t a_i(s)ds, \; i = 1,2,\cdots,n-1 \\
&\rho_1[t,T] = R_1[t,T], \\
&\rho_i[t,T] = \int_T^t \alpha_{i-2}[t,s]a_{i-1}(s)R_i[s,T]ds, \qquad i = 2,3,\cdots,n-1.
\end{aligned}$$

We shall also denote by $\sigma(t) = \min\{g_1(t), g_2(t), \cdots, g_m(t)\}$ and $\tau(t) = \max\{g_1(t), g_2(t), \cdots, g_m(t)\}$.

The following theorem is concerned with the oscillatory and asymptotic behavior of the equation $(E_{102};1)$, when the functions F_i, $i = 1,2,\cdots,m$ are locally of bounded variation, i.e. F_i, $i = 1,2,\cdots,m$ are not necessarily monotonic.

Theorem 2.19.3. Let $\gamma > 0$, conditions (2.19.1) and (C_1) hold, and let (G_i, H_i) be a pair of continuous components of F_i, $i = 1,2,\cdots,m$. Moreover, suppose that

$$\frac{H_i(x_i)}{x_i} \geq c_i > 0 \quad \text{for} \quad x_i \neq 0, \quad i = 1,2,\cdots,m \tag{2.19.4}$$

and for all sufficiently large T, every $\ell \in \{1,2,\cdots,n-1\}$ with $n+\ell$ odd and every $|k| \geq 1$

$$\liminf_{t\to\infty} \int_t^{\sigma(t)} \gamma_\ell[s,T] \int_s^\infty \beta_\ell[u,s] \sum_{i=1}^m c_i q_i(u) G_i(k, \alpha[g_i(u),T])duds > \frac{1}{e}, \tag{2.19.5;ℓ}$$

then for n even, equation $(E_{102};1)$ is oscillatory. If in addition,

$$L_i\eta(t) \to 0 \quad \text{as} \quad t \to \infty, \quad i = 1,2,\cdots,n-1 \tag{2.19.6}$$

and

$$\int^{\infty} \beta_0[u,T]\sum_{i=1}^{m} q_i(u)du = \infty \quad \text{for all large } T, \tag{2.19.7}$$

then for n odd, every solution $x(t)$ of the equation $(E_{102};1)$ is either oscillatory or $L_ix(t) \to 0$ as $t \to \infty$, $i = 1,2,\cdots,n-1$.

Proof. Let $x(t)$ be a nonoscillatory solution of the equation $(E_{102};1)$. Without any loss of generality we assume that $x(t) \neq 0$ for all $t \geq t_0 \geq \gamma$. Furthermore, we suppose that $x(t) > 0$ and $x[g_i(t)] > 0$ for $t \geq t_0 \geq \gamma$, $i = 1,2,\cdots,m$, since the substitution $u = -x$ transforms equation $(E_{102};1)$ into a equation of the same form subject to the assumptions of the theorem.

Consider the function $x(t) = y(t) + \eta(t)$, then from the equation $(E_{102};1)$

$$L_ny(t) = -f(t,x[g_1(t)],\cdots,x[g_m(t)]) < 0 \quad \text{for all} \quad t \geq t_0 \tag{2.19.8}$$

so that L_ny is eventually negative for all $t \geq t_0$. Hence, $L_iy(t)$, $i = 0,1,\cdots,n-1$ are monotone and one-signed for all large t, say $t \geq t_0$.

Now, if $y(t) < 0$ for $t \geq t_0$, then $y(t) + \eta(t) > 0$ implies $\eta(t) > -y(t) > 0$, a contradiction to the oscillatory character of $\eta(t)$. Hence, $y(t) > 0$ for $t \geq t_0$.

By Lemma 2.17.1, there exists a $t_1 \geq t_0$ and $\ell \in \{0,1,\cdots,n-1\}$ with $n+\ell$ odd such that

$$\begin{cases} L_iy(t) > 0 \quad \text{on} \quad [t_1,\infty) \quad \text{for} \quad 0 \leq i \leq \ell \\ (-1)^{i-\ell}L_iy(t) > 0 \quad \text{on} \quad [t_1,\infty) \quad \text{for} \quad \ell \leq i \leq n. \end{cases} \tag{2.19.9}$$

Suppose $\ell \in \{1,2,\cdots,n-1\}$. Then, from (2.19.9) we have

$$y'(t) > 0 \quad \text{and} \quad L_{n-1}y(t) > 0 \quad \text{for} \quad t \geq t_1. \tag{2.19.10}$$

Recall that $x[g_i(t)] = y[g_i(t)] + \eta[g_i(t)]$. Since $y(t)$ is positive, increasing and $\eta(t) \to 0$ as $t \to \infty$, there exists a $t_2 \geq t_1$ and a constant λ, $0 < \lambda < 1$ so that

$$x[g_i(t)] \geq \lambda y[g_i(t)] \quad \text{for} \quad t \geq t_2, \quad i = 1,2,\cdots,m. \tag{2.19.11}$$

Since $L_{n-1}y(t)$ is positive and nonincreasing, there exists $k_1 > 0$ such that $L_{n-1}y(t) \leq k_1$ for $t \geq t_2$. By successive integration from t_2

to t, we conclude that there exist a $t_3 \geq t_2$ and $k_2 > 0$ such that $y(t) \leq k_2\alpha[t, t_2]$ for $t \geq t_3$. Choose $t_4 \geq t_3$ so that $g_i(t) \geq t_3$ for $t \geq t_4$, then

$$y[g_i(t)] \leq k_2\alpha[g_i(t), t_2] \quad \text{for} \quad t \geq t_4, \quad i = 1, 2, \cdots, m.$$

Using condition (2.19.1), there exist $t_5 \geq t_4$ and $k \geq 1$ such that

$$x[g_i(t)] \leq k\alpha[g_i(t), t_2] \quad \text{for} \quad t \geq t_5, \quad i = 1, 2, \cdots, m. \tag{2.19.12}$$

From (C_1), (2.19.4) and (2.19.11), for $t \geq t_5$ we obtain

$$\begin{aligned} f(t, x[g_1(t)], \cdots, x[g_m(t)]) &\geq \sum_{i=1}^{m} q_i(t)F_i(x[g_i(t)]) \\ &= \sum_{i=1}^{m} q_i(t)G_i(x[g_i(t)])H_i(x[g_i(t)]) \\ &\geq \sum_{i=1}^{m} c_i q_i(t)G_i(k\alpha[g_i(t), t_2])x[g_i(t)] \\ &\geq \lambda\sum_{i=1}^{m} c_i q_i(t)G_i(k\alpha[g_i(t), t_2])y[g_i(t)]. \end{aligned}$$

From (2.19.10) and the definition of the function $\sigma(t)$, there exists a $t_6 \geq t_5$ so that

$$y[g_i(t)] \geq y[\sigma(t)] \quad \text{for} \quad t \geq t_6, \quad i = 1, 2, \cdots, m. \tag{2.19.13}$$

Thus, for $t \geq t_6$

$$f(t, x[g_1(t)], \cdots, x[g_m(t)]) \geq \lambda\sum_{i=1}^{m} c_i q_i(t)G_i(k\alpha[g_i(t), t_2])y[\sigma(t)].$$

Now equation (2.19.8) becomes

$$L_n y(t) + \lambda\left(\sum_{i=1}^{m} c_i q_i(t)G_i(k\alpha[g_i(t), t_2])\right)y[\sigma(t)] \leq 0 \quad \text{for} \quad t \geq t_6. \tag{2.19.14}$$

Assume $\ell \in \{2, 3, \cdots, n-1\}$. Then, from Lemma 2.19.1(ii), for $t \geq s \geq t_6$ we obtain

$$\begin{aligned} L_\ell y(t) &= \sum_{j=\ell}^{n-1}(-1)^{j-\ell}I_{j-\ell}(s, t; a_j, \cdots, a_{\ell-1})L_j y(s) \\ &\quad +(-1)^{n-\ell}\int_t^s I_{n-\ell-1}(u, t; a_{n-1}, \cdots, a_{\ell+1})L_n y(u)du. \end{aligned}$$

Using (2.19.9) and (2.19.14) and letting $s \to \infty$, for $t \geq t_6$ we get

$$L_\ell y(t) \;\geq\; \lambda \int_t^\infty \beta_\ell[u,t] \sum_{i=1}^m c_i q_i(u) G_i(k\alpha[g_i(u), t_2]) y[\sigma(u)]du. \quad (2.19.15)$$

On the other hand, applying Lemma 2.19.1(i), we find

$$\begin{aligned} y'(t) \;&=\; a_1(t) \sum_{j=1}^{\ell-1} I_{j-1}(t, t_6; a_2, \cdots, a_j) L_j y(t_6) \\ &\quad + a_1(t) \int_{t_6}^t I_{\ell-2}(t, s; a_2, \cdots, a_{\ell-1}) a_\ell(s) L_\ell y(s) ds, \qquad t \geq t_6. \end{aligned}$$

From (2.19.9) and the fact that the function $L_\ell y(t)$ is nonincreasing for $t \geq t_6$, we obtain

$$\begin{aligned} y'(t) \;&\geq\; a_1(t) \int_{t_6}^t I_{\ell-2}(t, s; a_2, \cdots, a_{\ell-1}) a_\ell(s) ds \;\cdot\; L_\ell y(t) \\ &=\; a_1(t) I_{\ell-1}(t, t_6; a_2, \cdots, a_\ell) L_\ell y(t) \\ &=\; \gamma_\ell[t, t_6] L_\ell y(t) \quad \text{for} \quad t \geq t_6. \qquad (2.19.16) \end{aligned}$$

Combining (2.19.15) and (2.19.16), we get

$$y'(t) \;\geq\; \lambda \gamma_\ell[t, t_6] \int_t^\infty \beta_\ell[u,t] \sum_{i=1}^m c_i q_i(u) G_i(k\alpha[g_i(u), t_2]) y[\sigma(u)]du. \quad (2.19.17)$$

Inequality (2.19.17), in view of condition (2.19.5; ℓ) and Lemma 2.19.2 has no eventually positive solution, a contradiction to the fact that $y(t) > 0$ for $t \geq t_1$.

Next, suppose $\ell = 1$. This is the case when $\delta = 1$ and n is even; from Lemma 2.19.1(ii), we obtain

$$\begin{aligned} y'(t) \;&=\; a_1(t) L_1 y(t) \\ &=\; a_1(t) \sum_{j=1}^{n-1} (-1)^{j-1} I_{j-1}(s, t; a_j, \cdots, a_2) L_j y(s) \\ &\quad + (-1)^{n-1} a_1(t) \int_t^s I_{n-2}(u, t; a_{n-1}, \cdots, a_2) L_n y(u) du. \end{aligned}$$

Thus,

$$\begin{aligned} y'(t) &\geq \lambda a_1(t) \int_t^\infty I_{n-2}(u, t; a_{n-1}, \cdots, a_2) \sum_{i=1}^m c_i q_i(u) G(k\alpha[g_i(u), t_2]) y[\sigma(u)]du \\ &= \lambda a_1(t) \int_t^\infty \beta_1[u,t] \sum_{i=1}^m c_i q_i(u) G(k\alpha[g_i(u), t_2]) y[\sigma(u)]du. \qquad (2.19.18) \end{aligned}$$

From condition $(2.19.5;\ell)$ and Lemma 2.19.2, it follows that $y(t)$ must be eventually negative for $t \geq t_6$, which is a contradiction.

Finally, let $\ell = 0$. Then, $\delta = 1$ and n is odd. Now, it follows from (2.19.9) that

$$(-1)^j L_j y(t) > 0 \quad \text{for} \quad t \geq t_1, \quad j = 0, 1, \cdots, n. \qquad (2.19.19)$$

Since $y'(t) < 0$ on $[t_1, \infty)$, $y(t) \to d \geq 0$ as $t \to \infty$. If $d > 0$, from (2.19.1), (C_1) and the continuity of F_i, $i = 1, 2, \cdots, m$ there exists a $t_2 \geq t_1$ such that

$$F_i(x[g_i(t)]) = F_i(y[g_i(t)] + \eta[g_i(t)]) \geq \frac{F_i(d)}{2} \quad \text{for} \quad t \geq t_2, \quad i = 1, \cdots, m.$$

By Lemma 2.19.1(ii), we get

$$\begin{aligned} y(t_2) &= \sum_{j=0}^{n-1} (-1)^j I_j(s, t_2; a_j, \cdots, a_1) L_j y(s) \\ &\quad + (-1)^n \int_{t_2}^{s} I_{n-1}(u, t_2; a_{n-1}, \cdots, a_1) L_n y(u) du. \end{aligned}$$

Using (C_1), (2.19.4), (2.19.8) and (2.19.19), we have

$$y(t_2) \geq \nu \int_{t_2}^{s} \beta_0[u, t_2] \sum_{i=1}^{m} q_i(u) du \to \infty \quad \text{as} \quad s \to \infty,$$

where $\nu = \{c_1 F_1(d)/2, \cdots, c_m F_m(d)/2\}$, which is a contradiction. Thus, $d = 0$. ∎

If condition (2.19.4) is replaced by

$$\frac{F_i(x_i)}{x_i} \geq c_i > 0 \quad \text{for} \quad x_i \neq 0, \quad i = 1, 2, \cdots, m \qquad (2.19.20)$$

then condition $(2.19.5;\ell)$ takes the form

$$\liminf_{t \to \infty} \int_t^{\sigma(t)} \gamma_\ell[s, T] \int_s^{\infty} \beta_\ell[u, s] \left(\sum_{i=1}^{m} c_i q_i(u) \right) du ds > \frac{1}{e}, \qquad (2.19.21;\ell)$$

for $\ell \in \{1, 2, \cdots, n-1\}$ and all sufficiently large T. Thus, we have the following result.

Theorem 2.19.4. Let conditions (2.19.1), (C_1) and (2.19.20) hold. A sufficient condition for the equation $(E_{102};\delta)$ to be almost oscillatory is that

(i) when $\delta = 1$ and n is even, condition $(2.19.21;\ell)$, $\ell = 1, 3, \cdots, n-1$,

(ii) when $\delta = 1$ and n is odd, conditions (2.19.21; ℓ), $\ell = 2, 4, \cdots, n-1$, (2.19.6) and (2.19.7),

(iii) when $\delta = -1$ and n is even, conditions (2.19.21; ℓ), $\ell = 2, 4, \cdots, n-2$, (2.19.6) and (2.19.7), and

$$\int^{\infty} \sum_{i=1}^{m} c_i q_i(s) \alpha_{n-1}[g_i(s), T] ds = \infty \quad \text{for all large } T, \tag{2.19.22}$$

(iv) when $\delta = -1$ and n is odd, conditions (2.19.21; ℓ), $\ell = 1, 3, \cdots, n-2$, and (2.19.22).

Proof. Let $x(t)$ be a nonoscillatory solution of the equation $(E_{102}; \delta)$. Assume $x(t) > 0$ and $x[g_i(t)] > 0$ for $t \geq t_0$, $i = 1, 2, \cdots, m$. Furthermore, we consider the function y defined as in the proof of Theorem 2.19.3 and then for $t \geq t_1$, we obtain

$$L_n y(t) = -\delta f(t, x[g_1(t)], \cdots, x[g_m(t)]), \quad t \geq t_0. \tag{2.19.23}$$

Thus, $L_n y(t)$ is of constant sign, eventually negative if $\delta = 1$ and eventually positive if $\delta = -1$. In both cases, $L_i y(t)$ are monotone and are of one sign on $[t_0, \infty)$, and as in the proof of Theorem 2.19.3, we see that $y(t) > 0$ for $t \geq t_1$ and (2.19.9) holds on $[t_1, \infty)$ with $n + \ell$ odd if $\delta = 1$ and $n + \ell$ even if $\delta = -1$. Proceeding as in the proof of Theorem 2.19.3 and letting $G(x) = 1$, we see that the cases $\ell = 0$, $\ell = 1$ and $\ell \in \{2, 3, \cdots, n-1\}$ are impossible. Thus, it remains to consider the case $\ell = n$.

Suppose $\ell = n$. Clearly, this is the case when $\delta = -1$ and n is odd or n is even. From (2.19.9), we obtain

$$L_j y(t) > 0 \quad \text{for} \quad t \geq t_1, \quad j = 0, 1, \cdots, n. \tag{2.19.24}$$

On the other hand, by L'Hospital's rule

$$\lim_{t \to \infty} \frac{y(t)}{\alpha_{n-1}[t, t_1]} = \lim_{t \to \infty} L_{n-1} y(t) > 0.$$

Since $\lim_{t \to \infty} g_i(t) = \infty$, $i = 1, 2, \cdots, m$ and $\lim_{t \to \infty} \eta(t) = 0$, there exists a constant $b > 0$ and a $t_2 \geq t_1$ such that

$$x[g_i(t)] = y[g_i(t)] + \eta[g_i(t)] \geq b\alpha_{n-1}[g_i(t), t_1] \quad \text{for } t \geq t_2, \quad i = 1, 2, \cdots, m. \tag{2.19.25}$$

Now integrating (2.19.23) from t_2 to t and using (2.19.25), we find

$$L_{n-1} y(t) = L_{n-1} y(t_2) + \int_{t_2}^{t} \sum_{i=1}^{m} b c_i q_i(s) \alpha_{n-1}[g_i(s), t_1] ds \to \infty \quad \text{as } t \to \infty.$$

Thus, $L_{n-1} x(t) \to \infty$ as $t \to \infty$ and consequently $L_j x(t) \to \infty$ as $t \to \infty$, $j = 0, 1, \cdots, n-1$. ∎

Remark 2.19.1. In the literature advanced differential equations have been treated like ordinary differential equations. However, this is not always the case. The differential equation

$$\left(\frac{1}{t}x'(t)\right)' + \frac{\sqrt{6}}{4t^3}x[c^*t] = 0, \quad t \geq 1 \qquad (E_{103})$$

is oscillatory by Theorem 2.19.3 for all $c^* > \exp(4\sqrt{6}/3e)$, while the corresponding ordinary differential equation

$$\left(\frac{1}{t}x'(t)\right)' + \frac{\sqrt{6}}{4t^3}x(t) = 0, \quad t \geq 1 \qquad (E_{104})$$

has a nonoscillatory solution $x(t) = \sqrt{t}$. Thus, the advanced argument in (E_{103}) generates oscillations.

On the other hand, when the function f in the equation $(E_{102};\delta)$ is locally of bounded variation, we see that the advanced arguments in some equations of the type $(E_{102};\delta)$ may damage the oscillatory character of the analogous ordinary differential equations. The ordinary differential equation

$$x''(t) + \frac{1}{4t}\left(1 + \frac{1}{t^4}\right)\frac{x(t)}{1 + x^2(t)} = 0, \quad t \geq 1 \qquad (E_{105})$$

is oscillatory by Theorem 6 in [210], while the advanced equation

$$x''(t) + \frac{1}{4t}\left(1 + \frac{1}{t^4}\right)\frac{x[t^5]}{1 + x^2[t^5]} = 0, \quad t \geq 1 \qquad (E_{106})$$

has a nonoscillatory solution $x(t) = \sqrt{t}$. That is, the advanced argument in equation (E_{106}) disrupts oscillations.

Theorems 2.19.3 and 2.19.4 when specialized to the equation

$$x^{(n)}(t) + \delta f(t, x[g_1(t)], \cdots, x[g_m(t)]) = 0, \qquad (E_{102};\delta)^*$$

where $\delta = \pm 1$, f and g_i, $i = 1, 2, \cdots, m$ are as in equation $(E_{102};\delta)$ can be stated as the following corollaries.

Corollary 2.19.5. Let conditions $(2.19.5;\ell)$, $(2.19.6)$ and $(2.19.7)$ of Theorem 2.19.3 be replaced by

$$\liminf_{t\to\infty} \int_t^{\sigma(t)} s^{\ell-1} \int_s^{\infty} (u-s)^{n-\ell-1} \sum_{i=1}^{m} c_i q_i(u) G\left(kg^{n-1}(u)\right) du ds$$
$$> \frac{(\ell-1)!\,(n-\ell-1)!}{e} \qquad (2.19.5;\ell)^*$$

for every $|k| \geq 1$ and $\ell \in \{1, 2, \cdots, n-1\}$,

$$\eta^{(i)}(t) \to 0 \quad \text{as} \quad t \to \infty, \quad i = 1, 2, \cdots, n-1 \tag{2.19.6)*}$$

and

$$\int^{\infty} u^{n-1} \sum_{i=1}^{m} q_i(u) du = \infty, \tag{2.19.7)*}$$

respectively. Then, the conclusion of Theorem 2.19.3 holds for the equation $(E_{102}; \delta)^*$.

Corollary 2.19.6. If conditions (2.19.21; ℓ) and (2.19.22) of Theorem 2.19.4 are replaced by

$$\liminf_{t\to\infty} \int_t^{\sigma(t)} s^{\ell-1} \int_s^{\infty} (u-s)^{n-\ell-1} \sum_{i=1}^{m} c_i q_i(u) du ds > \frac{(\ell-1)!\,(n-\ell-1)!}{e} \tag{2.19.21; ℓ)*}$$

for every $\ell \in \{1, 2, \cdots, n-1\}$, and

$$\int^{\infty} \sum_{i=1}^{m} c_i q_i(u)(g_i(u))^{n-1} du = \infty, \tag{2.19.22)*}$$

respectively, and also conditions (2.19.6) and (2.19.7) are replaced respectively by (2.19.6)* and (2.19.7)*, then the conclusion of Theorem 2.19.4 holds for the equation $(E_{102}; \delta)^*$.

Proof. It suffices to note that for $(E_{102}; \delta)^*$

$$\alpha_i[t,s] = \frac{(t-s)^i}{i!}, \quad \alpha[t,T] = \frac{(t-T)^{n-1}}{(n-1)!}, \quad \beta_i[t,s] = \frac{(t-s)^{n-i-1}}{(n-i-1)!}$$

and

$$\gamma_i[t,s] = \frac{(t-s)^{i-1}}{(i-1)!}, \quad i = 1, 2, \cdots, n-1. \quad \blacksquare$$

Example 2.19.1. For the differential equation

$$x''(t) + t^{9/4}\left(\frac{x[t^2]}{1 + x^2[t^2]}\right) = \left(\frac{\sin t}{t}\right)'', \quad t \geq 1$$

all the conditions of Corollary 2.19.5 are satisfied and hence it is oscillatory.

Theorem 2.19.7. Let $\gamma > 0$, conditions (2.19.1) and (C_1) hold, and let (G_i, H_i) be a pair of continuous components of F_i, $i = 1, 2, \cdots, m$ and H_i, $i = 1, 2, \cdots, m$ satisfy condition (2.19.4). In addition, suppose that

$$\liminf_{t\to\infty} \int_t^{\tau(t)} a_\ell(s) \int_s^{\infty} \frac{\beta_\ell[u,s]}{R_\ell[\tau(u),T]} \left(\sum_{i=1}^{m} c_i q_i(u) G_i(k\alpha[g_i(u),T]) \rho_\ell[g_i(u),T]\right) du ds$$

$$> \frac{1}{e}, \tag{2.19.26; ℓ)}$$

for every $\ell \in \{1, 2, \cdots, n-1\}$ with $n + \ell$ odd, every $|k| \geq 1$ and all large T. Then, the equation $(E_{102}; 1)$ is oscillatory provided n is even.

Moreover, if conditions (2.19.6) and (2.19.7) hold, and n is odd, then every solution $x(t)$ of the equation $(E_{102}; 1)$ is either oscillatory, or $L_i x(t) \to 0$ as $t \to \infty$, $i = 0, 1, \cdots, n-1$.

Proof. Let $x(t)$ be a nonoscillatory solution of the equation $(E_{102}; 1)$, say $x(t) > 0$ for $t \geq t_0 \geq \gamma$. Furthermore, we consider the function $y(t)$ defined as in the proof of Theorem 2.19.3 and then for $t \geq t_6$, we obtain (2.19.9), (2.19.12) and (2.19.14).

Now, suppose $\ell \in \{2, 3, \cdots, n-1\}$. Then, from Lemma 2.19.1(ii), (2.19.9) and (C_2), we obtain

$$L_\ell y(t) \geq \lambda \int_t^\infty \beta_\ell[u, t] \sum_{i=1}^m c_i q_i(u) G_i(k\alpha[g_i(u), t_2]) y[g_i(u)] du, \quad t \geq t_6. \tag{2.19.27}$$

Again, from Lemma 2.19.1(i), we get

$$\begin{aligned} y(t) &= a_1(t) \sum_{j=1}^{\ell-2} I_j(t, t_6; a_2, \cdots, a_j) L_j y(t_6) \\ &\quad + a_1(t) \int_{t_6}^t I_{\ell-2}(t, u; a_1, \cdots, a_{\ell-2}) a_{\ell-1}(u) L_{\ell-1} y(u) du, \quad t \geq t_6. \end{aligned}$$

From (2.19.9), we find

$$y[g_i(t)] \geq \int_{t_6}^{g_i(t)} \alpha_{\ell-2}[g_i(t), u] a_{\ell-1}(u) L_{\ell-1} y(u) du, \quad i = 1, 2, \cdots, m. \tag{2.19.28}$$

Using the fact that $L_\ell y(t)$ is nonincreasing for $t \geq t_1$, we obtain

$$L_{\ell-1} y(t) \geq R_\ell[t, t_1] L_\ell y(t) \quad \text{for} \quad t \geq t_1$$

and the function

$$\frac{L_{\ell-1} y(t)}{R_\ell[t, t_1]} \quad \text{is nonincreasing on} \quad [t_1, \infty). \tag{2.19.29}$$

Thus, the inequality (2.19.27) takes the form

$$\begin{aligned} y[g_i(t)] &\geq \frac{L_{\ell-1} y[\tau(t)]}{R_\ell[\tau(t), t_1]} \int_{t_6}^{g_i(t)} \alpha_{\ell-2}[g_i(t), s] a_{\ell-1}(s) R_\ell[s, t_1] ds \\ &= \rho_\ell[g_i(t), t_6] \frac{R_{\ell-1} y[\tau(t)]}{R_\ell[\tau(t), t_1]} \quad \text{for } t \geq t_6, \ i = 1, \cdots, m. \end{aligned} \tag{2.19.30}$$

Combining (2.19.27) and (2.19.30), we get

$$(L_{\ell-1}x(t))' \geq a_\ell(t)\int_t^\infty \frac{\beta_\ell[u,t]}{R_\ell[\tau(u),t_1]}\left(\sum_{i=1}^m \lambda c_i q_i(u) G_i(k\alpha[g_i(u),t_2])\right.$$

$$\left.\times\ \rho_\ell[g_i(u),t_6]\right) L_{\ell-1}y[\tau(u)]du.$$

From (2.19.6; ℓ), $\ell = 2,3,\cdots,n-1$ and Lemma 2.19.2, it follows that $L_{\ell-1}x(t)$ is eventually negative for $t \geq t_6$, which is a contradiction.

Next, suppose $\ell = 1$. This is the case when n is even, and from Lemma 2.19.1(ii) and (2.19.29), we obtain

$$x'(t) \geq a_1(t)\int_t^\infty \frac{\beta_1[u,t]}{R_1[\tau(t),t_1]}\sum_{i=1}^m \lambda c_i q_i(u) G_i(k\alpha[g_i(u),t_2])$$

$$\times\ \rho_1[g_i(u),t_6]x[\tau(u)]du, \quad t \geq t_6.$$

From condition (2.19.26; 1), we obtain a contradiction to the fact that $x(t) > 0$ for $t \geq t_0$.

A similar proof as in Theorem 2.19.3 covers the case $\ell = 0$. ∎

When condiiton (2.19.4) is replaced by condiiton (2.19.20), then condiiton (2.19.26; ℓ) will take the form

$$\liminf_{t\to\infty}\int_t^{\tau(t)} a_\ell(s)\int_s^\infty \frac{\beta_\ell[u,s]}{R_\ell[\tau(u),T]}\left(\sum_{i=1}^m c_i q_i(u)\rho_\ell[g_i(u),T]\right) du ds > \frac{1}{e}, \tag{2.19.31; ℓ}$$

for every $\ell \in \{1,2,\cdots,n-1\}$ and all large T. Thus, we obtain the following criterion.

Theorem 2.19.8. Let conditions (2.19.1), (2.19.20) and (C_1) hold. The equation (E_{102}; δ) is almost oscillatory if

(i) for $\delta = 1$ and n even, condition (2.19.31; ℓ), $\ell = 1,3,\cdots,n-1$ holds,

(ii) for $\delta = 1$ and n odd, conditions (2.19.31; ℓ), $\ell = 2,4,\cdots,n-1$, (2.19.6) and (2.19.7) hold,

(iii) for $\delta = -1$ and n even, conditions (2.19.31; ℓ), $\ell = 2,4,\cdots,n-2$, (2.19.6), (2.19.7) and (2.19.22) hold,

(iv) for $\delta = -1$ and n odd, conditions (2.19.31; ℓ), $\ell = 1,3,\cdots,n-2$ and (2.19.22) hold.

Proof. The proof is similar to that of Theorems 2.19.4 and 2.19.7. ∎

If $a_i(t) = 1$, $i = 1, 2, \cdots, n-1$, $\rho_i[g_i(t), T] = (g_i(t) - T)^i/i!$, $i = 1, 2, \cdots, n-1$, Theorems 2.19.7 and 2.19.8 when specialized to the equation $(E_{102}; \delta)^*$ take the following form:

Corollary 2.19.9. Let condition $(2.19.26; \ell)$ be replaced by

$$\liminf_{t\to\infty} \int_t^{\tau(t)} \int_s^{\infty} \frac{(u-s)^{n-\ell-1}}{\tau(u)} \sum_{i=1}^{m} c_i q_i(u) g_i^{\ell}(u) G\left(k g_i^{n-1}(u)\right) du ds$$

$$> \frac{(n-\ell-1)!\,\ell!}{e}, \qquad (2.19.26; \ell)^*$$

for every $|k| \geq 1$ and $\ell \in \{1, 2, \cdots, n-1\}$, and conditions (2.19.6) and (2.19.7) are replaced by $(2.19.6)^*$ and $(2.19.7)^*$ respectively. Then, the conclusion of Theorem 2.19.7 holds for the equation $(E_{102}; \delta)^*$.

Corollary 2.19.10. Let condition $(2.19.31; \ell)$ be replaced by

$$\liminf_{t\to\infty} \int_t^{\tau(t)} \int_s^{\infty} \frac{(u-s)^{n-\ell-1}}{\tau(u)} \sum_{i=1}^{m} c_i q_i(u) g_i^{\ell}(u) du ds > \frac{(n-\ell-1)!\,\ell!}{e} \qquad (2.19.31; \ell)^*$$

for every $\ell \in \{1, 2, \cdots, n-1\}$, and conditions (2.19.6), (2.19.7) and (2.19.22) are replaced respectively by $(2.19.6)^*$, $(2.19.7)^*$ and $(2.19.22)^*$, then the conclusion of Theorem 2.19.8 holds for the equation $(E_{102}; \delta)^*$.

Remark 2.19.2. We note that conditions $(2.19.5; \ell)$ and $(2.19.26; \ell)$ are independent. Also, conditions $(2.19.21; \ell)$ and $(2.19.31; \ell)$ are independent. To illustrate these cases we consider the following example.

Example 2.19.2. For $t \geq 1$ consider the differential equation

$$\left(\frac{1}{t}x'(t)\right)' + \frac{e}{(e+1)t^3}x[et] + \frac{1}{(e+1)t^3}x[e^2t] = \frac{2}{t^4}(\sin\ln t - 2\cos\ln t). \qquad (E_{107})$$

Since

$$a_1(t) = t, \quad q_1(t) = \frac{e}{(e+1)t^3}, \quad q_2(t) = \frac{1}{(e+1)t^3},$$

$$g_1(t) = et, \quad g_2(t) = e^2t \quad \text{and} \quad Q(t) = \frac{2}{t^4}(\sin\ln t - 2\cos\ln t),$$

we obtain

$$\sigma(t) = et, \quad \tau(t) = e^2t \quad \text{and} \quad \eta(t) = \frac{\sin\ln t}{t}.$$

It is easy to check that condition $(2.19.5; 1)$ (respectively condition $(2.19.21; 1)$) is satisfied, while condition $(2.19.26; 1)$ (respectively condition

(2.19.31; 1)) is violated. Thus, we conclude that equation (E_{107}) is oscillatory by Theorem 2.19.3 (or Theorem 2.19.4).

Next, for $t \geq 1$ we consider the differential equation

$$\left(\frac{1}{t}x'(t)\right)' + \frac{1}{3et^2}\left[x[t+1] + x[t+2] + x[t+3]\right] = \frac{2}{t^4}(\sin \ln t - 2\cos \ln t). \qquad (E_{108})$$

Here, we take

$$a_1(t) = t, \quad q_i(t) = \frac{1}{3et^2}, \quad i = 1,2,3, \quad g_1(t) = t+1, \quad g_2(t) = t+2$$

$$g_3(t) = t+3 \quad \text{and} \quad Q(t) = \frac{2}{t^4}(\sin \ln t - 2\cos \ln t),$$

and hence we have

$$\sigma(t) = t+1, \quad \tau(t) = t+3 \quad \text{and} \quad \eta(t) = \frac{\sin \ln t}{t}.$$

One can easily see that condition (2.19.26; 1) (respectively condition (2.19.31; 1)) is satisfied while condition (2.19.5; 1) (respectively condition (2.19.21; 1)) fails to apply. Therefore, we conclude that equation (E_{108}) is oscillatory by Theorem 2.19.7 (or Theorem 2.19.8).

Theorem 2.19.11. Let $\gamma > 0$, conditions (2.19.1) and (C_2) hold, and let (G_i, H_i) be a pair of continuous components of F_i, $i = 1, 2, \cdots, m$. Suppose that there exist nonnegative constants λ_i, $i = 1, 2, \cdots, m$ such that

$$H_i(|x_i|) \geq |x_i|^{\lambda_i} \quad \text{for} \quad x_1 x_i > 0, \quad i = 1, 2, \cdots, m \qquad (2.19.32)$$

and $\lambda_1 + \lambda_2 + \cdots + \lambda_m = 1$. Furthermore, assume that

$$\sum_{i=1}^{m} \lambda_i \liminf_{t\to\infty} \int_t^{g_i(t)} \gamma_\ell[s,T] \int_s^\infty \beta_\ell[u,s] \prod_{i=1}^{m} G_i(k\alpha[g_i(u),T])duds > \frac{1}{e}, \qquad (2.19.33;\ell)$$

for every $\ell \in \{1, 2, \cdots, n-1\}$ with $n + \ell$ odd, and every $|k| \geq 1$ and all large T. Then, for n even, the equation (E_{102}; 1) is oscillatory. If in addition, condition (2.19.6) holds, and

$$\int^\infty \beta_0[u,T]q(u)du = \infty \quad \text{for all large } T, \qquad (2.19.34)$$

then for n odd, every solution $x(t)$ of the equation (E_{102}; 1) is either oscillatory, or $L_j x(t) \to 0$ as $t \to \infty$, $i = 0, 1, \cdots, n-1$.

Proof. Let $x(t)$ be a nonoscillatory solution of the equation (E_{102}; 1), say $x(t) > 0$ for $t \geq t_0 \geq \gamma$. Considering the function $y(t) = x(t) - \eta(t)$

and proceeding as in Theorem 2.19.3, we obtain (2.19.9), (2.19.11) and (2.19.12). By Lemma 2.2.8 and conditions (2.19.32) and (C_2), for $t \geq t_5$ we have

$$\begin{aligned} f(t, x[g_1(t)], \cdots, x[g_m(t)]) &\geq q(t) \prod_{i=1}^{m} F_i(x[g_i(t)]) \\ &\geq q(t) \prod_{i=1}^{m} G_i(x[g_i(t)]) \prod_{i=1}^{m} (x[g_i(t)])^{\lambda_i} \\ &\geq \lambda q(t) \prod_{i=1}^{m} G_i(k\alpha[g_i(t), t_2]) \prod_{i=1}^{m} (y[g_i(t)])^{\lambda_i}. \end{aligned}$$

Thus, the equation (2.19.8) takes the form

$$L_n y(t) + \lambda q(t) \prod_{i=1}^{m} G_i(k\alpha[g_i(t), t_2]) \prod_{i=1}^{m} (y[g_i(t)])^{\lambda_i} \leq 0 \quad \text{for} \quad t \geq t_5. \tag{2.19.35}$$

Suppose $\ell \in \{2, 3, \cdots, n-1\}$. Then, from Lemma 2.19.1(ii) and (2.19.35), we obtain

$$L_\ell y(t) \geq \lambda \int_t^\infty \beta_\ell[u, t] q(u) \prod_{i=1}^{m} g_i(k\alpha[g_i(u), t_2]) \prod_{i=1}^{m} (y[g_i(u)])^{\lambda_i} du, \quad t \geq t_5. \tag{2.19.36}$$

Next, by Lemma 2.19.1(i) and the fact that $L_\ell y(t)$ is nonincreasing on $[t_1, \infty)$, we have

$$y'(t) \geq \gamma_\ell[t, t_5] L_\ell y(t) \quad \text{for} \quad t \geq t_5. \tag{2.19.37}$$

Combining (2.19.36) and (2.19.37), for $t \geq t_5$ we get

$$y'(t) \geq \lambda \gamma_\ell[t, t_5] \int_t^\infty \beta_\ell[u, t] q(u) \prod_{i=1}^{m} G_i(k\alpha[g_i(u), t_2]) \prod_{i=1}^{m} (y[g_i(u)])^{\lambda_i} du. \tag{2.19.38}$$

Inequality (2.19.38), in view of condition (2.19.33; ℓ) and Lemma 2.19.2 has eventually negative solutions, which is a contradiction.

Now, assume $\ell = 1$. Then, n must be even and from Lemma 2.19.1(ii) and (2.19.35), for $t \geq t_5$ we have

$$y'(t) \geq \lambda a_1(t) \int_t^\infty \beta_1[u, t] q(u) \prod_{i=1}^{m} G_i(k\alpha[g_i(u), t_2]) \prod_{i=1}^{m} (y[g_i(u)])^{\lambda_i} du. \tag{2.19.39}$$

By Lemma 2.19.2 and condition (2.19.33; ℓ) inequality (2.19.39) has no eventually positive solutions, which is a contradiction.

The proof of the case $\ell = 0$ is similar to that of as in Theorem 2.19.3 and hence omitted. ∎

We can replace condition (2.19.32) with

$$\prod_{i=1}^{m} F_i(|x_i|) \geq \prod_{i=1}^{m} |x_i|^{\lambda_i} \quad \text{for} \quad x_1 x_i > 0, \quad i = 1, 2, \cdots, m \qquad (2.19.40)$$

and $\lambda_1 + \lambda_2 + \cdots + \lambda_m = 1$.

But, then condition (2.19.33; ℓ) takes the from

$$\sum_{i=1}^{m} \lambda_i \liminf_{t\to\infty} \int_t^{g_i(t)} \gamma_\ell[s,T] \int_s^\infty \beta_\ell[u,s] q(u) du ds > \frac{1}{e} \qquad (2.19.41;\ell)$$

for $\ell \in \{1, 2, \cdots, n-1\}$ and all sufficiently large T.

Now we have the following result.

Theorem 2.19.12. Let conditions (2.19.1), (C_2) and (2.19.40) hold. Then, $(E_{102}; \delta)$ is almost oscillatory if

(i) when $\delta = 1$ and n is even, condition (2.19.41; ℓ), $\ell = 1, 3, \cdots, n-1$ holds,

(ii) when $\delta = 1$ and n is odd, conditions (2.19.41; ℓ), $\ell = 2, 4, \cdots, n-1$, (2.19.6) and (2.19.34) hold,

(iii) when $\delta = -1$ and n is even, conditions (2.19.41; ℓ), $\ell = 2, 4, \cdots, n-2$, (2.19.6), (2.19.34), and

$$\int^\infty q(u) \prod_{i=1}^{m} (\alpha_{n-1}[g_i(u), T])^{\lambda_i} du = \infty \quad \text{for all large} \quad T \qquad (2.19.42)$$

hold,

(iv) when $\delta = -1$ and n is odd, conditions (2.19.41; ℓ), $\ell = 1, 3, \cdots, n-2$, and (2.19.42) hold.

Proof. The proof is clear from Theorems 2.19.11 and 2.19.4. ∎

The following criteria are concerned with the oscillatory and asymptotic behavior of the equation $(E_{102}; -1)$.

In what follows we let $\xi_i : \mathbb{R}_0 \to \mathbb{R}_0$, $i = 1, 2, \cdots, m$ be continuous functions such that

$$t \leq \xi_i(t) \leq g_i(t), \quad i = 1, 2, \cdots, m, \quad t \geq 0.$$

We also let

$$\xi(t) = \min\{\xi_1(t), \xi_2(t), \cdots, \xi_m(t)\}, \quad t \geq 0.$$

Theorem 2.19.13. Let $n \geq 3$, and conditions (2.19.1), (C_1) and (2.19.20) hold. Suppose that for every $\ell \in \{1, 2, \cdots, n-2\}$ with $n+\ell$ even and for all large T, condition (2.19.21; ℓ) (or (2.19.31; ℓ)) holds, and

$$\liminf_{t\to\infty} \int_t^{\xi(t)} \sum_{i=1}^m c_i q_i(s)\alpha_{n-1}[g_i(s), \xi_i(s)]ds > \frac{1}{e}. \qquad (2.19.43)$$

Then, for n odd, equation $(E_{102}; -1)$ is oscillatory. If in addition, conditions (2.19.6) and (2.19.7) hold, then for n even, every solution $x(t)$ of the equation $(E_{102}; -1)$ is either oscillatory or $L_j x(t) \to 0$ as $t \to \infty$, $j = 0, 1, \cdots, n-1$.

Theorem 2.19.14. Suppose that $n \geq 3$, conditions (2.19.1), (C_2) and (2.19.40) hold. Furthermore, assume that condition (2.19.41; ℓ) holds for $\ell = 1, 2, \cdots, n-2$ with $n+\ell$ even and all large T, and

$$\sum_{j=1}^m \lambda_j \liminf_{t\to\infty} \int_t^{\xi(t)} q(s) \prod_{i=1}^m (\alpha_{n-1}[g_i(s), \xi(s)])^{\lambda_i} ds > \frac{1}{e}. \qquad (2.19.44)$$

Then, every solution of the equation $(E_{102}; -1)$ is oscillatory provided n is odd. Moreover, let conditions (2.19.6) and (2.19.34) hold. If n is even, then every solution $x(t)$ of the equation $(E_{102}; -1)$ is either oscillatory or $L_i x(t) \to 0$ as $t \to \infty$, $i = 0, 1, \cdots, n-1$.

Proofs of Theorems 2.19.13 and 2.19.14. Let $x(t)$ be a nonoscillatory solution of the equation $(E_{102}; -1)$. Assume $x(t) > 0$ for $t \geq t_0$. Consider the function $y(t) = x(t) - \eta(t)$ and proceed as in the proof of Theorem 2.19.3, we see that $L_i y(t)$, $i = 0, 1, \cdots, n$ are of fixed signs, $y(t) > 0$ for $t \geq t_1$ and (2.19.9) holds with $n+\ell$ even. Also, for the case $\ell \geq 1$, the inequality (2.19.11) holds. Since the proofs of the cases $\ell = 0$, $\ell = 1$, and $\ell \in \{2, 3, \cdots, n-1\}$ are similar to those of earlier theorems we delete them, and consider only the case $\ell = n$. From Lemma 2.19.1(ii), we have

$$y(t) = \sum_{j=0}^{n-1} I_j(t, s; a_1, \cdots, a_j) L_j y(t) + \int_s^t I_{n-1}(t, u; a_1, \cdots, a_{n-1}) L_n y(u) du.$$

Using (2.19.9) when $\ell = n$, we get

$$y(t) \geq \alpha_{n-1}[t, s] L_{n-1} y(s) \quad \text{for} \quad t \geq s \geq t_2,$$

which gives for $i = 1, 2, \cdots, m$

$$y[g_i(t)] \geq \alpha_{n-1}[g_i(t), \xi_i(t)] L_{n-1} y[\xi_i(t)] \quad \text{for} \quad t \geq t_2.$$

Using (2.19.11), we obtain

$$x[g_i(t)] \geq \lambda \alpha_{n-1}[g_i(t), \xi_i(t)] L_{n-1} y[\xi_i(t)] \quad \text{for} \quad t \geq t_2. \qquad (2.19.45)$$

Therefore, from the equation $(E_{102}; -1)$, the substitution $y(t) = x(t) - \eta(t)$, and conditions (C_1), (C_2), (2.19.20) and (2.19.40), and the inequality (2.19.45), we get

$$\begin{aligned} L_n y(t) &= f(t, x[g_1(t)], \cdots, x[g_m(t)]) \\ &\geq \lambda \sum_{i=1}^{m} c_i q_i(t) \alpha_{n-1}[g_i(t), \xi_i(t)] L_{n-1} y[\xi(t)] \end{aligned}$$

and

$$L_n y(t) \geq \lambda q(t) \prod_{i=1}^{m} (\alpha_{n-1}[g_i(t), \xi_i(t)] L_{n-1} y[\xi_i(t)])^{\lambda_i}.$$

From the above inequality, in view of conditions (2.19.43) and (2.19.44) and Theorems 3 and 4 in [300], it follows that $L_{n-1}y(t)$ is eventually of negative sign on $[t_2, \infty)$, which is a contradiction. ∎

To illustrate how the above results can be used in practice we consider the following examples.

Example 2.19.3. Consider the differential equation

$$L_n x(t) + t^{4-2n} \frac{x[t^2]}{1 + |x[t^2]|^{1/(2n-2)}} = 0, \quad t \geq 1 \qquad (E_{109})$$

where

$$L_0 x(t) = x(t), \quad L_k x(t) = \frac{1}{t}(L_{k-1} x(t)), \quad k = 1, 2, \cdots, n.$$

Since,

$$a_i(t) = t, \quad i = 1, 2, \cdots, n-1, \quad g(t) = \sigma(t) = t^2, \quad F(x) = x$$

and

$$G(x) = \frac{1}{1 + |x[t^2]|^{1/(2n-2)}}$$

all the conditions of Theorem 2.19.4 (respectively Theorem 2.19.12) are satisfied and hence the equation (E_{109}) is almost oscillatory.

Example 2.19.4. Consider the differential equation

$$L_n x(t) - t^{2-2n} x[2t] = 0, \quad t \geq 1 \qquad (E_{110})$$

where L_n is as in Example 2.19.3. If we take $\xi(t) = (3/2)t$, then all hypotheses of Theorems 2.19.13 and 2.19.14 are satisfied. Hence we conclude that

(i) for $n \geq 3$ odd, equation (E_{110}) is oscillatory,

(ii) for $n \geq 3$ even, every solution of the equation (E_{110}) is either oscillatory or $L_j x(t) \to 0$ as $t \to \infty$, $i = 0, 1, \cdots, n-1$.

2.20. Systems of Higher Order Functional Differential Equations

Consider the systems of delay differential equations

$$(-1)^{m+1} y_i^{(m)}(t) + \sum_{j=1}^{n} q_{ij}(t) y_j[g_{jj}(t)] \;=\; 0, \quad i, j = 1, \cdots, n \qquad (E_{111})$$

and

$$y_i^{(m)}(t) + \sum_{j=1}^{n} q_{ij}(t) y_j[\tau(t)] \;=\; 0, \quad i, j = 1, \cdots, n \qquad (E_{112})$$

where τ, g_{jj}, $q_{ij} \in C[\mathbb{R}_0, \mathbb{R}]$, $\tau(t) \leq t$, $\lim_{t\to\infty} \tau(t) = \infty$ and for every $j = 1, \cdots, n$, g_{jj} is increasing, $g_{jj}(t) \leq t$ and $\lim_{t\to\infty} g_{jj}(t) = \infty$.

We say that a solution $y(t) = (y_1(t), \cdots, y_n(t))^T$ of (E_{111}) (or (E_{112})) *oscillates* if it is eventually trivial, or if at least one component does not have eventually constant sign. Otherwise the solution is called *nonoscillatory*.

Here we shall follow Agarwal and Grace [7] to provide sufficient conditions for the oscillation of solutions of the systems (E_{111}) and (E_{112}) for $m \geq 1$. We shall also investigate the oscillatory behavior of the solutions of the neutral differential system

$$(-1)^{m+1}(y_i(t) + c y_i[\sigma(t)])^{(m)} + \sum_{j=1}^{n} q_{ij}(t) y_j[\tau(t)] \;=\; 0, \quad i = 1, \cdots, n \quad (E_{113})$$

where c is a real constant, $\sigma \in C[\mathbb{R}_0, \mathbb{R}]$ and $\lim_{t\to\infty} \sigma(t) = \infty$.

Our first result about the oscillation of all bounded solutions of (E_{111}) is embodied in the following:

Theorem 2.20.1. Assume that the following conditions hold:

(i) there exist a (componentwise) positive vector $u = (u_1, \cdots, u_n)^T$ and a function $q \in C[\mathbb{R}_0, \mathbb{R}]$ such that for all $j = 1, \cdots, n$

$$-u_j q_{jj}(t) + \sum_{i=1, i\neq j}^{n} u_i |q_{ij}(t)| \;\leq\; -q(t) u_j, \quad t \geq 0,$$

(ii) $\int^{\infty} s^{m-1}q(s)ds = \infty,$

(iii) $g(t) = \max_{1\le j\le n} g_{jj}(t)$, $g'(t) > 0$ and $\min_{1\le j\le n} g_{jj}^{-1}(t) > g^{-1}(t)$, $t \ge 0$, $j = 1,\cdots,n$,

(iv) there exists a function $Q \in C[\mathbb{R}_0, \mathbb{R}_0]$ such that

$$Q(t) = \min_{1\le j\le n} q[g_{jj}^{-1}(t)](g_{jj}^{-1}(t))', \quad t \ge 0,$$

where g^{-1} is the inverse function of g,

(v) every bounded solution of the differential equation

$$(-1)^{m+1}z^{(m)}(t) + Q[g(t)]g'(t)z[g(t)] = 0, \qquad (E_{114})$$

is oscillatory.

Then, every bounded solution of (E_{111}) is oscillatory.

Proof. Assume that (E_{111}) has a bounded and nonoscillatory solution $y(t) = (y_1(t),\cdots, y_n(t))^T$. Then, there exists a $t_0 \ge 0$ such that for all $i = 1,\cdots,n$ and for $t \ge t_0$

$$\delta_i = \text{sgn } y_i(t_0) = \text{sgn } y_i(t), \quad \delta_i y_i(t) = |y_i(t)|$$

and

$$\sum_{i=1}^{n} |y_i(t)| > 0 \quad \text{for } t \ge t_0.$$

Set $w(t) = \sum_{i=1}^{n} u_i\delta_i y_i(t)$, $t \ge t_0$. Then, $w(t) > 0$ and for all sufficiently large t,

$$\begin{aligned}
(-1)^{m+1}w^{(m)}(t) &= \sum_{i=1}^{n}(-1)^{m+1}u_i\delta_i y_i^{(m)}(t) \\
&= -\sum_{i=1}^{n} u_i\delta_i \sum_{j=1}^{n} q_{ij}(t)y_j[g_{jj}(t)] \\
&\le \sum_{j=1}^{n}[-u_j q_{jj}(t)y_j[g_{jj}(t)]\text{sgn } y_j[g_{jj}(t)]] \\
&\quad + \sum_{i=1, i\ne j}^{n} u_j|q_{ij}(t)||y_j[g_{jj}(t)]| \\
&\le (-q(t))\sum_{j=1}^{n} u_j|y_j[g_{jj}(t)]| \le 0. \qquad (2.20.1)
\end{aligned}$$

Now from the boundedness, nonoscillation and eventual positivity of $w(t)$ and by Lemma 2.2.1, we have

$$(-1)^k w^{(k)}(t) > 0 \quad \text{eventually}, \quad k = 0,1,\cdots,m. \qquad (2.20.2)$$

Thus, $w(t)$ is a decreasing function for $t \geq t_1$, $t_1(\geq t_0)$ sufficiently large. Hence, we conclude that $|y_j(t)|$ converges to zero as $t \to \infty$, $j = 1, \cdots, n$. We let

$$\lim_{t\to\infty} |y_j(t)| \;\geq\; \alpha_j \;>\; 0, \qquad j = 1, \cdots, n.$$

We claim that $\alpha_j = 0$, $j = 1, \cdots, n$. Suppose not, then there exists a $t_2 \geq t_1$ so large that

$$|y_j[g_{jj}(t)]| \;\geq\; \frac{\alpha_j}{2} \quad \text{for} \quad t \geq t_2. \tag{2.20.3}$$

From (2.20.2) and (2.20.3), we get

$$(-1)^{m+1} w^{(m)}(t) \;\leq\; -\frac{1}{2} q(t) \sum_{j=1}^{n} u_j \alpha_j, \quad t \geq t_2. \tag{2.20.4}$$

We multiply (2.20.4) by t^{m-1} and integrate (by parts) from t_2 to t, to obtain

$$(-1)^{m+1} \left[t^{m-1} w^{(m-1)}(t) - \int_{t_2}^{t} (m-1) s^{m-2} w^{(m-1)}(s) \right]$$

$$\leq C - \beta \int_{t_2}^{t} s^{m-1} q(s) ds, \tag{2.20.5}$$

where C is a constant and $\beta = -(1/2)\sum_{j=1}^{n} u_j \alpha_j$. Now, using (iii) in (2.20.5) we conclude that

$$\lim_{t\to\infty} (-1)^{m+1} \left[t^{m-1} w^{(m-1)}(t) - \int_{t_2}^{t} (m-1) s^{m-2} w^{(m-1)}(s) ds \right] \;=\; -\infty.$$

Let us define

$$u(t) \;=\; \int_{t_2}^{t} s^{m-2} w^{(m-1)}(s) ds,$$

then $u'(t) = t^{m-2} w^{(m-1)}(t)$. Hence,

$$\lim_{t\to\infty} (-1)^{m+1} \left[t u'(t) - (m-1) u(t) \right] \;=\; -\infty.$$

By Lemma 2.2.7, we have $\lim_{t\to\infty} u(t) = \pm\infty$. Therefore, since

$$w^{(m-1)}(t) \;>\; 0 \text{ if } m \text{ is odd} \quad \text{and} \quad w^{(m-1)}(t) \;<\; 0 \text{ if } m \text{ is even},$$

we have

$$\lim_{t\to\infty} \int_{t_2}^{t} s^{m-2} w^{(m-1)}(s) ds \;=\; \infty \quad \text{if } m \text{ is odd}$$

and

$$\lim_{t\to\infty}\int_{t_2}^{t} s^{m-2}w^{(m-1)}(s)ds \;=\; -\infty \quad \text{if } m \text{ is even.}$$

We continue this process of integration by parts until we obtain $\int_{t_2}^{\infty} w'(s)ds = -\infty$, which is a contradiction to the positiveness of $w(t)$.

Thus, we have $\sum_{j=1}^{n} u_j\alpha_j = 0$ and hence $\alpha_j = 0$ for $j = 1,\cdots,n$, $\lim_{t\to\infty} y_j(t) = 0$, $j = 1,\cdots,n$ and $\lim_{t\to\infty} w(t) = 0$. Next, integrating both sides of (2.20.1) m–times from t to v, letting $v \to \infty$, and using (2.20.2), to get

$$\begin{aligned}
w(t) \;&\geq\; \int_t^\infty \frac{(s-t)^{m-1}}{(m-1)!}q(s)\sum_{j=1}^{n} u_j|y_j[g_{jj}(s)]|ds \\
&=\; \sum_{j=1}^{n}\int_t^\infty \frac{(s-t)^{m-1}}{(m-1)!}q(s)u_j|y_j[g_{jj}(s)]|ds \\
&=\; \sum_{j=1}^{n}\int_{g_{jj}(t)}^\infty \frac{(g_{jj}^{-1}(\xi)-t)^{m-1}}{(m-1)!}q[g_{jj}^{-1}(\xi)](g_{jj}^{-1}(\xi))'u_j|y_j(\xi)|d\xi \\
&\geq\; \sum_{j=1}^{n}\int_{g(t)}^\infty \frac{(g_{jj}^{-1}(\xi)-t)^{m-1}}{(m-1)!}Q(\xi)u_j|y_j(\xi)|d\xi \\
&=\; \sum_{j=1}^{n}\int_t^\infty \frac{(\eta-t)^{m-1}}{(m-1)!}Q[g(\eta)]g'(\eta)u_j|y_j[g(\eta)]|d\eta \\
&=\; \int_t^\infty \frac{(\eta-t)^{m-1}}{(m-1)!}(Q[g(\eta)]g'(\eta))w[g(\eta)]d\eta, \quad t \geq t_2.
\end{aligned}$$

The function w is positive and strictly decreasing on $[t_2,\infty)$. Thus, Theorem 2.5.7 ensures the existence of a positive solution z of (E_{114}) with $\lim_{t\to\infty} z(t) = 0$, which is a contradiction to the assumptions of our theorem. ∎

The following two results deal with the oscillation of all solutions of (E_{112}).

Theorem 2.20.2. Let m be even, and condition (i) holds. If the equation

$$z^{(m)}(t) + q(t)z[\tau(t)] \;=\; 0 \qquad (E_{115})$$

is oscillatory, then (E_{112}) is oscillatory.

Proof. Let $y(t) = (y_1(t),\cdots,y_n(t))^T$ be a nonoscillatory solution of (E_{112}). As in the proof of Theorem 2.20.1, inequality (2.20.1) takes the form

$$w^{(m)}(t) \;\leq\; -q(t)w[\tau(t)] \quad \text{for all large } t$$

or

$$w^{(m)}(t) + q(t)w[\tau(t)] \ \le \ 0 \tag{2.20.6}$$

has an eventually positive solution. Hence, by Lemma 2.5.1 the equation (E_{115}) also has an eventually positive solution. This contradicts the hypothesis and the proof is complete. ∎

Theorem 2.20.3. Let m be odd, condition (i) holds, and

$$\int^{\infty} q(s)ds \ = \ \infty. \tag{2.20.7}$$

If the equation (E_{115}) is oscillatory, then (E_{112}) is oscillatory.

Proof. Let $y(t) = (y_1(t), \cdots, y_n(t))^T$ be a nonoscillatory solution of (E_{112}). As in the proof of Theorem 2.20.2, we obtain the inequality (2.20.6), which in view of Theorem 2.5.17 implies that equation (E_{115}) has an eventually positive solution. This contradiction proves the assertion of the theorem. ∎

Next, we shall consider the neutral system (E_{113}) and prove the following results:

Theorem 2.20.4. Let condition (i) hold. If either

(I) $\sigma(t) \ge t$, $t \ge t_0$, $c \in (0,1)$, and every bounded solution of the equation

$$(-1)^{m+1}v^{(m)}(t) + (1-c)q(t)v[\tau(t)] \ = \ 0 \tag{E_{116}}$$

is oscillatory,

(II) $\sigma(t)$ is strictly increasing, $\sigma(t) \le t$, $\sigma^{-1} \circ \tau(t) \le t$ for $t \ge t_0$, $c \in (1,\infty)$, and every bounded solution of the equation

$$(-1)^{m+1}V^{(m)}(t) + \left(\frac{c-1}{c^2}\right) q(t)V[\sigma^{-1} \circ \tau(t)] \ = \ 0 \tag{E_{117}}$$

is oscillatory, or

(III) $\sigma(t) \le t$, $c^* = -c \in (0,1]$, and every bounded solution of the equation

$$(-1)^{m+1}W^{(m)}(t) + q(t)W[\sigma(t)] \ = \ 0 \tag{E_{118}}$$

is oscillatory, then every bounded solution of (E_{113}) is oscillatory.

Proof. Let $y(t) = (y_1(t), \cdots, y_n(t))^T$ be a bounded and nonoscillatory solution of (E_{113}). There exists a $t_0 \ge 0$ such that for all $i = 1, 2, \cdots, n$ and $t \ge t_0$

$$\delta_i \ = \ \text{sgn}\ y_i(t_0) \ = \ \text{sgn}\ y_i(t), \quad \delta_i y_i(t) \ = \ |y_i(t)|$$

and

$$\sum_{i=1}^{n} |y_i(t)| > 0 \quad \text{for} \quad t \geq t_0.$$

Set $w(t) = \sum_{i=1}^{n} u_i \delta_i y_i(t)$, and

$$z(t) = w(t) + cw[\tau(t)] \quad \text{for} \quad t \geq t_0. \tag{2.20.8}$$

Then, $w(t) > 0$ and $z(t) > 0$ if $c \in (0,1)$ or $c \in (1,\infty)$ and as in the proof of Theorem 2.20.1, we obtain for all sufficiently large t

$$(-1)^{m+1} z^{(m)}(t) \leq -q(t)w[\tau(t)] \leq 0. \tag{2.20.9}$$

From (2.20.9), and Lemma 2.2.1, we have

$$(-1)^k z^{(k)}(t) > 0 \quad \text{for} \quad k = 0, 1, \cdots, m \quad \text{and all large } t. \tag{2.20.10}$$

That is, $z(t)$ is a decreasing function for all large t. Using this fact, we see from (2.20.8) that if $c \in (0,1)$ and $\sigma(t) \geq t$, thus

$$w(t) = z(t) - cw[\sigma(t)] = z(t) - c\left[z[\sigma(t)] - cw[\sigma \circ \sigma(t)]\right] \geq (1-c)z(t) \tag{2.20.11}$$

for all large t, and if $c \in (1,\infty)$ and $\sigma(t) \leq t$, then

$$\begin{aligned} w(t) &= (1/c)\left[z[\sigma^{-1}(t)] - w[\sigma^{-1}(t)]\right] \\ &= (1/c)z[\sigma^{-1}(t)] - (1/c^2)\left[z[\sigma^{-1} \circ \sigma^{-1}(t)] - w[\sigma^{-1} \circ \sigma^{-1}(t)]\right] \\ &\geq \left(\frac{c-1}{c^2}\right) z[\sigma^{-1}(t)], \end{aligned} \tag{2.20.12}$$

for all large t.

Finally, if $c^* = -c \in (0,1]$. From (2.20.9), we see that $z(t)$ is either eventually negative or else eventually positive. We claim that $z(t) < 0$ is impossible. Otherwise, since $w(t)$ is bounded, there is a sequence $\{t_j\}$ such that

$$\lim_{j\to\infty} t_j = \infty, \quad \lim_{j\to\infty} w(t_j) = \limsup_{t\to\infty} w(t).$$

Without loss of generality, we assume that $\{w[\tau(t_j)]\}$ is convergent. Then,

$$0 > \lim_{j\to\infty} z(t_j) \geq \limsup_{t\to\infty} w(t)(1-c^*) \geq 0,$$

which is a contradiction. Thus, $z(t) > 0$ and hence (2.20.10) holds, and

$$w(t) \geq z(t) \quad \text{for all large } t. \tag{2.20.13}$$

1. Suppose (I) holds. From (2.20.9) and (2.20.11), we obtain

$$(-1)^{m+1} z^{(m)}(t) + (1-c)q(t)z[\tau(t)] \leq 0 \quad \text{for all large } t.$$

2. Suppose (II) holds. From (2.20.9) and (2.20.12), we have

$$(-1)^{m+1}z^{(m)}(t)+\left(\frac{c-1}{c^2}\right)q(t)z[\sigma^{-1}\circ\tau(t)]\ \le\ 0\quad\text{for all large } t.$$

3. Suppose (III) holds. From (2.20.9) and (2.20.13), we get

$$(-1)^{m+1}z^{(m)}(t)+q(t)z[\tau(t)]\ \le\ 0\quad\text{for all large } t.$$

The rest of the proof is similar to that of Theorem 2.20.1. ■

Next, we shall consider the neutral system

$$(y_i(t)+cy_i[\sigma(t)])^{(m)}+\sum_{j=1}^{n}q_{ij}(t)y_j[\tau(t)]\ =\ 0,\quad m \text{ is even}\qquad (E_{119})$$

where c, σ, τ and q_{ij} are as in (E_{113}), and establish the following oscillation criterion:

Theorem 2.20.5. Let condition (i) hold. If either

(1) $\sigma(t)\le t$, $t\ge t_0$, $c\in(0,1)$, and the equation

$$\nu^{(m)}(t)+(1-c)q(t)\nu[\tau(t)]\ =\ 0\qquad (E_{120})$$

is oscillatory,

(2) $\sigma(t)$ is strictly increasing, $\sigma(t)\ge t$, $\sigma^{-1}\circ\tau(t)\le t$ for $t\ge t_0$, and the equation

$$V^{(m)}(t)+\left(\frac{c-1}{c^2}\right)q(t)V[\sigma^{-1}\circ\tau(t)]\ =\ 0\qquad (E_{121})$$

is oscillatory, or

(3) $\sigma(t)\le t$, $c^*=-c\in(0,1]$, and the equation (E_{115}) is oscillatory,

then (E_{119}) is oscillatory.

Proof. Let $y(t)=(y_1(t),\cdots,y_n(t))^T$ be a nonoscillatory solution of (E_{119}). Next, we define the functions w and z as in the proof of Theorem 2.20.4, then inequality (2.20.9) takes the form

$$z^{(m)}(t)\ \le\ -q(t)w[\tau(t)]\ \le\ 0\quad\text{for all large } t.\qquad (2.20.14)$$

Clearly the functions w and z are positive eventually for the two cases $c\in(0,1)$ and $c\in(1,\infty)$. Now, from (2.20.14) and the fact that m is even, $z(t)$ is a nondecreasing function for all large t, and hence for $c\in(0,1)$

inequality (2.20.11) holds, and for $c \in (1, \infty)$ inequality (2.20.12) holds. In view of (2.20.14) and (2.20.11), we have

$$z^{(m)}(t) + (1-c)q(t)z[\tau(t)] \leq 0 \quad \text{for all large } t \tag{2.20.15}$$

and from (2.20.14) and (2.20.12), we obtain

$$z^{(m)}(t) + \left(\frac{c-1}{c^2}\right) q(t)z[\sigma^{-1} \circ \tau(t)] \leq 0, \quad \text{for all large } t. \tag{2.20.16}$$

Finally, if $c^* = -c \in (0, 1]$. From (2.20.14), we see that $z(t)$ is either negative or positive eventually, and as in the proof of Theorem 2.20.4(III), we see that if w is bounded, then $z(t) < 0$ is impossible. Now, we claim that if $w(t)$ is unbounded, then $z(t) < 0$ is also impossible. Otherwise, if $w(t)$ is unbounded, there exists a sequence $\{t_k\}$ and $\lim_{k\to\infty} t_k = \infty$ and $w(t_k) = \max_{t \leq t_k} w(t)$. Then,

$$0 > z(t_k) = w(t_k) - c^* w[\tau(t_k)] \geq w(t_k)(1 - c^*) \geq 0,$$

which is a contradiction, and hence $z(t) > 0$ eventually and (2.20.13) holds. In view of (2.20.14) and (2.20.13), we have

$$z^{(m)}(t) + q(t)z[\tau(t)] \leq 0 \quad \text{for all large } t. \tag{2.20.17}$$

The rest of the proof is similar to that of Theorem 2.20.2. ∎

Remark 2.20.1. From Theorems 2.4.17(v) and 2.4.18(v), we see that all bounded solutions of the equation (E_{114}) are oscillatory if the following condition holds

$$\limsup_{t\to\infty} \int_{g(t)}^{t} \frac{(s - g(t))^{m-i-1}}{(m-i-1)!} \frac{(g(t) - g(s))}{i!} Q[g(s)]g'(s)ds > 1 \tag{2.20.18}$$

for some $i = 0, 1, \cdots, m-1$.

Now we obtain the following oscillation criterion for all bounded solutions of (E_{111}).

Corollary 2.20.6. Let conditions (i) – (iv) and (2.20.18) hold, then all bounded solutions of (E_{111}) are oscillatory.

Remark 2.20.2. For the special case when $q_{ij}(t) = q_{ij}$ and $g_{jj}(t) = t - \tau_{jj}$ where q_{ij} and τ_{jj}, $i, j = 1, 2, \cdots, n$ are real numbers and $\tau_{jj} > 0$. We observe that conditions (i) – (iv), and the equation (E_{114}) in Theorem 2.20.1 can be replaced, respectively, by

$$q = \min_{1 \leq i \leq n} \left[q_{ii} - \sum_{j=1, j\neq i}^{n} |q_{ij}| \right] > 0 \tag{2.20.19}$$

and

$$(-1)^{m+1}z^{(m)}(t)+qz(t-\tau) \;=\; 0, \quad \tau=\min_{1\le j\le n}\tau_{jj}. \qquad (E_{114})'$$

Therefore, one can easily formulate a result similar to that of Theorem 2.20.1 for such a special case. Also, note that we may use condition (2.20.18), or

$$q^{1/m}\frac{\tau}{m}e \;>\; 1 \qquad (2.20.20)$$

(see [198]), to investigate the oscillation of all bounded solutions of the equation $(E_{114})'$, and a similar result to Corollary 2.20.6 can be stated.

Remark 2.20.3. From the proof of Theorem 2.20.1 and the observation that if $y(t)=(y_1(t),\cdots,\,y_n(t))^T$ is a bounded nonoscillatory solution of (E_{111}), then $\lim_{t\to\infty}y_i(t)=0,\ i=1,2,\cdots,n,$ Theorems 2.20.1 and 2.20.4 can be easily extended to nonlinear systems

$$(-1)^{m+1}y_i^{(m)}(t)+\sum_{j=1}^{n}q_{ij}(t)f_j(y_j[g_{jj}(t)]) \;=\; 0 \qquad (E_{122})$$

and

$$(-1)^{m+1}(y_i(t)+cy_i[\sigma(t)])^{(m)}+\sum_{j=1}^{n}q_{ij}(t)f_j(y_j[\tau(t)]) \;=\; 0, \qquad (E_{123})$$

where $c,\ \sigma,\ \tau,\ q_{ij}$ and g_{jj} are as in (E_{111}) and (E_{113}), and each f_j satisfies

$$f_j\in C[\mathbb{R},\mathbb{R}],\ \ uf_j(u)>0 \text{ for } u\neq 0 \text{ and } \lim_{u\to\infty}\frac{f_j(u)}{u}=1,\ \ j=1,\cdots,n.$$

Here, we refer to the results in Sections 10.6 – 10.8 of [129] and omit the details.

Remark 2.20.4. It would be interesting to obtain criteria similar to those in Theorems 2.20.1 and 2.20.4 to investigate the oscillatory properties of all solutions of (E_{111}) and (E_{113}). Also, to obtain criteria similar to Theorems 2.20.2 – 2.20.5 for the systems of the form

$$y_i^{(m)}(t)+\sum_{j=1}^{n}q_{ij}(t)y_j[g_{jj}(t)] \;=\; 0 \qquad (E_{124})$$

and

$$(y_i(t)+p_i(t)y_i[\sigma_i(t)])^{(m)}+\sum_{j=1}^{n}q_{ij}(t)y_j[g_{jj}(t)] \;=\; 0, \qquad (E_{125})$$

where q_{ij} and g_{jj} are as in (E_{111}), $p_i,\ \sigma_i\in C[\mathbb{R}_0,\mathbb{R}]$ and $\lim_{t\to\infty}\sigma_i(t)=\infty,\ \ i=1,2,\cdots,n.$

References

1. **R.P. Agarwal,** Oscillation and asymptotic behavior of solutions of differential equations with nested arguments, *Bulletin UMI, Serie IV* **1–C**(1982), 137–146. (Special issue on Analisi Funzionale E Applicazioni).
2. **R.P. Agarwal,** *Difference Equations and Inequalities, Marcel Dekker*, New York, 1992.
3. **R.P. Agarwal and S.R. Grace,** The oscillation of higher–order differential equations with deviating arguments, *Computers Math. Applic.* **38**(3–4)(1999), 185–199.
4. **R.P. Agarwal and S.R. Grace,** Oscillation of certain functional differential equations, *Computers Math. Applic.* **38**(5–6)(1999), 143–153.
5. **R.P. Agarwal and S.R. Grace,** Oscillation of certain difference equations, *Mathl. Comput. Modelling* **29**(1999), 1–8.
6. **R.P. Agarwal and S.R. Grace,** The oscillation of certain difference equations, *Mathl. Comput. Modelling* **30**(1999), 53–66.
7. **R.P. Agarwal and S.R. Grace,** Oscillation of certain systems of functional differential equations, *Dynamic Systems Applic.* **8**(1999), 45–52.
8. **R.P. Agarwal and S.R. Grace,** Oscillation of higher order difference equations, *Applied Math. Letters*, to appear.
9. **R.P. Agarwal and S.R. Grace,** Oscillation of higher order nonlinear difference equations of neutral type, *Applied Math. Letters*, to appear.
10. **R.P. Agarwal and S.R. Grace,** On the oscillation of systems of difference equations, *Applied Math. Letters*, to appear.
11. **R.P. Agarwal and S.R. Grace,** Oscillation theorems for certain neutral functional differential equations, *Computers Math. Applic.*, to appear.
12. **R.P. Agarwal and S.R. Grace,** Oscillation theorems for certain difference equations, *Dynamic Systems Applic.*, to appear.
13. **R.P. Agarwal and S.R. Grace,** Forced oscillation of nth order nonlinear differential equations, to appear.
14. **R.P. Agarwal and S.R. Grace,** Oscillation of certain third order difference equations, to appear.
15. **R.P. Agarwal and S.R. Grace,** Oscillations of forced functional differential equations generated by advanced arguments, to appear.
16. **R.P. Agarwal and S.R. Grace,** On the oscillation of perturbed functional differential equations, to appear.
17. **R.P. Agarwal and S.R. Grace,** On the oscillation of certain second order differential equations, to appear.
18. **R.P. Agarwal, W.–C. Lian and C.–C. Yeh,** Levin's comparison theorems for nonlinear second order differential equations, *Applied Math. Letters* **9**(6)(1996), 29–35.

19. **R.P. Agarwal and J. Popenda,** On the oscillation of recurrence equations, *Nonlinear Analysis* **36**(1999), 231–268.
20. **R.P. Agarwal, S.–H. Shieh and C.–C. Yeh,** Oscillation criteria for second–order retarded differential equations, *Mathl. Comput. Modelling* **26**(4)(1997), 1–11.
21. **R.P. Agarwal, M.M. Susai Manuel and E. Thandapani,** Oscillatory and nonoscillatory behavior of second order neutral delay difference equations, *Mathl. Comput. Modelling* **24**(1)(1996), 5–11.
22. **R.P. Agarwal, M.M. Susai Manuel and E. Thandapani,** Oscillatory and nonoscillatory behavior of second order neutral delay difference equations II, *Appl. Math. Letters* **10**(2)(1997), 103–109.
23. **R.P. Agarwal and E. Thandapani,** Asymptotic behavior and oscillation of solutions of differential equations with deviating arguments, *Bulletin UMI,* **17–B**(1980), 82–93.
24. **R.P. Agarwal, E. Thandapani and P.J.Y. Wong,** Oscillation of higher order neutral difference equations, *Appl. Math. Letters* **10**(1)(1997), 71–78.
25. **R.P. Agarwal and P.J.Y. Wong,** *Advanced Topics in Difference Equations*, *Kluwer*, Dordrecht, 1997.
26. **R.P. Agarwal and P.J.Y. Wong,** On the oscillation of second order nonlinear difference equations, *Mathl. Ineq. Applic.* **1**(1998), 349–365.
27. **R.P. Agarwal and Y. Zhou,** Oscillation of partial difference equations with continuous variables, *Mathl. Comput. Modelling*, to appear.
28. **C.D. Ahlbrandt and A.C. Peterson,** *Discrete Hamiltonian Systems: Difference Equations, Continued Fractions and Riccati Equations*, Kluwer, Dordrecht, 1996.
29. **G.V. Ananeva and B.I. Balaganskii,** Oscillation of the solutions of certain differential equations of high order, *Uspehi Mat. Nauk.* **14**(1959), 135–140.
30. **F.V. Atkinson,** On second order nonlinear oscillations, *Pacific J. Math.* **5**(1955), 643–647.
31. **D. Bainov and D.P. Mishev**, *Oscillation Theory for Neutral Differential Equations with Delay*, *Adam Hilger*, New York, 1991.
32. **D. Bainov and V. Petrov,** On some conjectures on the nonoscillatory solutions of neutral differential equations, *J. Math. Anal. Appl.* **191**(1995), 168–179.
33. **Š. Belohorec,** On some properties of the equation $y''(x) + f(x)y^{\alpha}(x) = 0,\ 0 < \alpha < 1$, *Mat. Fyz. Časopis Sloven Akad. Veid.* **17**(1967), 10–19.
34. **N.P. Bhatia,** Some oscillation theorems for second order differential equations, *J. Math. Anal. Appl.* **15**(1966), 442–446.
35. **I. Bihari,** Oscillation and monotonity theorems concerning nonlinear differential equations of the second order, *Acta Math. Sci. Hungarica* **9**(1958), 83–104.
36. **L.E. Bobisud,** Oscillation of nonlinear differential equations with small nonlinear damping, *SIAM J. Appl. Math.* **18**(1970), 74–76.

37. **G.A. Bogar,** Oscillation of nth order differential equations with retarded argument, *SIAM J. Math. Anal.* **5**(1974), 473–481.
38. **J.S. Bradley,** Oscillation theorems for a second order delay equation, *J. Differential Equations* **8**(1970), 394–403.
39. **F. Burkowski,** Nonlinear oscillation of a second order sublinear functional differential equation, *SIAM J. Appl. Math.* **21**(1971), 486–490.
40. **T.A. Burton and R. Grimmer,** Oscillation, continuation, and uniqueness of solutions of retarded differential equations, *Trans. Amer. Math. Soc.* **179**(1973), 193–209.
41. **G.J. Butler,** The oscillatory behavior of a second order nonlinear differential equations with damping, *J. Math. Anal. Appl.* **57**(1977), 273–289.
42. **T.A. Chanturia,** Some comparison theorems for higher order ordinary differential equations, *Bull. Acad. Polon. Sci. Ser. Sci. Math. Astronom. Phys.* **25**(1977), 749–756 (Russian).
43. **T.A. Chanturia,** Integral tests for the oscillation of the solutions of higher order linear differential equations, *Differencial'nye Uravnenija* **16**(1980), 470–482.
44. **Kuo–Liang Chiou,** Oscillation and nonoscillation theorems for second order functional differential equations, *J. Math. Anal. Appl.* **45**(1974), 382–403.
45. **C.V. Coffman and J.S.W. Wong,** Oscillation and nonoscillation of solutions of generalized Emden–Fowler equations, *Trans. Amer. Math. Soc.* **167**(1972), 399–434.
46. **W.J. Coles,** Oscillation criterion for nonlinear second order equations, *Ann. Mat. Pura Appl.* **82**(1969), 123–133.
47. **R. Cristescu,** *Ordered Vector Spaces and Linear Operator*, *Abacus Press, Tunbridge Wells*, Kent, 1976.
48. **R.S. Dahiya and B. Singh,** On the oscillatory behavior of even order delay equations, *J. Math. Anal. Appl.* **42**(1973), 183–190.
49. **Y. Domshlak,** On the oscillation of solutions of vector differential equations, *Soviet Math. Dokl.* **11**(1970), 21–23.
50. **Y. Domshlak,** Oscillatory properties of linear difference equations with continuous time, *Differential Equations and Dynamical Systems* **1**(1993), 311–324.
51. **S. Elaydi,** *An Introduction to Difference Equations*, *Springer–Verlag*, New York, 1996.
52. **A. Elbert and T. Kusano,** Oscillation and nonoscillation theorems for a class of second order quasilinear differential equations, *Acta Mathematica Hungarica* **56**(1990), 325–336.
53. **L. Erbe,** Oscillation theorems for second order nonlinear differential equations, *Proc. Amer. Math. Soc.* **24**(1970), 811–814.
54. **L. Erbe, Q.K. Kong and B.G. Zhang,** *Oscillation Theory for Functional Differential Equations*, *Marcel Dekker*, New York, 1995.
55. **J.M. Ferreira and A.M. Pedro,** Oscillations of delay difference systems, *J. Math. Anal. Appl.* **221**(1998), 364–383.

56. **W.B. Fite,** Concerning the zeros of the solutions of certain differential equations, *Trans. Amer. Math. Soc.* **19**(1918), 341–352.
57. **K. Foster,** Oscillation of forced sublinear differential equations of even order, *J. Math. Anal. Appl.* **55**(1976), 634–643.
58. **K. Foster,** Criteria for oscillation and growth of nonoscillatory solutions of forced differential equations of even order, *J. Differential Equations* **20**(1976), 115–132.
59. **K. Foster and R.C. Grimmer,** Nonoscillatory solutions of higher order differential equations, *J. Math. Anal. Appl.* **71**(1979), 1–17.
60. **K. Foster and R.C. Grimmer,** Nonoscillatory solutions of higher order delay differential equations, *J. Math. Anal. Appl.* **77**(1980), 150–164.
61. **K. Gopalsamy,** *Stability and Oscillation in Delay Differential Equations of Population Dynamics*, Kluwer, Dordrecht, 1992.
62. **K. Gopalsamy, S.R. Grace and B.S. Lalli,** Oscillation of even order neutral differential equations, *Indian J. Math.* **35**(1993), 9–25.
63. **K. Gopalsamy, B.S. Lalli and B.G. Zhang,** Oscillation of odd order neutral differential equations, *Czech. Math. J.* **42**(1992), 313–323.
64. **S.R. Grace,** Oscillation of even order nonlinear functional differential equations with deviating arguments, *Funkcialaj Ekvac.* **32**(1989), 265–272.
65. **S.R. Grace,** On the oscillatory and asymptotic behavior of even order nonlinear differential equations with retarded arguments, *J. Math. Anal. Appl.* **137**(1989), 528–540.
66. **S.R. Grace,** Comparison theorems for forced functional differential equations, *J. Math. Anal. Appl.* **144**(1989), 168–182.
67. **S.R. Grace,** Oscillation theorems for second order nonlinear differential equations with damping, *Math. Nachr.* **141**(1989), 117–127.
68. **S.R. Grace,** Oscillation criteria for second order differential equations with damping, *J. Austral. Math. Soc. Ser. A* **49**(1990), 43–54.
69. **S.R. Grace,** Oscillation criteria for forced functional differential equations with deviating arguments, *J. Math. Anal. Appl.* **145**(1990), 63–88.
70. **S.R. Grace,** Oscillation of functional differential equations with deviating arguments, *J. Math. Anal. Appl.* **149**(1990), 558–575.
71. **S.R. Grace,** Oscillatory properties of functional differential equations, *J. Math. Anal. Appl.* **160**(1991), 60–78.
72. **S.R. Grace,** Oscillatory and asymptotic behavior of certain functional differential equations, *J. Math. Anal. Appl.* **162**(1991), 177–188.
73. **S.R. Grace,** Oscillation of even order nonlinear functional differential equations with deviating arguments, *Math. Slovaca* **41**(1991), 189–204.
74. **S.R. Grace,** Oscillatory and asymptotic behavior of delay differential equations with a nonlinear damping term, *J. Math. Anal. Appl.* **168**(1992), 306–318.
75. **S.R. Grace,** Oscillation theorems for nonlinear differential equations of second order, *J. Math. Anal. Appl.* **171**(1992), 220–241.
76. **S.R. Grace,** Oscillation theorems for damped functional differential equations, *Funkcialaj Ekvacioj* **35**(1992), 261–278.

77. **S.R. Grace,** Oscillation theorems of comparison type of delay differential equations with a nonlinear damping term, *Math. Slovaca* **44**(1994), 303–314.
78. **G.R. Grace,** Oscillation criteria for nth order neutral functional differential equations, *J. Math. Anal. Appl.* **184**(1994), 44–55.
79. **S.R. Grace,** Oscillation theorems for certain functional differential equations, *J. Math. Anal. Appl.* **184**(1994), 100–111.
80. **S.R. Grace,** Oscillation of mixed neutral functional differential equations, *Appl. Math. Comput.* **68**(1995), 1–13.
81. **S.R. Grace,** An oscillation criterion for functional differential equations with deviating arguments, *J. Math. Anal. Appl.* **192**(1995), 371–380.
82. **S.R. Grace,** On the oscillations of mixed neutral equations, *J. Math. Anal. Appl.* **194**(1995), 377–388.
83. **S.R. Grace,** Oscillation theorems of comparison type for neutral nonlinear functional differential equations, *Czech. Math. J.* **45**(1995), 609–626.
84. **S.R. Grace,** Oscillation criteria for retarded differential equations with a nonlinear damping term, *Aequationes Math.* **51**(1996), 68–82.
85. **S.R. Grace,** Oscillation of higher order nonlinear functional differential equations of neutral type, *Dynam. Systems Appl.* **5**(1996), 399–406.
86. **S.R. Grace,** On the oscillation of certain forced functional differential equations, *J. Math. Anal. Appl.* **202**(1996), 555–577.
87. **S.R. Grace,** Oscillation of certain neutral difference equations of mixed type, *J. Math. Anal. Appl.* **224**(1998), 241–254.
88. **S.R. Grace and G.G. Hamadani,** On the oscillation of functional differential equations, *Math. Nachr.* **203**(1999), 111–123.
89. **S.R. Grace and B.S. Lalli,** An oscillation criterion for nth order nonlinear differential equations with functional arguments, *Canad. Math. Bull.* **26**(1983), 35–40.
90. **S.R. Grace and B.S. Lalli,** On oscillation of solutions of nth order delay differential equations, *J. Math. Anal. Appl.* **91**(1983), 328–339.
91. **S.R. Grace and B.S. Lalli,** Oscillation theorems for nth order delay differential equations, *J. Math. Anal. Appl.* **91**(1983), 352–366.
92. **S.R. Grace and B.S. Lalli,** Oscillation theorems for nth order nonlinear functional differential equations, *J. Math. Anal. Appl.* **94**(1983), 509–524.
93. **S.R. Grace and B.S. Lalli,** Oscillation theorems for nth order nonlinear differential equations with deviating arguments, *Proc. Amer. Math. Soc.* **90**(1984), 65–70.
94. **S.R. Grace and B.S. Lalli,** An oscillation criterion for nth order nonlinear differential equations with retarded arguments, *Czech. Math. J.* **34**(1984), 18–21.
95. **S.R. Grace and B.S. Lalli,** Oscillation theorems for damped differential equations of even order with deviating arguments, *SIAM J. Math. Anal.* **15**(1984), 308–316.
96. **S.R. Grace and B.S. Lalli,** Oscillatory and asymptotic behavior of solutions of differential equations with deviating arguments, *J. Math. Anal. Appl.* **104**(1984), 79–94.

97. **S.R. Grace and B.S. Lalli,** Oscillation of odd order nonlinear functional differential equations with deviating arguments, *Applicable Analysis* **20**(1985), 189–199.
98. **S.R. Grace and B.S. Lalli,** Oscillation theorems for nonlinear second order functional differential equations with damping, *Bull. Inst. Math. Acad. Sinica* **13**(1985), 183–192.
99. **S.R. Grace and B.S. Lalli,** Oscillatory behavior of nonlinear differential equations with deviating arguments, *Bull. Austral. Math. Soc.* **31**(1985), 127–136.
100. **S.R. Grace and B.S. Lalli,** A comparison theorem for general nonlinear ordinary differential equations, *J. Math. Anal. Appl.* **120**(1986), 39–43.
101. **S.R. Grace and B.S. Lalli,** Oscillation theorems for certain nonlinear differential equations with deviating arguments, *Czech. Math. J.* **36**(1986), 268–274.
102. **S.R. Grace and B.S. Lalli,** Oscillation theorems for certain neutral differential equations, *Czech. Math. J.* **38**(1988), 745–753.
103. **S.R. Grace and B.S. Lalli,** Oscillation theorems for damped–forced nth order nonlinear differential equations with deviating arguments, *J. Math. Anal. Appl.* **136**(1988), 54–65.
104. **S.R. Grace and B.S. Lalli,** Oscillation theorems for damped–forced nth order nonlinear differential equations with deviating arguments, *Math. Nachr.* **138**(1988), 255–262.
105. **S.R. Grace and B.S. Lalli,** Comparison and oscillation theorems for functional differential equations with deviating arguments, *Math. Nachr.* **144**(1989), 65–79.
106. **S.R. Grace and B.S. Lalli,** Oscillations of functional differential equations generated by advanced arguments, *Differential Integral Equations* **2**(1989), 21–36.
107. **S.R. Grace and B.S. Lalli,** Oscillation of even order differential equations with deviating arguments, *J. Math. Anal. Appl.* **147**(1990), 569–579.
108. **S.R. Grace and B.S. Lalli,** Integral averaging techniques for the oscillation of second order nonlinear differential equations, *J. Math. Anal. Appl.* **149**(1990), 277–311.
109. **S.R. Grace and B.S. Lalli,** Oscillation theorems for second order neutral functional differential equations, *Appl. Math. Comput.* **51**(1992), 119–133.
110. **S.R. Grace and B.S. Lalli,** On the oscillation of certain neutral functional differential equations, *Funkcialaj Ekvac.* **36**(1993), 303–310.
111. **S.R. Grace and B.S. Lalli,** On the oscillation of certain higher order functional differential equations of neutral type, *Funkcialaj Ekvac.* **37**(1994), 211–220.
112. **S.R. Grace and B.S. Lalli,** Oscillation theorems for certain neutral functional differential equations with periodic coefficients, *Dynam. Systems Appl.* **3**(1994), 85–93.
113. **S.R. Grace and B.S. Lalli,** Comparison theorems for forced functional differential equations with deviating arguments, *J. Mathl. Phyl. Sci.* **28**(1994), 23–42.

114. **S.R. Grace and B.S. Lalli,** Oscillation theorems for second order neutral difference equations, *Appl. Math. Comput.* **62**(1994), 47–60.
115. **S.R. Grace and B.S. Lalli,** Oscillation theorems for second order delay and neutral difference equations, *Utilitas Math.* **45**(1994), 197–211.
116. **S.R. Grace and B.S. Lalli,** Oscillation theorems for forced neutral difference equations, *J. Math. Anal. Appl.* **187**(1994), 91–106.
117. **J.R. Graef,** Some nonoscillation criteria for higher order nonlinear differential equations, *Pacific J. Math.* **66**(1976), 125–129.
118. **J.R. Graef and P.W. Spikes,** Asymptotic properties of solutions of functional differential equations of arbitrary order, *J. Math. Anal. Appl.* **60**(1977), 339–348.
119. **J.R. Graef, P.W. Spikes and M.K. Grammatikopoulos,** Asymptotic behavior of nonoscillatory solutions of neutral delay differential equations of arbitrary order, *Nonlinear Analysis* **21**(1993), 23–42.
120. **M.K. Grammatikopoulos,** Oscillatory and asymptotic behavior of differential equations with deviating arguments, *Hiroshima Math. J.* **6**(1976), 31–53.
121. **M.K. Grammatikopoulos, E.A. Grove and G. Ladas,** Oscillation and asymptotic behavior of neutral differential equations with deviating arguments, *Appl. Anal.* **22**(1986), 1–19.
122. **M.K. Grammatikopoulos, G. Ladas and A. Meimaridou,** Oscillation and asymptotic behavior of higher order neutral equations with variable coefficients, *Chinese Ann. Math. Ser. B.* **9**(1988), 322–338.
123. **M.K. Grammatikopoulos, G. Ladas and Y.G. Sficas,** Oscillation and asymptotic behavior of neutral equations with variable coefficients, *Radovi Mat.* **2**(1986), 279–303.
124. **M.K. Grammatikopoulos, Y.G. Sficas and V.A. Staikos,** Oscillatory properties of strongly superlinear differential equations with deviating arguments, *J. Math. Anal. Appl.* **67**(1979), 171–187.
125. **M.K. Grammatikopoulos, Y.G. Sficas and I.P. Stavroulakis,** Necessary and sufficient conditions for oscillations of neutral equations with several coefficients, *J. Differential Equations* **76**(1988), 294–311.
126. **R. Grimmer,** On nonoscillatory solutions of a nonlinear differential equation, *Proc. Amer. Math. Soc.* **34**(1972), 118–120.
127. **R. Grimmer,** Oscillation criteria and growth of nonoscillatory solutions of even order ordinary and delay differential equations, *Trans. Amer. Math. Soc.* **198**(1974), 215–228.
128. **I. Györi and G. Ladas,** Oscillation of systems of neutral differential equations, *Differential Integral Equations* **1**(1988), 281–286.
129. **I. Györi and G. Ladas,** *Oscillation Theory of Differential Equations with Applications*, *Clarendon Press*, Oxford, 1991.
130. **M.E. Hammett,** Nonoscillation properties of a nonlinear differential equation, *Proc. Amer. Math. Soc.* **30**(1971), 92–96.
131. **G.H. Hardy, J.E. Littlewood and G. Polya,** *Inequalities*, 2nd edition, *Cambridge University Press*, Cambridge, 1988.
132. **J.W. Heidel,** The existence of the oscillatory solutions for a nonlinear odd order differential equation, *Czech. Math. J.* **20**(1970), 93–97.

133. **J.W Hooker and W.T. Patula,** A second order nonlinear difference equation: Oscillation and asymptotic behavior, *J. Math. Anal. Appl.* **91**(1983), 9–29.
134. **J.W Hooker and W.T. Patula,** Growth and oscillation properties of solutions of a fourth order linear difference equations, *J. Austral. Math. Soc. Ser. B* **26**(1985), 310–328.
135. **A.F. Ivanov, Y. Kitamura, T. Kusano and V.N. Shevelo,** Oscillatory solutions of functional differential equations generated by deviation of arguments of mixed type, *Hiroshima Math. J.* **12**(1982), 645–655.
136. **A.J. Jerri,** *Linear Difference Equations with Discrete Transform Methods*, *Kluwer*, Dordrecht, 1996.
137. **J. Jeroš and T. Kusano,** On oscillation of linear neutral differential equations of higher order, *Hiroshima Math. J.* **20**(1990), 407–419.
138. **J. Jeroš and T. Kusano,** Oscillation properties of first order nonlinear functional differential equations of neutral type, *Differential and Integral Equations* **4**(1991), 425–436.
139. **I.V. Kamenev,** On the oscillation of solutions of a nonlinear equation of higher order, *Differencial'nye Uravnenija* **7**(1971), 927–929.
140. **I.V. Kamenev,** On integral criteria for nonoscillation, *Mat. Zametki* **13**(1973), 51–54.
141. **A.G. Kartsatos,** Criteria for oscillation of solutions of differential equations of arbitrary order, *Proc. Japan Akad.* **44**(1968), 599–602.
142. **A.G. Kartsatos,** On oscillations of even order nonlinear differential equations, *J. Differential Equations* **6**(1969), 232–237.
143. **A.G. Kartsatos,** Contributions to the research of the oscillation and asymptotic behavior of solutions of ordinary differential equations, *Bull. Soc. Math. Greece* **10**(1969), 1–48 (Greek).
144. **A.G. Kartsatos,** Oscillation properties of solutions of even order differential equations, *Bull. Fac. Sci. Ibaraki Univ. Ser. A* **2**(1969), 9–14.
145. **A.G. Kartsatos,** On the maintenance of oscillation of nth order equations under the effect of a small forcing term, *J. Differential Equations* **10**(1971), 355–363.
146. **A.G. Kartsatos,** Maintenance of oscillation under the effect of a periodic forcing term, *Proc. Amer. Math. Soc.* **33**(1972), 377–383.
147. **A.G. Kartsatos,** On positive solutions of perturbed nonlinear differential equations, *J. Math. Anal. Appl.* **47**(1974), 58–68.
148. **A.G. Kartsatos,** On nth order differential inequalities, *J. Math. Anal. Appl.* **52**(1975), 1–9.
149. **A.G. Kartsatos,** Oscillation and existence of unique positive solutions for nonlinear nth order equations with forcing term, *Hiroshima Math. J.* **6**(1976), 1–6.
150. **A.G. Kartsatos,** Recent results on oscillation of solutions of forced and perturbed nonlinear differential equations of even order, in *Proc. NSF–CBM Reg Conference, Stability of Dynamical Systems*, ed. J.R. Graef, Marcel Dekker, 1977, 17–72.
151. **A.G. Kartsatos,** Oscillation of nth order equations with perturbations, *J. Math. Anal. Appl.* **57**(1977), 161–169.

152. **A.G. Kartsatos,** Oscillation and nonoscillation for perturbed differential equations, *Hiroshima Math. J.* **8**(1978), 1–10.
153. **A.G. Kartsatos,** The oscillation of a forced equation implies the oscillation of the unforced equations–small forcing, *J. Math. Anal. Appl.* **76**(1980), 98–106.
154. **A.G. Kartsatos and M.N. Manougian,** Perturbations causing oscillations of functional differential equations, *Proc. Amer. Math. Soc.* **43**(1974), 111–117.
155. **A.G. Kartsatos and M.N. Manougian,** Further results on oscillation of functional differential equations, *J. Math. Anal. Appl.* **53**(1976), 28–37.
156. **A.G. Kartsatos and H. Onose,** On the maintenance of oscillations under the effect of a small nonlinear damping, *Bull. Fac. Sci. Ibaraki Univ. Ser. A* **4**(1972), 3–11.
157. **A.G. Kartsatos and J. Toro,** Comparison and oscillation theorems for equations with middle term of order $(n-1)$, *J. Math. Anal. Appl.* **66**(1978), 297–312.
158. **A.G. Kartsatos and J. Toro,** Oscillation and asymptotic behavior of forced nonlinear equations, *SIAM J. Math. Anal.* **10**(1979), 86–95.
159. **W.G. Kelley and A.C. Peterson,** *Difference Equations: An Introduction with Applications*, *Academic Press*, New York, 1991.
160. **I.T. Kiguradze,** On oscillatory solutions of some ordinary differential equations, *Soviet Math. Dokl.* **144**(1962), 33–36.
161. **I.T. Kiguradze,** On oscillatory solutions of the equation $\frac{d^m u}{dt^m} + a(t)|u|^m \times \operatorname{sgn} u = 0$, *Mat. Sbornik* **65**(1964), 172–187.
162. **I.T. Kiguradze,** On the oscillatory and monotone solutions of ordinary differential equations, *Arch. Math.* **XIV**(1978), 21–44.
163. **W.J. Kim,** Nonoscillatory solutions of a class of nth order linear differential equations, *J. Differential Equations* **27**(1978), 19–27.
164. **W.J. Kim,** Asymptotic properties of nonoscillatory solutions of higher order differential equations, *Pacific J. Math.* **92**(1981), 107–114.
165. **Y. Kitamura,** Oscillation of functional differential equations with general deviating arguments, *Hiroshima Math. J.* **15**(1985), 445–491.
166. **Y. Kitamura and T. Kusano,** Nonlinear oscillation of higher order functional differential equations with deviating arguments, *J. Math. Anal. Appl.* **77**(1980), 79–90.
167. **V.L. Kocic and G. Ladas,** *Global Behavior of Nonlinear Difference Equations of Higher Order with Applications*, *Kluwer*, Dordrecht, 1993.
168. **R.G. Koplatadze and T.A. Chanturiya,** On the oscillatory properties of differential equations with a deviating argument, *Tblisi Univ. Press*, Tblisi, 1977 (Russian).
169. **R.G. Koplatadze and T.A. Chanturia,** On oscillatory and monotone solutions of first order differential equations with deviating arguments, *Differencial'nye Uravnenija* **18**(1982), 1463–1465.
170. **W.A. Kosmala,** Oscillation and asymptotic behavior of nonlinear equations with middle terms of order $n-1$ and forcings, *Nonlinear Analysis* **6**(1982), 1115–1133.

171. **W.A. Kosmala,** Oscillation theorems for higher order delay equations, *Appl. Anal.* **21**(1986), 43–53.

172. **W.A. Kosmala,** On positive solutions of higher order equations, *Appl. Anal.* **22**(1986), 87–101.

173. **K. Kreith,** PDE generalization of Sturm comparison theorem, *Memoris Amer. Math. Soc.* **48**(1984), 31–46.

174. **K. Kreith and T. Kusano,** Oscillation theorems for nonlinear ordinary differential equations of even order, *Canad. Math. Bull.* **24**(1981), 409–413.

175. **M.R.S. Kulenovic, G. Ladas and Y.G. Sficas,** Comparison results for oscillations of delay equations, *Annal. Math. Pura. Appl.* **CLVI**(1990), 1–14.

176. **T. Kusano,** On strong oscillation of even order differential equations with advanced arguments, *Hiroshima Math. J.* **11**(1981), 617–620.

177. **T. Kusano,** On even order functional differential equations with advanced and retarded arguments, *J. Differential Equations* **45**(1982), 75–84.

178. **T. Kusano,** Oscillation of the even order linear differential equations with deviating arguments of mixed type, *J. Math. Anal. Appl.* **98**(1984), 341–347.

179. **T. Kusano and B.S. Lalli,** On oscillation of half–linear functional differential equations with deviating arguments, *Hiroshima Math. J.* **24**(1994), 549–563.

180. **T. Kusano and M. Naito,** Comparison theorems for functional differential equations with deviating arguments, *J. Math. Soc. Japan* **33**(1981), 509–532.

181. **T. Kusano, Y. Naito and A. Ogata,** Strong oscillation and non- oscillation of quasilinear differential equations of second order, Differential Equations and Dynamical Systems **2**(1994), 1–10.

182. **T. Kusano and H. Onose,** Nonlinear oscillation of a sublinear delay equations of arbitrary order, *Proc. Amer. Math. Soc.* **40**(1973), 219–224.

183. **T. Kusano and H. Onose,** Oscillations of functional differential equations with retarded arguments, *J. Differential Equations* **15**(1974), 269–277.

184. **T. Kusano and H. Onose,** Remarks on the oscillatory behavior of solutions of functional differential equations with deviating argument, *Hiroshima Math. J.* **6**(1976), 183–189.

185. **T. Kusano and N. Yoshida,** Nonoscillation theorems for a class of quasilinear differential equations of second order, *J. Math. Anal. Appl.* **189**(1995), 115–127.

186. **M.K. Kwong and J.S.W. Wong,** Linearization of second order nonlinear oscillation theorems, *Trans. Amer. Math. Soc.* **279**(1983), 705–722.

187. **G. Ladas,** Oscillation and asymptotic behavior of solutions of differential equations with retarded arguments, *J. Differential Equations* **10**(1971), 281–290.

188. **G. Ladas,** Oscillatory effects of retarded actions, *J. Math. Anal. Appl.* **60**(1977), 410–416.

189. **G. Ladas**, Explicit conditions for the oscillation of difference equations, *J. Math. Anal. Appl.* **153**(1990), 276–287.

190. **G. Ladas, G. Ladde and J.S. Papadakis**, Oscillations of functional differential equations generated by delays, *J. Differential Equations* **12**(1972), 385–395.

191. **G. Ladas and V. Lakshmikantham**, Oscillations caused by retarded actions, *Appl. Anal.* **4**(1974), 9–15.

192. **G. Ladas and C. Qian**, Linearized oscillations for odd order neutral delay differential equations, *J. Differential Equations* **88**(1990), 238–247.

193. **G. Ladas and C. Qian**, Linearized oscillations of even order neutral differential equations, *J. Math. Anal. Appl.* **159**(1991), 237–250.

194. **G. Ladas and C. Qian**, Comparison results and linearized oscillations for higher order difference equations, *Internat. J. Math. & Math. Sci.* **15**(1992), 129–142.

195. **G. Ladas and Y.G. Sficas**, Oscillation of neutral delay differential equations, *Canad. Math. Bull.* **29**(1986), 438–445.

196. **G. Ladas and Y.G. Sficas**, Oscillations of higher order neutral equations, *J. Austral. Math. Soc. Series B* **27**(1986), 502–511.

197. **G. Ladas and I.P. Stavroulakis**, On delay differential inequalities of first order, *Funkcialaj Ekvacioj* **25**(1982), 105–113.

198. **G. Ladas and I.P. Stavroulakis**, On delay differential inequalities of higher order, *Canad. Math. Bull.* **25**(1982), 348–354.

199. **G. Ladas and I.P. Stavroulakis**, Oscillation caused by several retarded and advanced arguments, *J. Differential Equations* **44**(1982), 134–152.

200. **G.S. Ladde, V. lakshmikantham and B.G. Zhang**, *Oscillation Theory of Differential Equations with Deviating Arguments, Marcel Dekker*, New York, 1987.

201. **V. Lakshmikantham, S. Sivasundaram and B. Kaymakcalan**, *Dynamic Systems on Measure Chains, Kluwer*, Dordrecht, 1996.

202. **V. Lakshmikantham and D. Trigiante**, *Difference Equations with Applications to Numerical Analysis, Academic Press*, New York, 1988.

203. **B.S. Lalli and S.R. Grace**, Oscillation theorems for second order neutral difference equations, *Appl. Math. Comput.* **62**(1994), 47–60.

204. **B.S. Lalli, B.G. Zhang and Jan Jhao**, On oscillation and existence of positive solutions of neutral difference equations, *J. Math. Anal. Appl.* **158**(1991), 213–233.

205. **I. Licko and M. Svec**, Le caractere oscillatoris des solutions del' equation $y^{(n)} + f(x)y^{\alpha} = 0,\ n > 1$, *Czech. Math. J.* **13**(1963), 481–489.

206. **X. Liu and X. Fu**, Nonlinear differential inequalities with distributed deviating arguments and applications, *Nonlinear World* **1**(19 94), 409–427.

207. **D.L. Lovelady**, An asymptotic analysis of an odd order linear differential equation, *Pacific J. Math.* **57**(1975), 475–480.

208. **D.L. Lovelady**, Oscillation and even order linear differential equations, *Rocky Mount. J. Math.* **6**(1976), 299–304.

209. **D.L. Lovelady**, Oscillation and a class of linear delay differential equations, *Trans. Amer. Math. Soc.* **228**(1977), 345–364.

210. **W.E. Mahfoud,** Oscillation and asymptotic behavior of solutions of nth order nonlinear delay differential equations, *J. Differential Equations* **24**(1977), 75–98.
211. **W.E. Mahfoud,** Characterization of oscillation of solutions of the equation $x^{(n)}(t) + a(t)f(x[g(t)]) = 0$, *J. Differential Equations* **28**(1978), 437–451.
212. **W.E. Mahfoud,** Comparison theorems for delay differential equations, *Pacific J. Math.* **83**(1979), 187–197.
213. **P. Marusiak,** Note on the Ladas' paper: On oscillation and asymptotic behavior of solutions of differential equations with retarded argument, *J. Differential Equations* **13**(1973), 150–156.
214. **R.E. Mickens,** *Difference Equations: Theory and Applications*, *Van Nostrand Reinhold*, New York, 2nd edition, 1990.
215. **R.E. Mickens,** *Nonstandard Finite Difference Models of Differential Equations*, *World Scientific*, Singapore, 1994.
216. **J.G. Mikusinski,** On Fite's oscillation theorems, *Colloq. Math.* **2**(1949), 34–39.
217. **D.S. Mitrinovic, J.E. Pecaric and A.M. Fink,** *Classical and New Inequalities in Analysis*, *Kluwer*, Dordrecht, 1993.
218. **K. Motohiko and T. Kusano,** On a class of second order quasilinear ordinary differential equations, *Hiroshima Math. J.* **25**(1995), 321–355.
219. **M. Naito,** Oscillation theorems for a damped nonlinear differential equations, *Proc. Japan Akad.* **50**(1974), 104–108.
220. **M. Naito,** Asymptotic behavior of solutions of second order differential equations with integrable coefficients, *Trans. Amer. Math. Soc.* **282**(1984), 577–588.
221. **M. Naito,** Nonoscillatory solutions of second order differential equations with integrable coefficients, *Proc. Amer. Math. Soc.* **109**(1990), 769–774.
222. **Z. Nehari,** A nonlinear oscillation problem, *J. Differential Equations* **5**(1969), 452–460.
223. **J. Ohriska,** Oscillation of second order delay and ordinary differential equations, *Czech. Math. J.* **34**(1984), 107–112.
224. **R. Oláh,** Note on oscillation of differential equation with advanced argument, *Math. Slovaca* **33**(1983), 241–248.
225. **H. Onose,** Oscillatory property of certain nonlinear ordinary differential equations, *Proc. Japan Akad.* **44**(1968), 110–113.
226. **H. Onose,** Oscillatory property of certain nonlinear ordinary differential equations II, *Proc. Japan Akad.* **44**(1968), 876–878.
227. **H. Onose,** Oscillatory property of ordinary differential equations of arbitrary order, *J. Differential Equations* **7**(1970), 454–458.
228. **H. Onose,** Oscillatory property of certain nonlinear ordinary differential equations, *SIAM J. Appl. Math.* **18**(1970), 715–719.
229. **H. Onose,** Oscillatory properties of solutions of even order differential equations, *Pacific J. Math.* **38**(1971), 747–757.
230. **H. Onose,** Some oscillation criteria for nth order nonlinear delay differential equations, *Hiroshima Math. J.* **1**(1971), 171–176.

231. **H. Onose,** On oscillation of solutions of nth order differential equations, *Proc. Amer. Math. Soc.* **33**(1972), 495–500.
232. **H. Onose,** Oscillation and asymptotic behavior of solutions of retarded differential equations of arbitrary order, *Hiroshima Math. J.* **3**(1973), 333–360.
233. **H. Onose,** A comparison theorem and the forced oscillation, *Bull. Austral. Math. Soc.* **13**(1975), 13–19.
234. **H. Onose,** Oscillatory properties of the first order nonlinear advanced and delayed differential inequalities, *Nonlinear Analysis* **8**(19 84), 171–180.
235. **W.T. Patula,** Growth and oscillation properties of second order linear difference equations, *SIAM J. Math. Anal.* **10**(1979), 55–61.
236. **H. Péics,** On the asymptotic behaviour of difference equations with continuous arguments, *Dynamics of Continuous, Discrete and Impulsive Systems*, to appear.
237. **Ch.G. Philos,** Oscillatory and asymptotic behavior of the bounded solutions of differential equations with deviating arguments, *Hiroshima Math. J.* **8**(1978), 31–48.
238. **Ch.G. Philos,** Oscillatory and asymptotic behavior of all solutions of differential equations with deviating arguments, *Proc. Royal Soc Edinburgh* **81 A**(1978), 195–210.
239. **Ch.G. Philos,** Bounded oscillation generated by retardation for differential equations of arbitrary order, *Utilitas Math.* **15**(1979), 161–182.
240. **Ch.G. Philos,** A comparison result for retarded differential equations, *Utilitas Math.* **17**(1980), 259–269.
241. **Ch.G. Philos,** On the existence of nonoscillatory solutions tending to zero at ∞ for differential equations with positive delays, *Archivum Mathematicum* **36**(1981), 168–178.
242. **Ch.G. Philos,** A new criterion for the oscillatory and asymptotic behavior of delay differential equations, *Bull. Acad. Pol. Sci. Ser. Sci. Mat.* **39**(1981), 61–64.
243. **Ch.G. Philos,** Oscillation and asymptotic behavior of linear retarded differential equations of arbitrary order, *University of Ioannina, Tech. Report No.* **57**, 1981.
244. **Ch.G. Philos,** Oscillation theorems for linear differential equations of second order, *Arch. Math.* **53**(1989), 482–492.
245. **Ch. G. Philos,** On oscillation of some difference equations, *Funkcialaj Ekvacioj* **34**(1991), 157–172.
246. **Ch.G. Philos,** Oscillation in a class of difference equations, *Appl. Math. Comput.* **48**(1992), 45–57.
247. **Ch.G. Philos and Y.G. Sficas,** Oscillation and asymptotic behavior of second and third order retarded differential equations, *Czech. Math. J.* **32**(1982), 169–182.
248. **Ch.G. Philos and Y.G. Sficas,** Positive solutions of difference equations, *Proc. Amer. Math. Soc.* **108**(1990), 107–115.
249. **Ch.G. Philos, Y.G. Sficas and V.A. Staikos,** Some results on the asymptotic behavior of nonoscillatory solutions of differential equations with deviating arguments, *J. Austral. Math. Soc. Ser. A.* **32**(1982), 295–317.

250. **J. Popenda,** On the system of finite difference equations, *Fasc. Math.* **15**(1984), 119–126.

251. **J. Popenda,** One expression for the solutions of second order difference equations, *Proc. Amer. Math. Soc.* **100**(1987), 87–93.

252. **J. Popenda,** Oscillation and nonoscillation theorems for second-order difference equations, *J. Math. Anal. Appl.* **121**(1987), 34–38.

253. **J. Popenda,** The oscillation of solutions of difference equations, *Computers Math. Applic.* **28**(1994), 271–279.

254. **J. Popenda and E. Schmeidel,** Nonoscillatory solutions of third order difference equations, *Portug. Math.* **49**(1992), 233–239.

255. **J. Popenda and B. Szmanda,** On the oscillation of solutions of certain difference equations, *Demonstr. Math.* **17**(1984), 153–164.

256. **S.M. Rankin,** Oscillation theorems for second order nonhomogeneous linear differential equations, *J. Math. Anal. Appl.* **53**(1976), 550–553.

257. **S.M. Rankin,** Oscillation of a forced second order nonlinear differential equations, *Proc. Amer. Math. Soc.* **59**(1976), 279–282.

258. **V.N. Ševelo,** *Oscillation of the Solutions of Differential Equations with Deviating Arguments*, Izdat. Naukova Dumka, Keiv, 1987 (Russian).

259. **V.N. Ševelo and V.N. Vareh,** On the oscillation of solutions of higher order linear differential equations with retarded arguments, *Ukrain Mat. Z.* **24**(1972), 513–520.

260. **V.N. Ševelo and V.N. Vareh,** On some properties of solutions of differential equations with delay, *Ukrain Mat. Z.* **24**(1972), 807–813.

261. **Y.G. Sficas,** On oscillation and asymptotic behavior of certain class of differential equations with retarded arguments, *Utilitas Math.* **3** (1973), 239–249.

262. **Y.G. Sficas,** The effect of the delay on the oscillatory and asymptotic behavior of nth order retarded differential equations, *J. Math. Anal. Appl.* **49**(1975), 748–757.

263. **Y.G. Sficas and V.A. Staikos,** Oscillation of retarded differential equations, *Proc. Camb. Phil. Soc.* **75**(1974), 95–101.

264. **Y.G. Sficas and V.A. Staikos,** The effect of retarded actions on nonlinear oscillations, *Proc. Amer. Math. Soc.* **46**(1974), 259–264.

265. **Y.G. Sficas and V.A. Staikos,** Oscillations of differential equations with deviating arguments, *Funkcialaj Ekvacioj* **19**(1976), 35–43.

266. **A.N. Sharkovsky, Y.L. Maistrenko and E.Y. Romanenko,** *Difference Equations and Their Applications*, Kluwer, Dordrecht, 1993.

267. **J.H. Shen,** Oscillation and comparison theorems of difference equations with continuous arguments and applications, *Chinese Sci. Bull.* **41**(1996), 1441–1444.

268. **B. Singh,** A necessary and sufficient condition for the oscillation of an even order nonlinear delay differential equation, *Canad. J. Math.* **25**(1973), 1078–1089.

269. **B. Singh,** Oscillation and nonoscillation of even order nonlinear delay differential equations, *Quart. Appl. Math.* **31**(1973), 343–349.

270. **B. Singh,** Asymptotic nature of nonoscillatory solutions of nth order retarded differential equations, *SIAM J. Math. Anal.* **6**(1975), 784–795.
271. **B. Singh,** Comparison theorems for even order differential equations, *Bull. Canad. Math. Soc.* **67**(1975), 23–28.
272. **B. Singh,** Impact of delays on oscillation in general functional equations, *Hiroshima Math. J.* **5**(1975), 351-361.
273. **B. Singh,** Forced oscillations in general ordinary differential equations with deviating arguments, *Hiroshima Math. J.* **6**(1976), 7–14.
274. **V.A. Staikos,** Oscillatory property of certain delay differential equations, *Bull. Soc. Math. Greece* **11**(1970), 1-5.
275. **V.A. Staikos,** Basic results on oscillation for differential equations, *Hiroshima Math. J.* **10**(1980), 495–516.
276. **V.A. Staikos and Ch.G. Philos,** On the asymptotic behavior of nonoscillatory solutions of differential equations with deviating arguments, *Hiroshima Math. J.* **7**(1977), 9–31.
277. **V.A. Staikos and Y.G. Sficas,** Oscillatory and asymptotic behavior of functional differential equations, *J. Differential Equations* **12**(1972), 426–437.
278. **V.A. Staikos and Y.G. Sficas,** Criteria for asymptotic and oscillatory character of functional differential equations of arbitrary order, *Boll. Un. Mat. Ital.* **6**(1972), 185–192.
279. **V.A. Staikos and Y.G. Sficas,** Oscillatory and asymptotic properties of differential equations with retarded arguments, *Appl. Anal.* **5**(1975), 141–148.
280. **V.A. Staikos and Y.G. Sficas,** Oscillatory and asymptotic characterization of the solutions of differential equations with deviating arguments, *J. London Math. Soc.* **10**(1975), 39–47.
281. **V.A. Staikos and Y.G. Sficas,** Forced oscillation for differential equations of arbitrary order, *J. Differential Equations* **17**(1975), 1–11.
282. **V.A. Staikos and I.P. Stavroulakis,** Bounded oscillations under the effect of retardations for differential equations of arbitrary order, *Proc. Roy. Soc. Edinburgh* **77 A**(1977), 129–136.
283. **I.P. Stavroulakis,** Oscillatory and asymptotic properties of differential equations with deviating arguments, *Atti. Accad. Naz. Lincei Rend. Cl. Sci. Fis. Mat. Natur.* **60**(1976), 611–622.
284. **M. Švec,** Monotone solutions of some differential equations, *Colloq. Math.* **18**(1967), 7-21.
285. **C.A. Swanson,** *Comparison and Oscillation Theory of Linear Differential Equations, Academic Press*, New York, 1968
286. **X.H. Tang and J.H. Shen,** Oscillation and existence of positive solutions in a class of higher order neutral equations, *J. Math. Anal. Appl.* **213**(1997), 662–680.
287. **R.D. Terry,** Some oscillation criteria for delay differential equations of even order, *SIAM J. Appl. Math.* **28**(1975), 319–334.
288. **R.D. Terry,** Delay differential equations of odd order satisfying property P_k , *J. Austral. Math. Soc. Ser. A* **20**(1975), 451–467.

289. **H. Teufel, Jr.,** Forced second order nonlinear oscillation, *J. Math. Anal. Appl.* **40**(1972), 148–152.
290. **E. Thandapani, I. Györi and B.S. Lalli,** An application of discrete inequality to second order nonlinear oscillation, *J. Math. Anal. Appl.* **186**(1994), 200–208.
291. **E. Thandapani and S. Pandian,** On the oscillatory behaviour of solutions of second order nonlinear difference equations, *Zeitschrift für Analysis und ihre Anwendungen* **13**(1994), 347–358.
292. **E. Thandapani, M.M. Susai Manuel and R.P. Agarwal,** Oscillation and nonoscillation theorems for second order quasilinear difference equations, *Facta Universitatis (Nis): Mathematics and Informatics* **11**(1996), 49–65.
293. **C.C. Travis,** Oscillation theorems for second order differential equations with functional arguments, *Proc. Amer. Math. Soc.* **31** (1972), 194–202.
294. **C.C. Travis,** A note on second order nonlinear oscillations, *Math. Japon.* **18** (1973), 261–264.
295. **W.F. Trench,** Oscillation properties of perturbed disconjugate equations, *Proc. Amer. Math. Soc.* **52**(1975), 147–155.
296. **W.F. Trench,** An oscillation condition for differential equations of arbitrary order, *Proc. Amer. Math. Soc.* **82**(1981), 548–552.
297. **E. True,** A comparison theorem for certain functional differential equations, *Proc. Amer. Math. Soc.* **47**(1975), 127–132.
298. **W.R. Utz,** A note on second order nonlinear differential equations, *Proc. Amer. Math. Soc.* **7**(1956), 1047–1048.
299. **P. Waltman,** An oscillation criterion for an equation with functional argument, *Canad. Math. Bull.* **11**(1968), 593–595.
300. **J. Werbowski,** Oscillation of first order differential inequalities with deviating arguments, *Ann. Mat. Pura. Appl.* **140**(1985), 383–392.
301. **J. Werbowski,** Oscillations of differential equations generated by advanced arguments, *Funkcial Ekvac.* **30**(1987), 69–79.
302. **J. Werbowski,** Oscillation of advanced differential inequalities, *J. Math. Anal. Appl.* **137**(1989), 193–206.
303. **D. Willett,** Classification of second order linear differential equations with respect to oscillation, *Advances in Math.* **3**(1969), 594–693.
304. **A. Wintner,** A criterion of oscillatory stability, *Quart. Appl. Math.* **7**(1949), 115–117.
305. **J.S.W. Wong,** On the second order nonlinear oscillations, *Funkcial Ekvac.* **11**(1968), 207–234.
306. **J.S.W. Wong,** On the generalized Emden–Fowler equation, *SIAM Review* **17**(1975), 339–360.
307. **J.S.W. Wong,** Oscillation theorems for second order nonlinear differential equations, *Bull. Inst. Math. Acad. Sinica* **3**(1975), 283–309.
308. **J.S.W. Wong,** Second order nonlinear forced oscillations, *SIAM J. Math. Anal.* **19**(1989), 667–675.
309. **P.J.Y. Wong and R.P. Agarwal,** Oscillation theorems and existence of positive monotone solutions for second order nonlinear difference equations, *Mathl. Comput. Modelling* **21**(3)(1995), 63–84.

310. **P.J.Y. Wong and R.P. Agarwal,** Oscillation theorems and existence criteria of asymptotically monotone solutions for second order differential equations, *Dynamic Systems and Appl.* **4**(1995), 477–496.
311. **P.J.Y. Wong and R.P. Agarwal,** On the oscillation and asymptotically monotone solutions of second order quasilinear differential equations, *Appl. Math. Comp.* **79**(1996), 207–237.
312. **P.J.Y. Wong and R.P. Agarwal,** Oscillatory behaviour of solutions of certain second order nonlinear differential equations, *J. Math. Anal. Appl.* **198**(1996), 337–354.
313. **P.J.Y. Wong and R.P. Agarwal,** Oscillation theorems for certain second order nonlinear difference equations, *J. Math. Anal. Appl.* **204**(1996), 813–829.
314. **P.J.Y. Wong and R.P. Agarwal,** Summation averages and the oscillation of second order nonlinear difference equations, *Mathl. Comput. Modelling* **24**(9)(1996), 21–35.
315. **P.J.Y. Wong and R.P. Agarwal,** Oscillation criteria for nonlinear partial difference equations with delays, *Computers Math. Applic.* **32**(6)(1996), 57–86.
316. **P.J.Y. Wong and R.P. Agarwal,** On the oscillation of an mth order perturbed nonlinear difference equation, *Archivum Mathematicum* **32**(1996), 13–27.
317. **P.J.Y. Wong and R.P. Agarwal,** Comparison theorems for the oscillation of higher order difference equations with deviating arguments, *Mathl. Comput. Modelling* **24**(12)(1996), 39–48.
318. **P.J.Y. Wong and R.P. Agarwal,** Oscillation and monotone solutions of a second order quasilinear difference equation, *Funkcialaj Ekvacioj* **39**(1996), 491–517.
319. **P.J.Y. Wong and R.P. Agarwal,** Nonexistence of unbounded nonoscillatory solutions of partial difference equations, *J. Math. Anal. Appl.* **214**(1997), 503–523.
320. **P.J.Y. Wong and R.P. Agarwal,** On the oscillation of partial difference equations generated by deviating arguments, *Acta Mathematica Hungarica* **79**(1998), 1–29.
321. **P.J.Y. Wong and R.P. Agarwal,** Asymptotic behavior of solutions of higher order difference and partial difference equations with distributed deviating arguments, *Appl. Math. Comp.* **97**(1998), 139–164.
322. **P.J.Y. Wong and R.P. Agarwal,** Oscillation and nonoscillation of half–linear difference equations generated by deviating arguments, In Advances in Difference Equations II, *Computers Math. Applic.* **36**(10–12)(1999), 11–26.
323. **P.J.Y. Wong and R.P. Agarwal,** Nonoscillatory solutions of functional difference equations involving quasi–differences, *Funkcialaj Ekvacioj*, to appear.
324. **J. Yu,** Oscillation of odd order neutral differential equations with an 'integrally small' coefficients, *Chinese Ann. Math. Ser. A* **16**(1995), 33–42.
325. **J. Yu, Z. Wang and C. Qian,** Oscillation of neutral delay differential equations, *Bull. Austral. Math. Soc.* **45**(1992), 195–200.

326. **J. Yu, Z. Wang and B.G. Zhang,** Oscillation of higher order neutral differential equations, *Rocky Mountain J. Math.* **25**(1995), 557–568.
327. **B.G. Zhang,** Oscillation criteria of partial difference equations with continuous variables, *Acta Math. SINICA* **42**(1999), 487–494 (in Chinese).
328. **B.G. Zhang, J. Yan and S.K. Choi,** Oscillation for difference equations with continuous variable, *Comput. Math. Applic.* **36**(9) (1998), 11–18.
329. **B.G. Zhang and J. Yu,** The existence of positive solutions of neutral differential equations, *Scientia Sinica Ser. A* **8**(1992), 785–790.
330. **Y. Zhang and J. Yan,** Oscillation criteria for difference equations with continuous arguments, *Acta Math. SINICA* **38**(1995), 406–411 (in Chinese).
331. **Z. Zhou and J. Yu,** Decaying solutions of difference equations with continuous arguments, *Annals of Differential Equations* **14**(1998), 576–582.
332. **Y. Zhou and Y.H. Yu,** The distribution of zeros of difference equations with continuous variable, *Chinese Ann. Math.* **20A**(1999), 295–300 (in Chinese).

Subject Index

FSC
www.fsc.org
MIX
Papier aus verantwortungsvollen Quellen
Paper from responsible sources
FSC® C105338

If you have any concerns about our products,
you can contact us on
ProductSafety@springernature.com

In case Publisher is established outside the EU,
the EU authorized representative is:
Springer Nature Customer Service Center GmbH
Europaplatz 3, 69115 Heidelberg, Germany

Printed by Libri Plureos GmbH
in Hamburg, Germany